W0051133

Polyoxometalate Chemistry
for Nano-Composite Design

Nanostructure Science and Technology

Series Editor: David J. Lockwood, FRSC
National Research Council of Canada
Ottawa, Ontario, Canada

Current volumes in this series:

Polyoxometalate Chemistry for Nano-Composite Design
Edited by Toshihiro Yamase and Michael T. Pope

Self-Assembled Nanostructures
Jin Zhang, Zhong-lin Wang, Jun Liu, Shaowei Chen, and Gang-yu Liu

A Continuation Order Plan is available for this series. A continuation order will bring delivery of each new volume immediately upon publication. Volumes are billed only upon actual shipment. For further information please contact the publisher.

Polyoxometalate Chemistry for Nano-Composite Design

Edited by

Toshihiro Yamase

Chemical Resources Laboratory
Tokyo Institute of Technology
Yokohama, Japan

and

Michael T. Pope

Department of Chemistry
Georgetown University
Washington, DC

Springer Science+Business Media, LLC

ISBN 978-1-4757-8718-4   ISBN 978-0-306-47933-5 (eBook)
DOI 10.1007/978-0-306-47933-5

© 2002 Springer Science+Business Media New York
Originally published by Kluwer Academic/Plenum Publishers, New York in 2002
Softcover reprint of the hardcover 1st edition 2002

http://www.wkap.nl/

10 9 8 7 6 5 4 3 2 1

A C.I.P. record for this book is available from the Library of Congress

All rights reserved

No part of this book may be reproduced, stored in a retrieval system, or transmitted in any form or by
any means, electronic, mechanical, photocopying, microfilming, recording, or otherwise, without written
permission from the Publisher, with the exception of any material supplied specifically for the purpose of
being entered and executed on a computer system, for exclusive use by the purchaser of the work.

PREFACE

Polyoxometalates are discrete early transition metal-oxide cluster anions and comprise a class of inorganic complexes of unrivaled versatility and structural variation in both symmetry and size, with applications in many fields of science. Recent findings of both electron-transfer processes and magnetic exchange-interactions in polyoxometalates with increasing nuclearities, topologies, and dimensionalities, and with combinations of different magnetic metal ions and/or organic moieties in the same lattice attract strong attention towards the design of nano-composites, since the assemblies of metal-oxide lattices ranging from insulators to superconductors form the basis of electronic devices and machines in present-day industries. The editors organized the symposium, "Polyoxometalate Chemistry for Nano-Composite Design" at the Pacifichem 2000 Congress, held in Honolulu on December 17–19, 2000. Chemists from several international polyoxometalate research groups discussed recent results, including: controlled self-organization processes for the preparation of nano-composites; electronic interactions in magnetic mixed-valence cryptands and coronands; synthesis of the novel polyoxometalates with topological or biological significance; systematic investigations in acid-base and/or redox catalysis for organic transformations; and electronic properties in materials science.

It became evident during the symposium that the rapidly growing field of polyoxometalates has important properties pertinent to nano-composites. It is therefore easy for polyoxometalate chemists to envisage a "bottom-up" approach for their design starting from individual small-size molecules and moieties which possess their own functionalities relevant to electronic/magnetic devices (ferromagnetism, semiconductivity, proton-conductivity, and display), medicine (antitumoral, antiviral, and antimicrobacterial activities), and catalysis. The resulting exchange of ideas in the symposium has served to stimulate progress in numerous interdisciplinary areas of research: crystal physics and chemistry, materials science, bioinorganic chemistry (biomineralization), and catalysis. Each participant who contributed to this text highlights some of the more interesting and important recent results and points out some of the directions and challenges of future research for the controlled linking of simple (molecular) building blocks, a reaction with which one can create mesoscopic cavities and display specifically desired properties. We believe that this volume provides an overview of recent progress relating to polyoxometalate chemistry, but we have deliberately chosen to exclude discussion of infinite metal oxide assemblies.

Acknowledgment. The editors would like to thank Nissan Chemical Industries, Ltd., Rigaku, and the Donors of the Petroleum Research Fund of the American Chemical Society for contributions towards the support of the Symposium.

Toshihiro Yamase
Michael T. Pope

CONTENTS

SELF-ASSEMBLY AND NANOSTRUCTURES

ORGANOMETALLIC OXIDES AND SOLUTION CHEMISTRY

MAGNETIC, BIOLOGICAL, AND CATALYTIC INTERACTIONS

CHEMISTRY WITH NANOPARTICLES:
LINKING OF RING- AND BALL-SHAPED SPECIES

P. Kögerler[1] and A. Müller[2] *

[1]Ames Laboratory
Iowa State University
Ames, IA 50011, USA
[2]Department of Chemistry
University of Bielefeld
33501 Bielefeld, Germany

INTRODUCTION

The fabrication of well-ordered arrays of well-defined nanoparticles or clusters is of fundamental and technological interest. As this is a difficult task, different techniques have been employed.[1] An elegant approach would be to link well-defined building blocks in a chemically straightforward procedure yielding a monodisperse or a completely homogeneous material. We succeeded now to cross-link assembled nanosized metal-oxide-based clusters/composites – novel supramolecular entities – under one-pot conditions.

Pertinent targets include the synthesis of materials with network structures that have desirable and predictable properties, such as mesoporosity[2] (due to well-defined cavities and channels), electronic and ionic transport,[3] ferro- as well as ferrielasticity, luminescence and catalytic activity.[4] The synthesis of solids from pre-organized linkable building blocks with well-defined geometries and chemical properties is, therefore, of special interest.[5] In this article, we will focus on the relationship between some polyoxomolybdate-based wheel- and ball-shaped clusters and network structures derived from these precursors.[6] Accordingly, a strategy will be presented that allows the intentional synthesis of solid-state materials, both by designing and utilizing known clusters that can be treated as synthon-based building blocks (and thus these synthons can be linked together), with preferred structure and function.

Polyoxometalate Chemistry for Nano-Composite Design
Edited by Yamase and Pope, Kluwer Academic/Plenum Publishers, 2002

BUILDING BLOCKS OF THE NANOPARTICLES

The basic cluster entities – the synthons – involved in this approach can furthermore be decomposed to characteristical *transferable building groups*.[7] For instance, building blocks containing 17 molybdenum atoms ({Mo_{17}} groups) can be given as an example of a generally repeated building block or synthon which can be considered to form anions consisting of two or three of these units. The resulting species are of the {Mo_{36}} type (e.g., $[(MoO_2)_2\{H_{12}Mo_{17}(NO)_2O_{58}(H_2O)_2\}_2]^{12-}$ **1**, a two-fragment cluster, $\equiv$ {Mo'_1}$_2$\{Mo_{17}\}$_2$) or of the {Mo_{57}} type (e.g., $[(VO(H_2O))_6(Mo_2(H_2O)_2(OH))_3\{Mo_{17}(NO)_2O_{58}(H_2O)_2\}_3]^{21-}$ **2**, a three-fragment cluster, $\equiv$ {V}$_6$\{Mo'_2\}$_3$\{Mo_{17}\}$_3$), see Figure 1.[8] It has now been well established that a solution containing {Mo_{17}}-type-based species can be reduced and acidified further to yield a mixed-valence wheel-shaped cluster (and derivatives thereof) {Mo_{154}}, $[Mo_{154}(NO)_{14}O_{434}(OH)_{14}(H_2O)_{70}]^{28-}$ **3** (due to inherent problems with the determination of the exact composition, the initially published formula[9] was flawed with regard to the reduction and protonation grade).[10] Formally, this cluster can be regarded as a tetradecamer with D_{7d} symmetry (if the hydrogen atoms are excluded) and structurally generated by linking 140 MoO_6 octahedra and 14 $MoO_6(NO)$ pentagonal bipyramids.

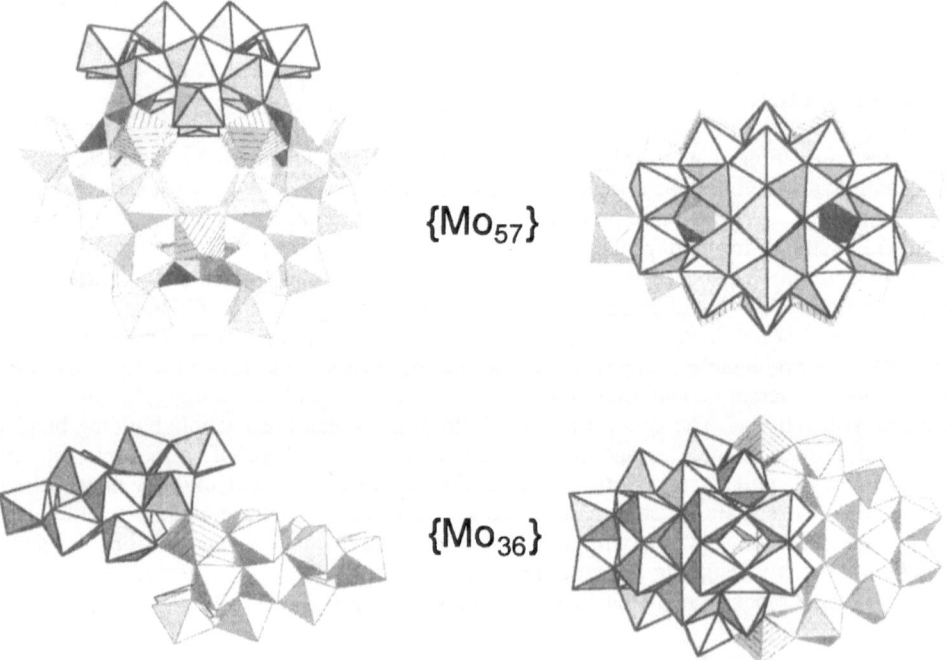

{Mo_{57}}

{Mo_{36}}

Figure 1. A structural comparison between the {$Mo_{57}V_6$}-type cluster **2** (upper half) and the {Mo_{36}}-type cluster **1** (lower half) along perpendicular views with a highlighted {Mo_{17}} building group ({Mo'_1} units: hatched, {Mo'_2} groups: dark gray, {V} units: hatched).

Using the general building block principle for this "classical" giant-wheel-type cluster the structural building blocks for other ring-shaped clusters can be deduced and expressed in terms of the three different building blocks as $[\{Mo_2\}_n\{Mo_8\}_n\{Mo_1\}_n]^{n-}$ ($n = 14$). The building blocks of the type {Mo_8}, {Mo_2}, and {Mo_1} are each present 14 times in the "original" cluster and the corresponding analogous (synthesized without the NO ligands) isopolyoxometalate cluster $[Mo_{154}O_{448}(OH)_{14}(H_2O)_{70}]^{14-}$ **4** (having 14 $[MoO]^{4+}$ instead of

14 $[MoNO]^{3+}$ groups) which turned out to comprise the prototype of the soluble molybdenum blue species.[10] Furthermore, a larger {Mo_{176}} "giant-wheel" cluster with D_{8d} symmetry can also be synthesized under similar conditions; the larger cluster geometrically results if two more of each of the three different types of building units are (formally) added to the "giant-wheel" {Mo_{154}} cluster.[11] This presents a hexadecameric ring structure, containing 16 ($n = 16$) instead of 14 of each of the three aforementioned building blocks (Figure 2).

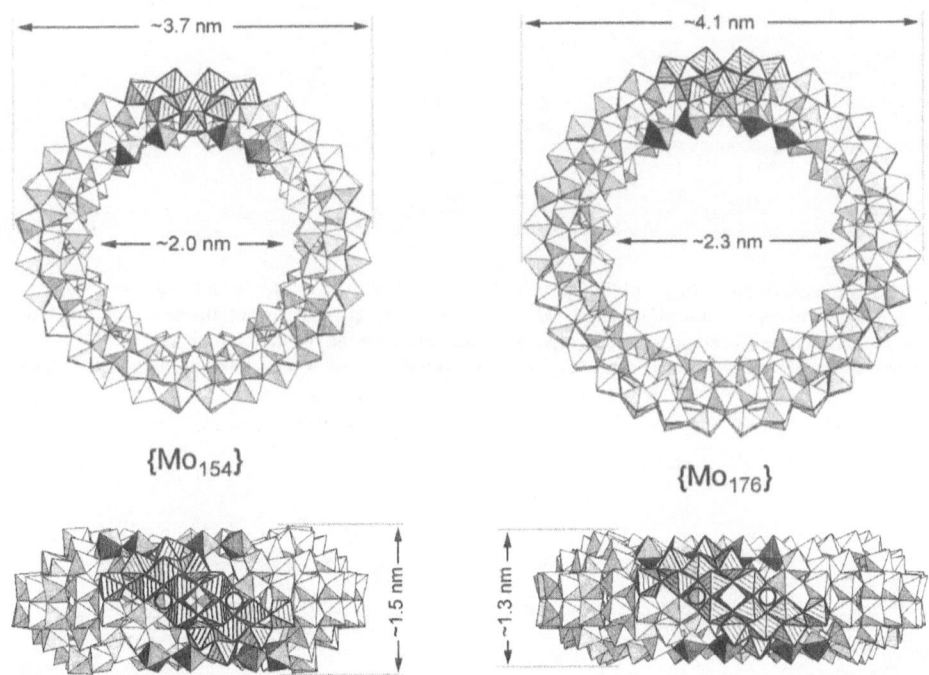

{Mo_{154}}   {Mo_{176}}

Figure 2. Polyhedral representation of a tetradecameric {Mo_{154}}-type (left) and a hexadecameric {Mo_{176}}-type cluster (right) along their C_7 and C_8 axes (upper half) and along their C_2 axes (lower half), respectively. Some individual building groups are highlighted hatched ({Mo_8}), dark gray ({Mo_2}), or encircled ({Mo_1}).

This consideration is interesting from the point of view that it is possible to express the architecture of these systems with a type of Aufbau principle. Furthermore, the C_{2v}-symmetrical {Mo_{17}} group can be subdivided again into one {Mo^*_1} and two {Mo_8} units (i.e. two {Mo_8}-type groups linked by an {Mo^*_1}-type unit). It is interesting to note that the {Mo_8} building blocks are found in many other large polyoxometalate structures and {Mo_8} itself can be divided into a (close-packed) pentagonal {$(Mo)Mo_5$} group – built up by a central MoO_7 pentagonal bipyramid sharing edges with five MoO_6 octahedra – and two more MoO_6 octahedra sharing corners with atoms of the pentagon (Figure 3). The mentioned pentagonal {$(Mo)Mo_5$} group comprises a necessary structural motif to construct spherical systems: while twelve edge-sharing (regular) pentagons form a dodecahedron $\equiv$ (pentagon)$_{12}$, the introduction of linkers in between the pentagons results in an extended structure (pentagon)$_{12}$(linker)$_{30}$ that preserves the I_h symmetry (Figure 4). In this so-called Keplerate-type structure the centers of the pentagons define the vertices of an icosahedron while the centers of the linker units define the vertices of an icosidodecahedron.

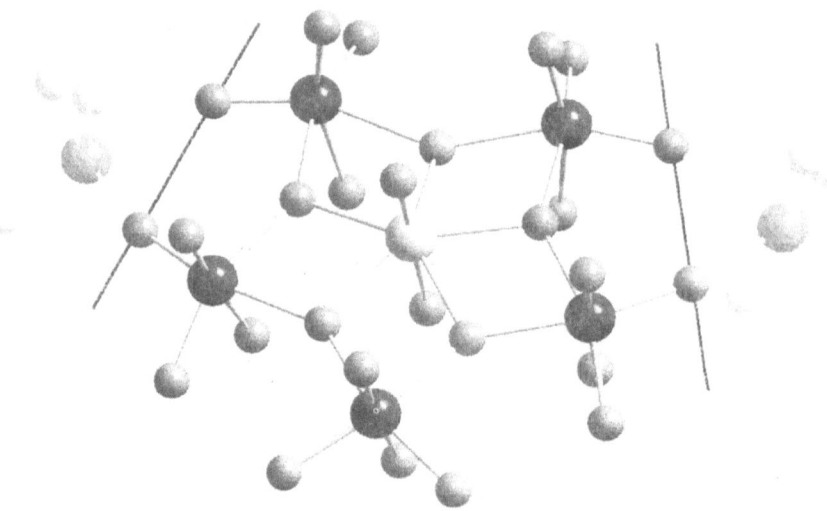

Figure 3. Structure of the {Mo$_8$} building group of the giant wheel-type cluster in ball-and-stick representation. One part of this group, the {(Mo)Mo$_5$} $\equiv$ [(Mo)Mo$_5$O$_{21}$]$^{6-}$ fragment, the basic building block of the spherical clusters is emphasized. The two gray lines illustrate the separation of the {(Mo)Mo$_5$} entity from the two adjoined MoO$_6$ octahedra. Mo: dark gray (with the central Mo position: bright gray), O: gray.

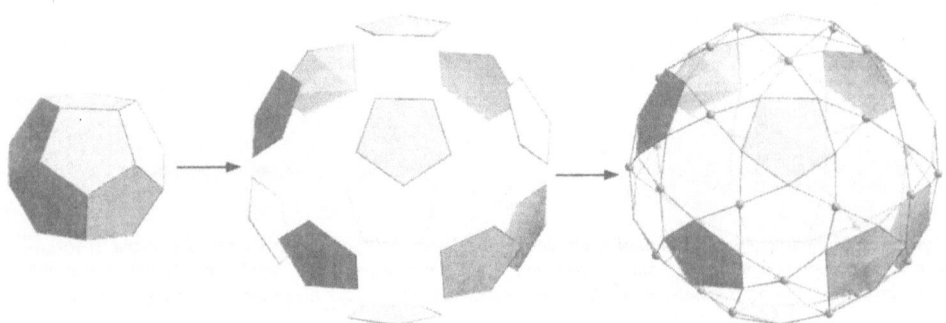

Figure 4. Schematic transition from a dodecahedron (left) via an "extended" polyhedron based on 12 separated pentagons to the final structure with interconnecting linker units (small spheres) while the I_h symmetry is pertained in both steps.

SPHERICAL NANOPARTICLES: SYNTHESES AND STRUCTURE

Recently we reported the first spherical nanostructured Keplerate cluster {Mo$_{132}$} $\equiv$ [{Mo$^V_2$O$_4$(CH$_3$COO)}$_{30}${(Mo)Mo$_5$O$_{21}$(H$_2$O)$_6$}$_{12}$]$^{42-}$ **5a**, as found in **5** $\equiv$ (NH$_4$)$_{42}$**5a** • 10 CH$_3$COONH$_4$ • 300 H$_2$O, with 12 pentagonal {(Mo)Mo$_5$O$_{21}$} groups defining the vertices of an icosahedron which are connected by 30 [Mo$^V_2$O$_4$(CH$_3$COO)]$^+$ linkers.[12] Subsequently we succeeded to substitute these linkers by 30 Fe^{3+} centers resulting in the formation of a relatively smaller icosahedral cluster $\{Mo_{72}Fe_{30}\}$ ≡ $[Mo_{72}Fe_{30}O_{252}(CH_3COO)_{12} \{Mo_2O_7(H_2O)\}_2\{H_2Mo_2O_8(H_2O)\}(H_2O)_{91}]$ **6a**, found in **6** ≡ **6a** • ca. 150 H_2O, in which 30 Fe^{3+} centers (the largest number of paramagnetic centers found in a discrete cluster until now) act as linkers or spacers between the 12 pentagonal $\{(Mo)Mo_5\}$ fragments, the essential building blocks for spherical species (Figure 5).[13] We could also obtain the "reactive" analogue $[Mo_{72}Fe_{30}O_{252}(CH_3COO)_{10}\{Mo_2O_7(H_2O)\} \{H_2Mo_2O_8(H_2O)\}_3 (H_2O)_{91}]$ • 140 H_2O **7** (≡ **7a** • 140 H_2O) of **6** which condenses to a layer framework occurring in $[H_4Mo_{72}Fe_{30}O_{254}(CH_3COO)_{10}\{Mo_2O_7(H_2O)\}\{H_2Mo_2O_8(H_2O)\}_3 (H_2O)_{87}]$ • ca. 80 H_2O **8** during the drying process of **7** at room temperature.[14] Furthermore we are able to follow this process and to characterize the relevant intermediate $[Mo_{72}Fe_{30}O_{252}(CH_3COO)_{10} \{Mo_2O_7(H_2O)\}\{H_2Mo_2O_8(H_2O)\}_3(H_2O)_{91}]$ **9a**, found in **9** ≡ **9a** • ca. 100 H_2O, showing the discrete ball-shaped units, too.[15] For further details on these compounds see Table 1.

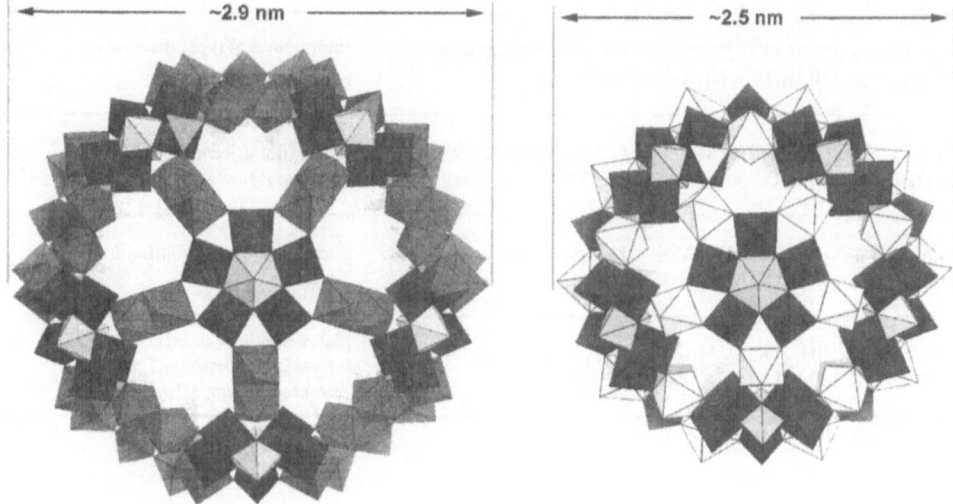

Figure 5. Structural comparison of the $\{Mo_{132}\}$- (left) and $\{Mo_{72}Fe_{30}\}$-type (right) clusters. Both consist of 12 $\{(Mo)Mo_5\}$ groups. The different linker groups L ($\{Mo_{132}\}$: L = $\{Mo^V_2\}$, medium grey; $\{Mo_{72}Fe_{30}\}$: L = $\{Fe\}$, light grey) can be used for sizing the spherical clusters.

A special preparation method, leading to the mixture of compound **6** and a similar one **7**, suggests that the clusters **6a** and **7a** (reactive) exist under equilibrium conditions with different capsule contents. The difference between **6a** and **7a** is that the latter species contains less acetate ligands and more dinuclear ligand units in its cavity corresponding to:

$$[6a] - 2\ CH_3COO^- - [Mo^{VI}_2O_7(H_2O)]^{2-} + 2\ [H_2Mo^{VI}_2O_8(H_2O)]^{2-} \equiv [7a].^{15}$$

Interestingly, **7** (and therefore **8** and **9** in principle) could be obtained by three different methods. As the ligands inside the cavity e.g. of **7a** are highly disordered the exact formula cannot be given. (Best descriptions: one $[Mo^{VI}_2O_7(H_2O)]^{2-}$, three $[H_2Mo^{VI}_2O_8(H_2O)]^{2-}$ groups and ten acetate ligands.) The same is of course true for the resulting condensation product **8** and intermediate compound **9**. The presence of the negatively charged $\{Mo^{VI}_2\}$-type dinuclear units in the icosahedral species is not surprising because the complete substitution of 30 $[Mo_2^V O_4(CH_3COO)]^+$ linkers of the anion **5a** (with rather high negative charge) by the 30 Fe^{III} centers alone would cause a positively charged cluster of the resulting relatively smaller spherical species. The source of the dinuclear $\{Mo^{VI}_2\}$ units, the presence of which leads to a neutral cluster compound, are the $[Mo_2^V O_4(CH_3COO)]^+$ linkers of **5a** which get air-oxidized during the substitution reaction.

Table 1. Selected data of Keplerate-based species

	Crystal habit / type of structure / space group	Ref.
$(NH_4)_{42}[\{Mo^V_2O_4(CH_3COO)\}_{30}\{(Mo)Mo_5O_{21}(H_2O)_6\}_{12}]$ · 10 CH_3COONH_4 · ca. 300 H_2O **5** ≡ $(NH_4)_{42}$**5a** · 10 CH_3COONH_4 · ca. 300 H_2O	distorted octahedra; discrete clusters; $Fm\bar{3}$	12
$[Mo_{72}Fe_{30}O_{252}(CH_3COO)_{12}\{Mo_2O_7(H_2O)\}_2\{H_2Mo_2O_8(H_2O)\}(H_2O)_{91}]$ · ca. 150 H_2O **6** ≡ **6a** · ca. 150 H_2O	rhombohedral type; discrete clusters; $R\bar{3}$	13
$[Mo_{72}Fe_{30}O_{252}(CH_3COO)_{10}\{Mo_2O_7(H_2O)\}\{H_2Mo_2O_8(H_2O)\}_3(H_2O)_{91}]$ · ca. 140 H_2O **7** ≡ **7a** · ca. 140 H_2O [a]	plates (not dried); discrete clusters; $P2_1/n$	15
$[H_4Mo_{72}Fe_{30}O_{254}(CH_3COO)_{10}\{Mo_2O_7(H_2O)\}\{H_2Mo_2O_8(H_2O)\}_3(H_2O)_{87}]$ · ca. 80 H_2O **8** ≡ **8a** · ca. 80 H_2O [a]	plates (dried); layer structure; $Cmca$	14
$[Mo_{72}Fe_{30}O_{252}(CH_3COO)_{10}\{Mo_2O_7(H_2O)\}\{H_2Mo_2O_8(H_2O)\}_3(H_2O)_{91}]$ · ca. 100 H_2O **9** ≡ **9a** · ca. 100 H_2O [a]	plates (medium dried), intermediate between **7** and **8**; discrete clusters; $P2_1/n$	15

[a] Regarding formula see text.

GETTING GIANT SPHERICAL CLUSTERS LINKED AND CROSS-LINKED

It turned out that the discrete clusters of the type $\{Mo_{72}Fe_{30}\}$ under the present packing conditions of **7**, which seems to be important, are reactive units even under room temperature and solid state conditions as the process finally results in the linking to form the layer structure of **8** (Figure 6). Remarkably the same process does not occur when rhombohedral crystals of **6** are dried; the linking is performed via Fe–O–Fe bridge formations between adjacent units which require deprotonation of the H_2O ligands at the Fe sites and subsequent condensation (see eq. 1). The initial step is dehydration, i.e. loss of crystal water (eq. 2) and the last step is the condensation reaction corresponding to equation (3).[16] We are not able to distinguish clearly whether the cluster unit or the crystal water molecules act as proton acceptors.

$$Fe(OH_2) + (H_2O)Fe \rightarrow Fe(OH) + (H_2O)Fe \rightarrow Fe–O–Fe \qquad (1)$$

$$[7] \xrightarrow{\text{- crystal water}} [9] \qquad (2)$$

$$[9] \xrightarrow{\text{deprotonation and condensation}} [8] \text{ (see eq. 1)} \qquad (3)$$

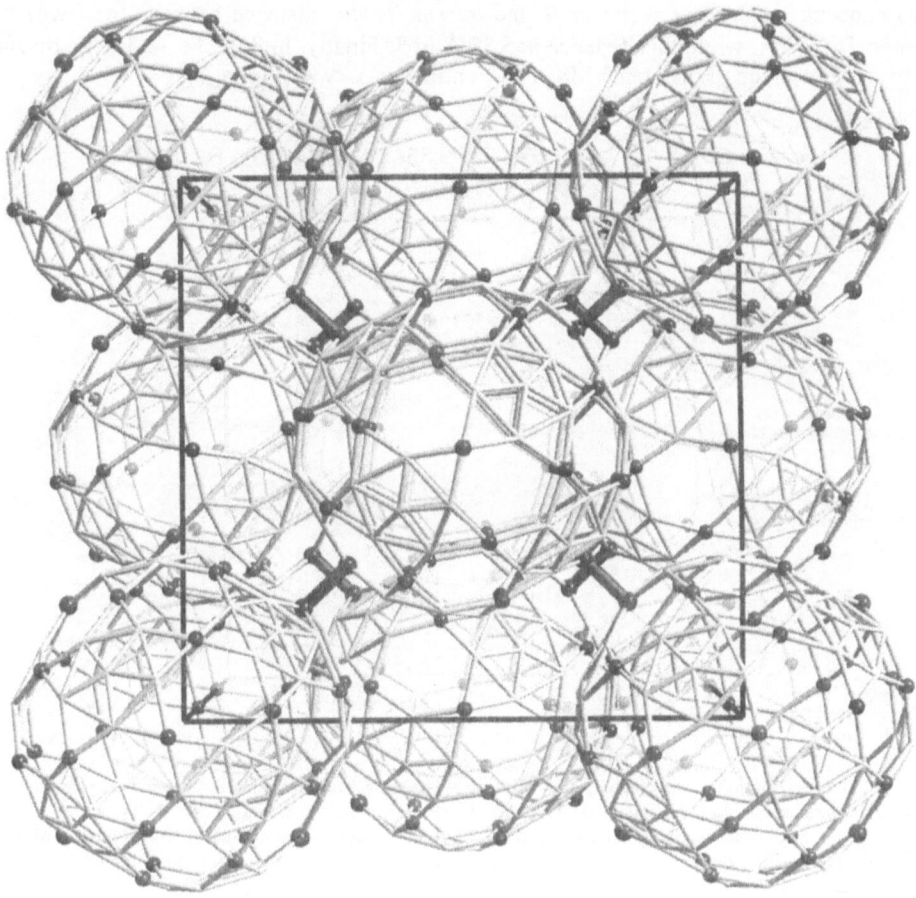

Figure 6. Wire frame representation of the layer structure of linked $\{Mo_{72}Fe_{30}\}$ units with emphasized Fe-O-Fe bridges along the a axis in **8** (only the metal atoms are shown; Fe centers, black spheres). The unit cell is also depicted, showing the approximate face-centered cubic packing of the cluster spheres.

Remarkably these consecutive processes can be detected from the determination of the crystal structures of **7**, **8** and **9**. The activity of the processes seems to be directly proportional to the rate of loss of crystal water molecules, i.e. the actual drying conditions.

The freshly precipitated (not dried) crystals of **7** but also the crystals of **9** contain the identical discrete spherical clusters of the $\{Mo_{72}Fe_{30}\}$ type. This is not surprising as reactions occur under solid state conditions. Each sphere contains 12 pentagonal fragments of the type $\{(Mo)Mo_5O_{21}\}$ with a central bipyramidal $\{MoO_7\}$ group which is linked by edge-sharing to five $\{MoO_6\}$ octahedra. These pentagonal fragments are connected by 30 $\{Fe^{III}(H_2O)_2\}$ linkers so that the overall shape of **7a** has approximately icosahedral symmetry. Each $\{Fe^{III}(H_2O)_2\}$ group is connected to oxygen atoms of two $\{MoO_6\}$ octahedra of two neighboring pentagons resulting in an $\{FeO_6\}$ octahedron. Interestingly, the 102 metal atoms (72 Mo^{VI} + 30 Fe^{III}) and their terminal ligands (O_{term} and H_2O) lie in two concentric spherical shells. In **7**, the *intermolecular* distance between the (two) Fe^{III} centers is 6.74 Å, while this distance is 5.35 Å in **9**. Finally, in **8** the Fe–(μ-O)–Fe distance between Fe centers of different $\{Mo_{72}Fe_{30}\}$ entities is 3.79 Å (Figure 7).

Fe···Fe: 6.74Å   →   Fe···Fe: 5.35Å   →   Fe-Fe: 3.79Å
                              (intermediate)

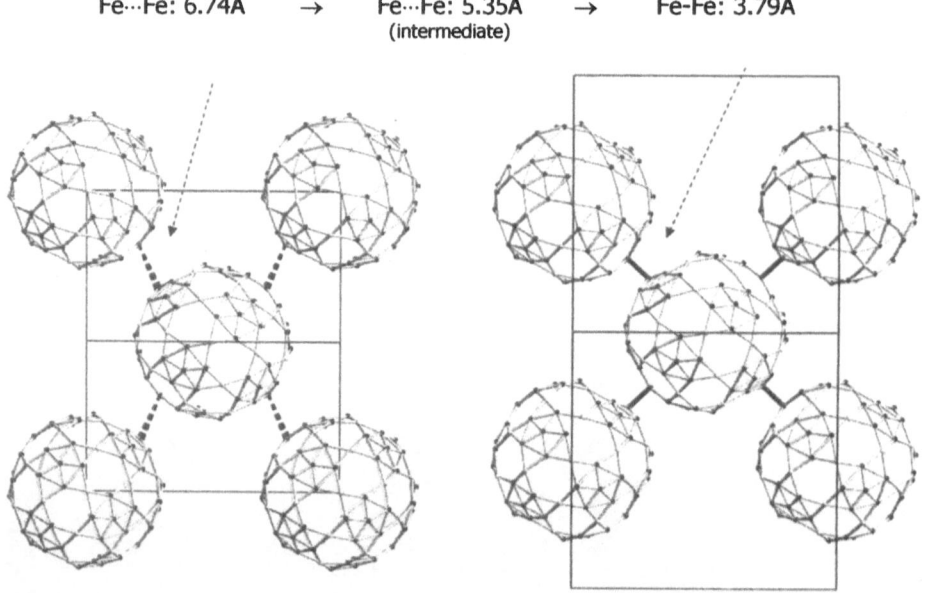

Figure 7. Representation of the condensation reaction proceeding from discrete cluster units (as present in freshly prepared monoclinic crystals of **7**; left side) via an intermediate **9**, in which discrete units are approached to each other reaching a minimum distance, to the layer-type compound **8**. The geometry of **9** is stabilized by hydrogen bonds between Fe-OH_2···H_2O-Fe groups which finally react to form the Fe-O-Fe linkers of **8**.

The value of $\chi_{mol}T$ = ca. 130 emu K mol^{-1} for compound **6** measured at room temperature corresponds nearly to 30 high spin Fe^{3+} centers ($\chi_{spin-only}T$ = 131.25 emu K mol^{-1}) whereas the corresponding value (113 emu K mol^{-1}) for compound **8** is consistent with 26 uncorrelated centers with s = 5/2 ($\chi_{spin-only}T$ = 113.75 emu K mol^{-1}). This clearly indicates that four of these 30 Fe^{3+} centers are strongly (antiferromagnetically) coupled.[14]

In the same manner, the supramolecular metal-oxide-based entity consisting of the icosahedral capsule of the type $\{Mo^{VI}_{72}Fe^{III}_{30}\}$ as host and the reduced Keggin cluster $[H_2PMo_{12}O_{40}]^{3-}$ (Keggin anion diameter ~ 14 Å) as nucleus (guest) can get linked (according to a modeling investigation before the synthetic approach it turned out that the Keggin anion just fits exactly into the capsule).[17]

In an acidified aqueous solution (pH 2) containing only polymolybdate, iron(II) chloride, and acetic acid as well as a relatively small amount of phosphate in the presence of air, a stepwise assembly process takes place leading to this new type of composite material, i.e. the neutral layer compound **10**.

$[H_2PMo^{VI}_{10}Mo^V_2O_{40} \subset H_4Mo^{VI}_{72}Fe^{III}_{30}(CH_3COO)_{15}O_{254}(H_2O)_{98}]$ · ca. 60 H_2O **10**
≡ **10a** · ca. 60 H_2O

While **10** can also be assembled by adding the normal Keggin anion $[PMo_{12}O_{40}]^{3-}$ directly to the aqueous reaction mixture according to our first approach, the other reaction (experimental section, method 1) corresponds to a molecular cascade with the formation of the Keggin ion as the initial step. Correspondingly, the reaction takes a different route (with no formation of **10**!) in the presence of larger amounts of phosphate, and adding the Keggin unit seems to accelerate the capsule formation as a template. It is important to start from Fe^{II} (which gets gradually oxidized) rather than from Fe^{III}, as the latter educt results immediately in a not well-defined precipitate.

The building block of each layer of **10** is the spherical icosahedral giant oxidized cluster cage of the $\{Mo_{72}Fe_{30}\}$ type but which now has a reduced metal-oxide-based cluster – the tetrahedral two-electron reduced Keggin $[H_2PMo_{12}O_{40}]^{3-}$ ion – as nucleus (Figure 8). Like in the layer compound **8**, each of the cluster-cluster composites is linked to four others via Fe–O–Fe bonds to form a layer structure.

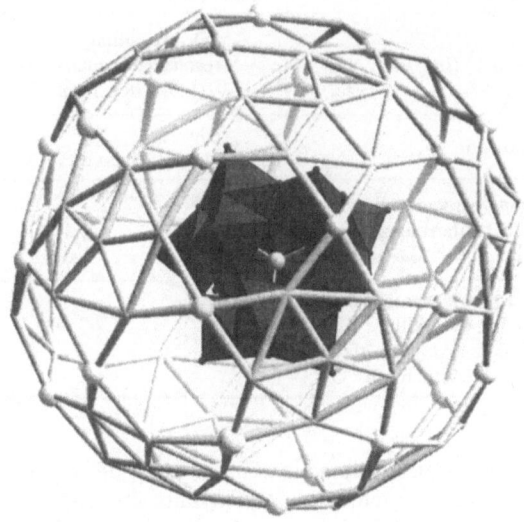

Figure 8. Demonstration of the structure of the building block of **10** with the capsule (host) and encapsulated Keggin-type cluster nucleus (guest): the metal $\{Mo_{72}Fe_{30}\}$ capsule is shown in wire frame representation – with the 30 Fe^{III} centers (highlighted as spheres) linking the 12 $\{(Mo)Mo_5\}$ pentagons – and the Keggin nucleus in polyhedral representation.

Selected physical properties of **10** are summarized in Table 2. They not only prove the existence of the two separate, non-covalently bonded parts of each supramolecular entity but show also its interesting topological, spectroscopic, electronic and magnetic properties. The reduced Keggin cluster can be identified nicely by means of the resonance Raman effect showing only the vibrational bands of this unit. The nanocapsules of the type $\{(Mo^{VI})Mo^{VI}_5\}_{12}Fe^{III}_{30}$ forming a system of magnetic dots (each individual discrete dot represents as yet the strongest known molecular paramagnet due to the presence of 30 Fe^{III} centers with 150 unpaired electrons) encapsulate the reduced nuclei (quantum dots) as guests which can be regarded as potential electron storage elements. It should be noted that the free Keggin cluster can be reduced by one electron, and further reduced in several two-electron steps in association with concomitant protonation thus keeping its charge constant.

Table 2. Selected properties of the composite compound **10**.[17]

| | Building blocks (capsule and nucleus) | | Interaction characteristics |
	Capsule (host)	Keggin nucleus (guest)	
Imposed Platonic / Archimedean solids	icosidodecahedron (Fe_{30}), icosahedron ($\{(Mo)Mo_5\}_{12}$)	cuboctahedron (Mo_{12}), rhombicuboctahedron (O_{24})	topological host-guest complementarity
Symmetry	icosahedral metal framework (I_h)	tetrahedral (T_d)	
Building units	90 octahedra ($Fe^{III}O_4(H_2O)_2$, $Mo^{VI}O_6$), 12 pentagonal bipyramids ($Mo^{VI}O_7$)	1 tetrahedron (P^VO_4), 12 octahedra ($Mo^{VI/V}O_6$)	
Electronic transitions [cm^{-1}]	$O \rightarrow Mo^{VI}$ charge transfer: ≈ 27000	IVCT ($Mo^V \rightarrow Mo^{VI}$): ≈ 11600 / ≈ 9500	nucleus $\rightarrow$ shell CT: ≈ 18100
Vibrations [cm^{-1}]	$\nu(MoO_t)$ (IR): 960	$\nu(O_{br})$-breathing (resonance Raman): 830 $\nu_{as}(PO_4)$ (IR): 1068	
Redox state	oxidized	reduced (electron reservoir)	
Magnetism	strong paramagnet corresponding to 26 antiferromagnetically coupled Fe^{III} ($s = 5/2$) centers		
Nuclei properties [$mm\ s^{-1}$]	^{57}Fe Mössbauer isomer shift / quadrupole splitting (180 K): $\delta = 0.48$ / $\Delta E_Q = 0.70$ (characteristic for $Fe^{III}O_6$)		

The non-covalent host-guest interactions are worthy of consideration as the reduced electron reservoir-type Keggin ion fits exactly into the capsule cavity (the shortest $O_{host} \cdots O_{guest}$ bond lengths, typical for hydrogen bonding, are of the order 2.6 Å). This type of composite/supramolecular entity with a reduced nucleus in an oxidized shell is unprecedented. The band observed at ~ 550 nm (~ 18.1×10^3 cm^{-1}) which contributes to the color can tentatively be assigned to a novel charge transfer transition of the type *reduced nucleus → oxidized shell*.

The knowledge of the chemistry of nanocapsules which are variable in size and linkable allows us the synthesis of new types of materials. It is even possible to open the capsules, exchange their contents, and close them again [18] which allows the fabrication of different types of cross-linked composites with core-shell topology. We refer to a new class of novel composite (a cluster encapsulated in a cluster) type material, in which the electronic/magnetic structure of the composite (quantum/magnetic dot) can in principle be tuned by changing the relevant properties of the constituents, for instance by changing the electron population of the nucleus (Keggin anion). Remarkably the nanoobjects can also get linked to chains in which the $\{Mo_{72}Fe_{30}\}$-type entities are linked via an Mo-O-Fe and an Fe-O-Mo bond to each nearest neighbor.[19]

GIANT RING-SHAPED CLUSTERS AS SYNTHONS FOR PERIODIC STRUCTURES

An extremely interesting observation is that it is possible to obtain "giant-wheel"-type species, which are structurally incomplete, comprising defects when compared to the parent isopolyoxomolybdate $\{Mo_{154}\}$-type cluster **4**.[10] These defects – which initiate linking in a way not well understood – manifest themselves as missing $\{Mo_2\}^{2+}$ units, but statistically these defects can sometimes be seen as under-occupied $\{Mo_2\}^{2+}$ units when the distribution is affected by rotational or translational disorder within the crystal structure. When wheel-type clusters with defects are considered the numbers of each type of building block are not identical as a fraction of the $\{Mo_2\}^{2+}$ groups have been removed. As a result the overall negative charge on the "giant-wheel" increases by two for each of the removed $\{Mo_2\}^{2+}$ groups. These compounds also can be expressed in terms of the "giant-wheel" architecture, with the general formula $[\{Mo_2\}_{n-x}\{Mo_8\}_n\{Mo_1\}_n]^{(n+2x)-}$ where the value x corresponds to the number of defects introduced into the system (in the case of the "giant-wheel" structures, only those that have $n = 14$ have been discovered with defects to date).

Important for linking of the $\{Mo_{154}\}$-type cluster is the increase of the nucleophilicity at special sites which can be realized either by removing several positively charged $\{Mo_2\}^{2+}$ groups with bidentate ligands like formate (that means via formation of defects),[20] or by placing electron-donating ligands like $H_2PO_2^-$ on the inner ring surfaces.[21] This leads to a linkage of the ring-shaped clusters via Mo-O-Mo bonds to form compounds with layers or chains (e.g., one derived from $\{Mo_{144}\}$ ring units with the formula $[Mo_{144}O_{437}H_{14}(H_2O)_{56}]^{24-}$ **11a** [22]) (Figure 9) according to a type of crystal engineering (see below). Single crystals of the chain-type compound exhibit interesting anisotropic electronic properties that represent promising fields for further research. In compounds of that type channels are present, the inner surfaces of which have basic properties in contrast to the acidic channels in zeolites.[23] The layer compound can take up small organic molecules such as formic acid, which according to the basicity of the system are partly deprotonated. The reduction of an aqueous solution of sodium molybdate by hypophosphorous (phosphinic) acid at low pH values (≈ 1) results in the formation of nanosized ring-shaped cluster units (defined above) which assemble to form layers of the compound $Na_{21}[Mo_{154}O_{462}H_{14}(H_2O)_{54}(H_2PO_2)_7] \cdot$ ca. 300 H_2O **12** (Figure 10).[21] The assembly is based on the synergetically induced complementarity of the amphiphilic $\{Mo_2\}$-type $O=Mo(H_2O)$ groups and corresponds to the replacement of H_2O ligands of rings by related terminal oxo groups also of the $\{Mo_2\}$-type $O=Mo(H_2O)$ units of other rings acting formally as ligands (and vice versa). The increased nucleophilicity of the relevant $O=Mo$ groups at the latter type ring is induced by coordinated $H_2PO_2^-$ ligands (Figure 11).

SUMMARY

By exploiting the concept of transferable building groups it is possible to deliberately generate highly symmetric nanometer-sized polyoxomolybdate-based clusters with the option to link them to 1-, 2-, and 3-dimensional networks. The building block concept even allows to size the ring- or sphere-shaped entities e.g. by using different linkers between the building groups (as in the case of the Keplerate-type clusters) or by varying the number of groups (as in the case of the tetradecameric and hexadecameric ring clusters). Inclusion of guest molecules in the spherical Keplerate-type clusters results in further functionalized supramolecular composite structures. The potential to attach different ligands to the prototypal cluster structures in order to alter the nucleophilicity of certain sites allows to control condensation processes that result in the formation of network structures including mesoporous systems. The assembly of well-ordered arrays of these nanosized molecules is therefore controllable to a great extent via a number of options.

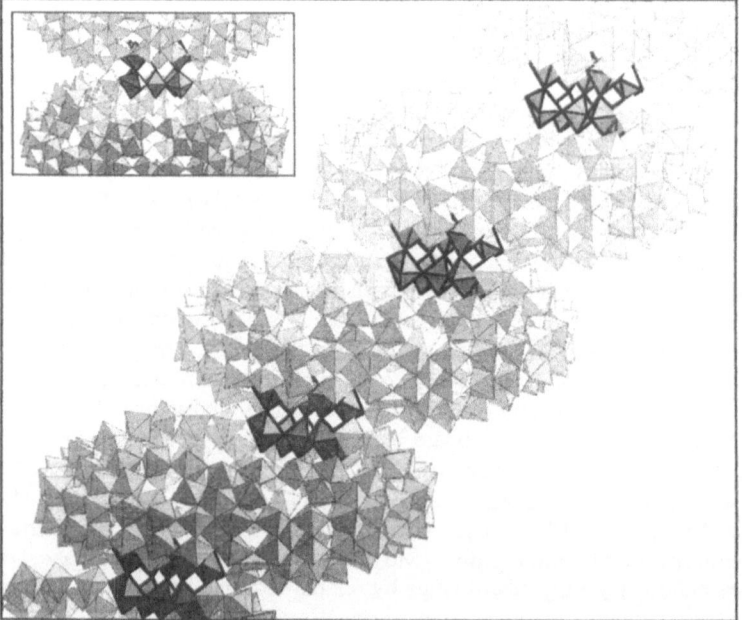

Figure 9. Polyhedral representation of {Mo$_{144}$} units **11a** linked to chains. The linking occurs via the {Mo$_2$} units – the polyhedra formed by the linked {Mo$_2$} units are shown within a black outline. A face-on representation of one of the linking units with a ring above and below the linked polyhedra (also highlighted with a black outline) is shown in the upper left corner for clarity.

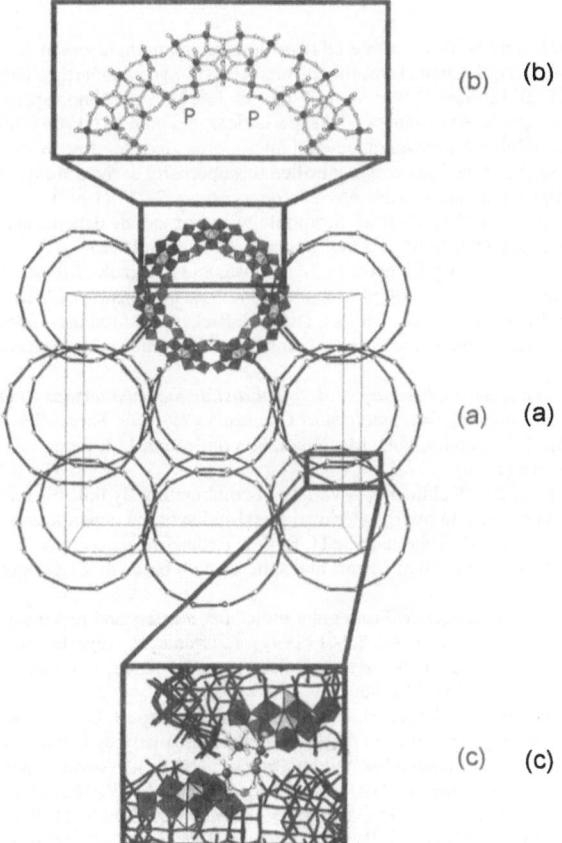

Figure 10. (a) Perspective view of the framework of $Na_{21}[Mo_{154}O_{462}H_{14}(H_2O)_{54}(H_2PO_2)_7] \cdot$ ca. 300 H_2O **12** along the crystallographic c axis, showing the abundance of nanotubes and cavities. For clarity, only one complete ring (without the P ligands) is shown in polyhedral representation. With respect to the other rings, only the centers of the $\{Mo_1\}$ units are given and connected. (b) Ball-and-stick representation of the upper half of a ring segment showing the principal positions of the $H_2PO_2^-$ ligands. (c) Detailed view [perpendicular to (a) and (b)] of the bridging region between two cluster rings emphasizing one $\{Mo_8\}$ and one $\{Mo_1\}$ unit (in polyhedral representation) as well as one $\{Mo_2\}^{2+}$ unit and one $H_2PO_2^-$ ligand (in ball-and-stick representation).[21]

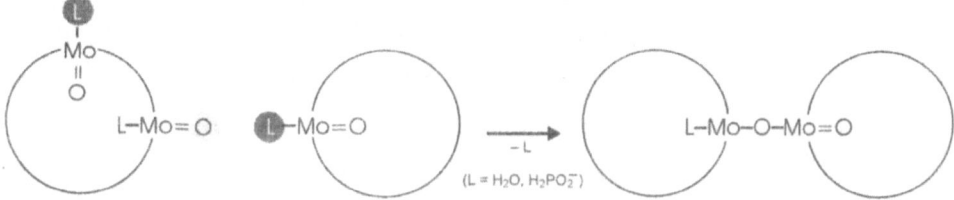

Figure 11. Schematic representation of the basic assembly principle of the "giant-wheel"-shaped cluster units forming layers of **12**. The formation is based on the synergetically induced functional complementarity of the $\{Mo_2\}$ units O=Mo(L) (L = H_2O, $H_2PO_2^-$) on their surfaces.

REFERENCES

1.   G. Schmid, M. Bäumle, and N. Beyer, Ordered two-dimensional monolayers of Au$_{55}$ clusters, *Angew. Chem. Int. Ed.* 39:181 (2000); S. Chen, Two-dimensional crosslinked nanoparticle networks, *Adv. Mater.* 12:186 (2000); S. Sun, C.B. Murray, D. Weller, L. Folks, and A. Moser, Monodisperse FePt nanoparticles and ferromagnetic FePt nanocrystal superlattices, *Science* 287:1989 (2000); C.L. Bowes and G.A. Ozin, Self-assembling frameworks: beyond microporous oxides, *Adv. Mater.* 8:13 (1996).

2.   K.A. Carrado and L.Q. Xu, Materials with controlled mesoporosity derived from synthetic polyvinylpyrrolidone-clay composites, *Micropor. Mesopor. Mater.* 27:87 (1999).

3.   N. Papageorgiou, C. Barbe, and M. Grätzel, Morphology and adsorbate dependence of ionic transport in dye sensitized mesoporous TiO$_2$ films, *J. Phys. Chem. B* 102:4156 (1998).

4.   C.K. Loong, P. Thiyagarajan, J.W. Richardson, M. Ozawa, and S. Suzuki, Microstructural evolution of zirconia nanoparticles caused by rare-earth modification and heat treatment, *J. Catal.* 171:498 (1997); T.R. Pauly, Y. Liu, T.J. Pinnavaia, S.Y.L. Bilinge, and T.P. Rieker, Textural mesoporosity and the catalytic activity of mesoporous molecular sieves with wormhole framework structures, *J. Am. Chem. Soc.* 121:8835 (1999).

5.   *Comprehensive Supramolecular Chemistry, Vol. 6, Solid-state Supramolecular Chemistry: Crystal Engineering*, and *Vol. 7, Solid-state Supramolecular Chemistry: Two and Three-dimensional Inorganic Networks*, J.L. Atwood, J.E.D. Davies, D.D. MacNicol, F. Vögtle, and J.M. Lehn, Eds., Pergamon/Elsevier, Oxford (1996).

6.   A. Müller, P. Kögerler, and C. Kuhlmann, A variety of combinatorially linkable units as disposition: from a giant icosahedral keplerate to multi-functional metal-oxide based network structures, *Chem. Comm.* 1347 (1999); A. Müller, P. Kögerler, and H. Bögge, Pythagorean harmony in the world of metal oxygen clusters of the {Mo$_{11}$} type: giant wheels and spheres both based on a pentagonal type unit, *Struct. Bond.* 96:203 (2000).

7.   A. Müller, D. Fenske, and P. Kögerler, From giant molecular clusters and precursors to solid-state structures, *Curr. Op. Solid State & Mat. Sci.* 3:141 (1999); L. Cronin, P. Kögerler, and A. Müller, Controlling growth of novel solid-state materials via discrete molybdenum-oxide-based building blocks as synthons, *J. Solid State Chem.* 152:57 (2000).

8.   A. Müller, E. Krickemeyer, S. Dillinger, H. Bögge, W. Plass, A. Proust, L. Dloczik, C. Menke, J. Meyer, and R. Rohlfing, New perspectives in polyoxometalate chemistry by isolation of compounds containing very large moieties as transferable building blocks: (NMe$_4$)$_5$[As$_2$Mo$_8$V$_4$AsO$_{40}$] · 3 H$_2$O, (NH$_4$)$_{21}$[H$_3$Mo$_{57}$V$_6$(NO)$_6$O$_{183}$(H$_2$O)$_{18}$] · 65 H$_2$O, (NH$_2$Me$_2$)$_{18}$(NH$_4$)$_6$[Mo$_{57}$V$_6$(NO)$_6$O$_{183}$(H$_2$O)$_{18}$] · 14 H$_2$O, and (NH$_4$)$_{12}$[Mo$_{36}$(NO)$_4$O$_{108}$(H$_2$O)$_{16}$] · 33 H$_2$O, *Z. Anorg. Allg. Chem.* 620:599 (1994).

9.   A. Müller, E. Krickemeyer, J. Meyer, H. Bögge, F. Peters, W. Plass, E. Diemann, S. Dillinger, F. Nonnenbruch, M. Randerath, and C. Menke, [Mo$_{154}$(NO)$_{14}$O$_{420}$(OH)$_{28}$ (H$_2$O)$_{70}$]$^{(25+5)-}$ – a water-soluble big wheel with more than 700 atoms and a relative molecular mass of about 24000, *Angew. Chem., Int. Ed. Engl.* 34:2122 (1995).

10.   A. Müller, C. Serain, Soluble molybdenum blues – "des Pudels Kern", *Acc. Chem. Res.* 33:2 (2000).

11.   A. Müller, E. Krickemeyer, H. Bögge, M. Schmidtmann, C. Beugholt, P. Kögerler, and C. Lu, Formation of a ring-shaped reduced "metal oxide" with the simple composition [(MoO$_3$)$_{176}$(H$_2$O)$_{80}$H$_{32}$], *Angew. Chem., Int. Ed.* 37:1220 (1998).

12.   A. Müller, E. Krickemeyer, H. Bögge, M. Schmidtmann, and F. Peters, Organizational forms of matter: an inorganic super fullerene and keplerate based on molybdenum oxide, *Angew. Chem., Int. Ed.* 37:3360 (1998).

13.   A. Müller, S. Sarkar, S.Q.N. Shah, H. Bögge, M. Schmidtmann, Sh. Sarkar, P. Kögerler, B. Hauptfleisch, A.X. Trautwein, and V. Schünemann, Archimedean synthesis and magic numbers: "sizing" giant molybdenum-oxide-based molecular spheres of the keplerate type, *Angew. Chem., Int. Ed.* 38:3238 (1999).

14.   A. Müller, E. Krickemeyer, S.K. Das, P. Kögerler, S. Sarkar, H. Bögge, M. Schmidtmann, Sh. Sarkar, Linking icosahedral, strong molecular magnets {Mo$^{VI}_{72}$Fe$^{III}_{30}$} to layers – a solid-state reaction at room temperature, *Angew. Chem., Int. Ed.* 39:1612 (2000).

15.   A. Müller, S.K. Das, E. Krickemeyer, P. Kögerler, H. Bögge, and M. Schmidtmann, Cross-linking nanostructured spherical capsules as building units by crystal engineering: related chemistry, *Solid State Sci.* 2:847 (2000).

16.   F.A. Cotton, G. Wilkinson, C.A. Murillo, and M. Bochmann: *Advanced Inorganic Chemistry*, 6th Ed., Wiley, New York (1999); W. Schneider, *Comments Inorg. Chem.* 3:204 (1984).

17.   A. Müller, S.K. Das, P. Kögerler, H. Bögge, M. Schmidtmann, A.X. Trautwein, V. Schünemann, E. Krickemeyer, and W. Preetz, A new type of supramolecular compound: molybdenum-oxide-based composites consisting of magnetic nanocapsules with encapsulated keggin-ion electron reservoirs cross-linked to a two-dimensional network, *Angew. Chem., Int. Ed.* 39:3413 (2000); see also J. Uppenbrink, A soupçon of phosphate, *Science* (Highlights, Editors' Choice) 290:411 (2000).

18. A. Müller, S. Polarz, S.K. Das, E. Krickemeyer, H. Bögge, M. Schmidtmann, and B. Hauptfleisch, "Open and shut" for guests in molybdenum oxide-based giant spheres, baskets, and rings containing the pentagon as a common structural element, *Angew. Chem., Int. Ed.* 38:3241 (1999).

19. A. Müller et al., in preparation.

20. A. Müller, S.K. Das, V.P. Fedin, E. Krickemeyer, C. Beugholt, H. Bögge, M. Schmidtmann, and B. Hauptfleisch, Rapid and simple isolation of the crystalline molybdenum-blue compounds with discrete and linked nanosized ring-shaped anions: $Na_{15}[Mo^{VI}_{126}Mo^{V}_{28}O_{462}H_{14}(H_2O)_{70}]_{0.5}$ $[Mo^{VI}_{124}Mo_{28}O_{457}H_{14}(H_2O)_{68}]_{0.5} \cdot$ ca. 400 H_2O and $Na_{22}[(Mo^{VI}_{118}Mo^{V}_{28}O_{442}H_{14}(H_2O)_{58}] \cdot$ ca. 250 H_2O, *Z. Anorg. Allg. Chem.* 625:1187 (1999).

21. A. Müller, S.K. Das, H. Bögge, C. Beugholt, and M. Schmidtmann, Assembling nanosized ring-shaped synthons to an anionic layer structure based on the synergetically induced functional complementarity of their surface-sites: $Na_{21}[Mo^{VI}_{126}Mo^{V}_{28}O_{462}H_{14}(H_2O)_{54}(H_2PO_2)_7] \cdot x\ H_2O$ ($x \approx 300$), *Chem. Comm.* 1035 (1999).

22. A. Müller, E. Krickemeyer, H. Bögge, M. Schmidtmann, F. Peters, C. Menke, and J. Meyer, An unusual polyoxomolybdate: giant wheels linked to chains, *Angew. Chem., Int. Ed.* 36:484 (1997).

23. A. Müller, E. Krickemeyer, H. Bögge, M. Schmidtmann, C. Beugholt, S.K. Das, F. Peters, and C. Lu, Giant ring-shaped building blocks linked to form a layered cluster network with nanosized channels: $[Mo^{VI}_{124}Mo^{V}_{28}O_{429}(\mu_3\text{-}O)_{28}H_{14}(H_2O)_{66.5}]^{16-}$, *Chem. Eur. J.* 5:1496 (1999).

PROSPECTS FOR RATIONAL ASSEMBLY OF COMPOSITE POLYOXOMETALATES

Nebebech Belai, Michael H. Dickman, Kee-Chan Kim,
Angelo Ostuni, Michael T. Pope*, Masahiro Sadakane,
Joseph L. Samonte, Gerta Sazani, and Knut Wassermann

Department of Chemistry, Box 571227
Georgetown University
Washington, DC 20057-1227

INTRODUCTION

Currently, polyoxometalates range in size from simple dimetalates such as $Mo_2O_7^{2-}$ to discrete water-soluble polyoxoanions containing more than 200 metal atoms, and species with relative molar masses of 40,000.[1] There is little doubt that even larger complexes can be synthesized, although complete characterization of these in the solid state and in solution will become increasingly challenging. There are important reasons for the development of the chemistry of such giant anions, which can be expected to exhibit both localized (molecular) and cooperative (solid state) properties. Controlled directed syntheses of ultra-large polyoxometalates with new structural frameworks can lead for example to further applications in catalysis, host-guest chemistry, and molecular recognition, as well as to new magnetic and optical materials.[2] Indeed the "emergence" of new or special properties resulting from increase of molecular size and complexity, exquisitely demonstrated by the structure and function of enzymes for example, is an additional powerful incentive.

We shall define a *composite polyoxometalate* as one containing two or more polyoxoanion "building blocks" and linker atoms or groups. Although the structures of such composite species suggest that they can be rationally synthesized by combination of building block with linker, in many cases the complete structure is formed from mononuclear components in a one-pot reaction .

The following examples, $[MnNb_{12}O_{38}]^{12-}$ (**1**),[3] $[Ln(XW_{11}O_{39})_2]^{n-}$ (**2**),[4] $[As_4W_{40}O_{140}]^{28-}$ (**3**),[5] $[B_3W_{39}O_{132}]^{21-}$ (**4**),[6] and $[NaSb_9W_{21}O_{86}]^{17-}$ (**5**)[7] illustrate some of the possibilities and complications.

Polyoxometalate Chemistry for Nano-Composite Design
Edited by Yamase and Pope, Kluwer Academic/Plenum Publishers, 2002

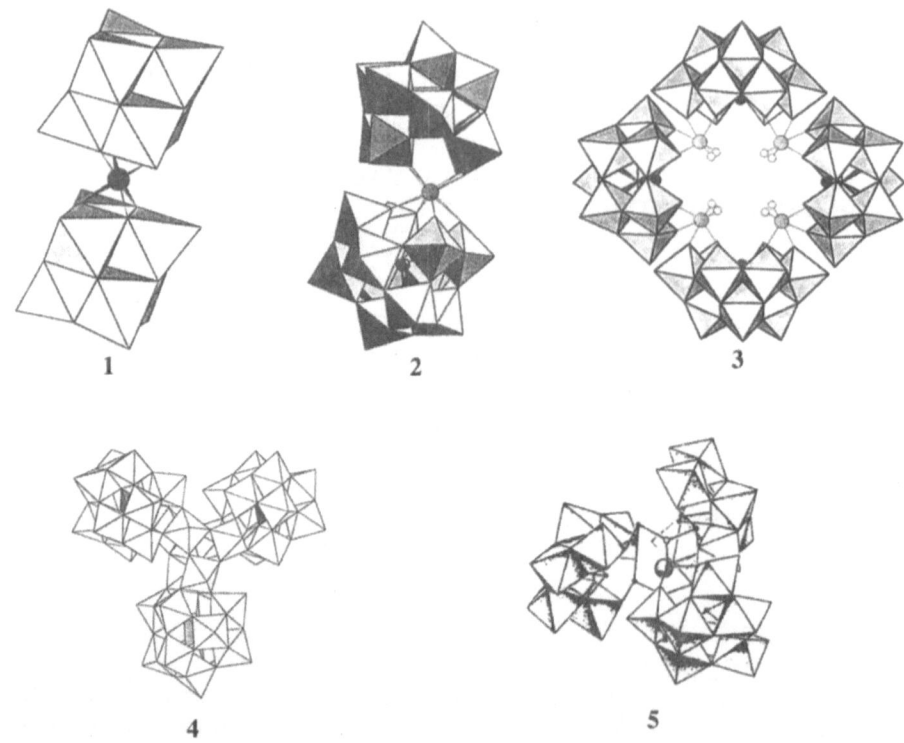

In **1**, **2**, **3** and **4** the building blocks are the independently stable and isolable anions $[Nb_6O_{19}]^{8-}$, $[XW_{11}O_{39}]^{n-}$, $[AsW_9O_{33}]^{9-}$, and $[BW_{11}O_{39}]^{9-}$ respectively. The linkers in **1** and **2** are six-coordinate Mn^{4+} and eight-coordinate Ln^{3+} cations, i.e. **1** and **2** can simply be egarded as oordination complexes, and can be synthesized by direct reaction of linker with building lock. The synthetic pathway for anions **3-5** becomes less clear-cut, for the linkers" must now be described as $\{cis\text{-}WO_2^{2+}\}$, $\{W_6O_{15}^{6+}\}$, and $\{Sb_6O_2^{14+}\}$ respectively. The "building block" in **5**, $\{SbW_7O_{28}^{11-}\}$, is an unknown lacunary derivative of a hypothetical[†] anion, $\beta\text{-}SbW_9O_{33}^{9-}$. Although **5** has been prepared starting with Sb^{3+} and the known α-isomer of $[SbW_9O_{33}]^{9-}$, (and an arsenate analog, $[Sb_6As_3W_{21}O_{86}]$ can be prepared using $\alpha\text{-}[AsW_9O_{33}]^{9-}$) any further insights into the mechanism of formation of **5** are lacking.

In the present paper we report some of our initial explorations of possible "rational" routes to composite polyoxometalates. Emphasis is placed on species that are hydrolytically stable in aqueous solution.

LANTHANIDE AND ACTINIDE CATION LINKERS

Mixed Ligand Peacock-Weakley Anions

Anions of type **2** above were first synthesized by Peacock and Weakley [8] in 1971, and the first reports of crystal structures of such (1:2) complexes $([U^{IV}(GeW_{11}O_{39})_2]^{12-}$ and

---

† This lacunary structure is anticipated to be metastable on the basis of the so-called Lipscomb criterion [W.N.Lipscomb, Paratungstate ion, *Inorg.Chem.* 4:132 (1965)]

$[Ce^{IV}(\alpha_2\text{-}P_2W_{17}O_{61})_2]^{16-}$) appeared about ten years later.[4,9] Recent investigations[10] have confirmed the earlier structures and provide more detailed metrical information. The intermediate 1:1 complexes, e.g. $[Ce(H_2O)_x(SiW_{11}O_{39})]^{7-}$, which have been characterized in solution (electrochemistry, NMR spectroscopy, luminescence lifetime measurements, etc),[11] often associate into "dimeric" or polymeric assemblies upon crystallization (Figure 1).[12]

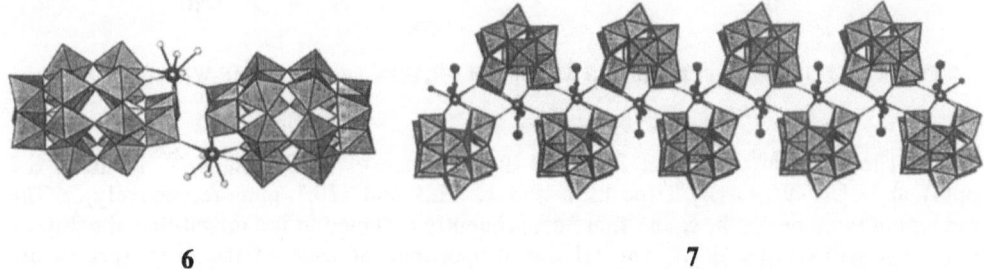

6   **7**

Figure 1.   Structures of $[Ce_2(H_2O)_8(\alpha_2\text{-}P_2W_{17}O_{61})_2]^{14-}$ (**6**) and $\{[Ce(H_2O)_3(PW_{11}O_{39})]^{5-}\}_x$ (**7**)

There are two common structural types of ligand building blocks observed in Peacock-Weakley anions, exemplified by $\alpha\text{-}[PW_{11}O_{39}]^{7-}$ and $\alpha_2\text{-}[P_2W_{17}O_{61}]^{10-}$. Both of these are stable in aqueous solution and this stability allows the straightforward determination of conditional formation constants, β_1 and β_2 for the 1:1 and 1:2 complexes respectively (see Table 1 for examples)

Table 1.   Conditional Formation Constants for Complexes of Ce(III) and Ce(IV).[13]

Ligand	Cation	$\log \beta_1$	$\log \beta_2$
$\alpha\text{-}[PW_{11}O_{39}]^{7-}$	Ce(III)	8.7	15.4
	Ce(IV)	22.9	33.5
$\alpha\text{-}[SiW_{11}O_{39}]^{8-}$	Ce(III)	9.4	15.6
	Ce(IV)	24.6	35
$\alpha_2\text{-}[P_2W_{17}O_{61}]^{10-}$	Ce(III)	8.8	14.8
	Ce(IV)	23.1	33.6

Some data[13] have been reported for complexes of the metastable α_1 isomer of $[P_2W_{17}O_{61}]^{10-}$ but these may be ambiguous in view of the facile $\alpha_1 \rightarrow \alpha_2$ isomerization in aqueous solution. Although 1:1 complexes of the "1 isomer with trivalent lanthanide cations have been convincingly characterized,[11(b)] isolated salts of presumed 1:2 complexes sometimes prove to contain both α_1 and α_2 ligands, e.g. $[Ce(\alpha_1\text{-}P_2W_{17}O_{61})(\alpha_2\text{-}P_2W_{17}O_{61})]^{17-}$ and $[U(\alpha_1\text{-}P_2W_{17}O_{61})(\alpha_2\text{-}P_2W_{17}O_{61})]^{16-}$.[10(b),14]

In an attempt to deliberately introduce different ligand building blocks into the same species we examined the interaction of U^{4+} with mixtures of the *stable* anions $\alpha\text{-}[PW_{11}O_{39}]^{7-}$ and $\alpha_2\text{-}[P_2W_{17}O_{61}]^{10-}$, see Figure 2.

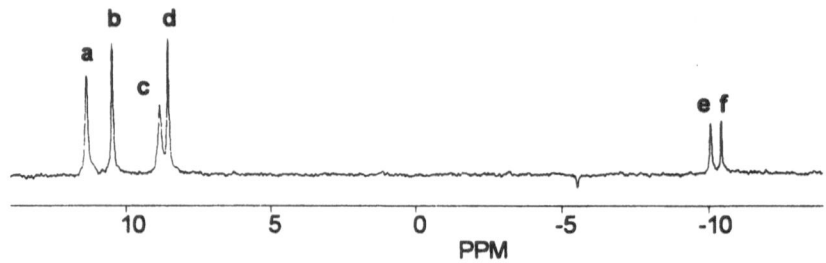

Figure 2.   P-31 NMR spectrum of a 1:1:1 mixture of U^{4+}, $[PW_{11}O_{39}]^{7-}$, and α_2-$[P_2W_{17}O_{61}]^{10-}$ at pH 5.

The spectrum in Figure 2 reveals the presence of $[U(PW_{11}O_{39})_2]^{10-}$ (peak **d**, 8.5 ppm) and $[U(P_2W_{17}O_{61})_2]^{16-}$ (peaks **a** and **e**, 11.3 and -10.1 ppm respectively). The remaining three peaks, **b**, **c**, and **f**, are consequently assigned to the mixed-ligand complex $[U(PW_{11}O_{39})(P_2W_{17}O_{61})]^{13-}$. The relative proportions of each of the three species are affected by the solution pH. For example, the ratio of $[U(PW_{11}O_{39})_2]^{10-}$: $[U(P_2W_{17}O_{61})_2]^{16-}$: $[U(PW_{11}O_{39})(P_2W_{17}O_{61})]^{13-}$ is *ca* 17:27:56 at pH 2.7 and 15:40:45 at pH 5.2

Pentatungstate Building Blocks

The decatungstolanthanide complexes $[LnW_{10}O_{36}]^{n-}$ or $[Ln(W_5O_{18})_2]^{n-}$, **8**, also first

8

characterized by Weakley [15] are clearly structurally analogous to those described above. In this case the ligand building blocks are lacunary versions of the hexatungstate anion $[W_6O_{19}]^{2-}$. However, unlike α-$[PW_{11}O_{39}]^{7-}$, α_2-$[P_2W_{17}O_{61}]^{10-}$, and related anions, $[W_5O_{18}]^{6-}$ has never been detected as a stable species in either aqueous or nonaqueous solution, or in the solid state‡. Solution chemistry of the hexatungstate anion is limited to non-hydrolytic solvents.

In addition to the several examples of **8** with different lanthanide and actinide central atoms, there exists a handful of other polyoxometalate structures that incorporate $\{W_5O_{18}\}$ groups.[1(b),16] Without exception all of these complexes have been synthesized in one-pot processes starting with WO_4^{2-} as the source of tungsten. As shown in Figures 3 and 4, appropriate mixtures of $[PW_{11}O_{39}]^{7-}$ (δ_P -10.5) and $[Ce(W_5O_{18})]^{9-}$ yield solutions containing the composite anion $[Ce(PW_{11}O_{39})(W_5O_{18})]^{10-}$ (δ_P -23.6).[17]

---

‡ The decatungstate anion, $[W_{10}O_{34}]^{4-}$, of D_{4h} symmetry, can be viewed as a condensed dimer of $\{W_5O_{18}\}$ units

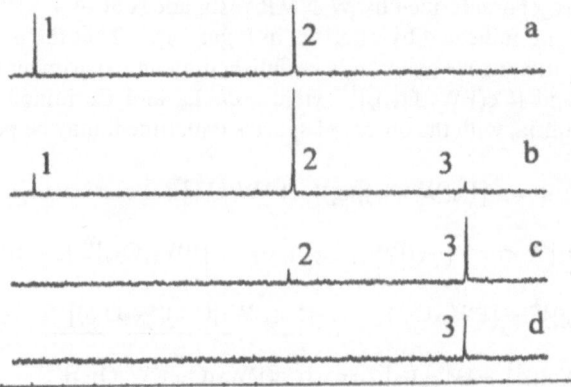

Figure 3. P-31 NMR spectra of mixtures of $[PW_{11}O_{39}]^{7-}$ (**A**) and $[Ce(W_5O_{18})_2]^{9-}$ (**B**). Mole ratio **A:B** = 3:1 (**a**), 2:1 (**b**), 1:2 (**c**), 1:3 (**d**). Signals shown correspond to $[PW_{11}O_{39}]^{7-}$ (**1**), $[Ce(PW_{11}O_{39})_2]^{11-}$ (**2**), and $[Ce(W_5O_{18})(PW_{11}O_{39})]^{10-}$ (**3**)

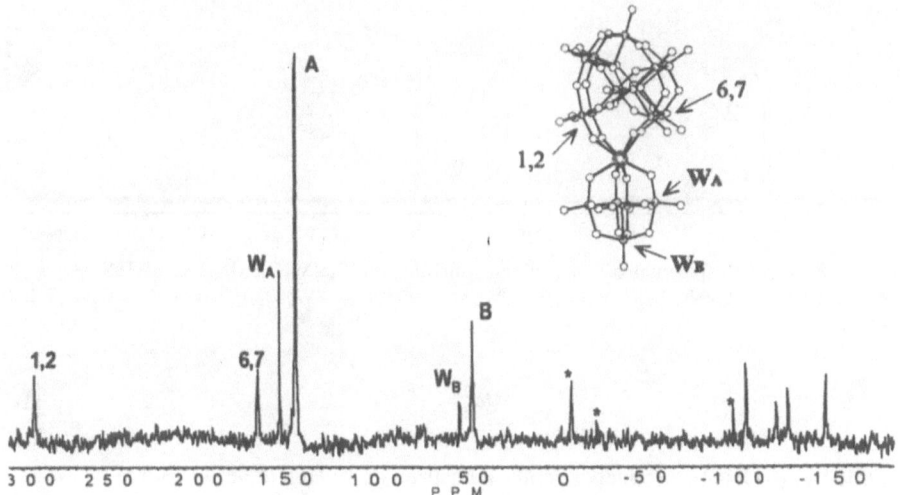

Figure4. W-183 NMR spectrum of solution shown in Figure 2(d). A and B are resonances of excess $[Ce(W_5O_{18})]^{9-}$; Other *labeled* resonances are from tungsten nuclei adjacent to the paramagnetic cerium(III) center in the mixed anion shown. Asterisks denote resonances from species resulting from the displaced $\{W_5O_{18}\}$ unit.

The formation of the mixed anion shown in Figures 3 and 4 can be described as a ligand displacement reaction,

$$[PW_{11}O_{39}]^{7-} + [Ce(W_5O_{18})_2]^{9-} ==> [Ce(PW_{11}O_{39})(W_5O_{18})]^{10-} + ``\{W_5O_{18}{}^{6-}\}"\qquad(1)$$

The displaced pentatungstate moiety ultimately is converted into one or more so-far unidentified species, characterized by W-NMR resonances at -4.4, -18.0, -92.9, and -179.8 ppm (three of these are indicated by asterisks in Figure 4).   That the pentatungstate ligand has kinetic stability in aqueous solution is established by an experiment in which a mixture of $[La(W_5O_{18})_2]^{9-}$ and $[Ce(PW_{11}O_{39})_2]^{11-}$ yield *both* La and Ce mixed anions (Figure 5). The following equilibria, with the observed species underlined, may be postulated.

$$[La(W_5O_{18})_2]^{9-} \Longleftrightarrow [La(W_5O_{18})(aq)]^{3-} + [W_5O_{18}]^{6-} \tag{2}$$

$$\underline{[Ce(PW_{11}O_{39})_2]^{11-}} \Longleftrightarrow [Ce(PW_{11}O_{39})(aq)]^{4-} + [PW_{11}O_{39}]^{7-} \quad K \sim 10^{-6.7} \text{ (ref 13)} \tag{3}$$

$$[La(W_5O_{18})(aq)]^{3-} + [PW_{11}O_{39}]^{7-} \Longrightarrow \underline{[La(PW_{11}O_{39})(W_5O_{18})]^{10-}} \tag{4}$$

$$[Ce(PW_{11}O_{39})(aq)]^{4-} + [W_5O_{18}]^{6-} \Longrightarrow \underline{[Ce(PW_{11}O_{39})(W_5O_{18})]^{10-}} \tag{5}$$

$$[La(PW_{11}O_{39})(W_5O_{18})]^{10-} + [PW_{11}O_{39}]^{7-} \Longrightarrow \underline{[La(PW_{11}O_{39})_2]^{11-}} \tag{6}$$

$$[Ce(PW_{11}O_{39})(W_5O_{18})]^{10-} + [W_5O_{18}]^{6-} \Longrightarrow [Ce(W_5O_{18})_2]^{9-} + [PW_{11}O_{39}]^{7-} \tag{7}$$

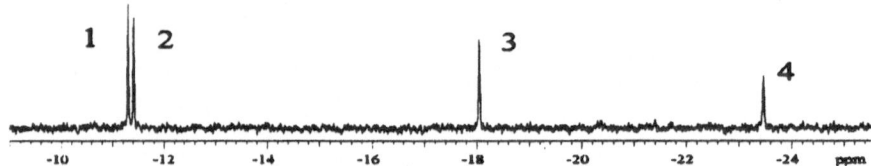

Figure 5. P-31 NMR spectrum of an equimolar mixture of $[La(W_5O_{18})_2]^{9-}$ and $[Ce(PW_{11}O_{39})_2]^{11-}$ **(3)** showing formation of $[La(W_5O_{18})(PW_{11}O_{39})]^{10-}$ **(2)** and $[Ce(W_5O_{18})(PW_{11}O_{39})]^{10-}$ **(4)**.   Resonance **1** is from $[La(PW_{11}O_{39})_2]^{11-}$.

Of these, (3) has been independently measured and (2) is the only source of $[W_5O_{18}]^{6-}$. The presence of unreacted $[La(W_5O_{18})_2]^{9-}$ and the formation of $[Ce(W_5O_{18})_2]^{9-}$ in the final solution is not yet established.

The use of lanthanide cations as linkers in anions like **2** and **6**, can be extended to the synthesis of even larger entities (**7**).   We have recently shown that the building blocks themselves may be composite polyanions.   Thus, reaction of **3** with lanthanide cations leads to "dimeric" (**9**) and "polymeric" (**10**) assemblies.

$$[As_4W_{40}O_{140}]^{28-} \text{ (3)} + 3 \text{ Ln}^{3+} \Longrightarrow [\{Ln(Ln_2OH)As_4W_{40}O_{140}\}_2]^{40-} \text{ (9)} \tag{8}$$

$$[As_4W_{40}O_{140}]^{28-} + 4 \text{ Ln}^{3+} \Longrightarrow [\{(Ln_2OH)_2As_4W_{40}O_{140}\}_x]^{18x-} \text{ (10)} \tag{9}$$

Equations (8) , Ln = Ce, Nd, Sm, Eu, Gd, and (9), Ln = La, Ce, Gd, do not tell the complete story.   Assembly of these large polyoxometalates requires high salt concentrations, typically 1-4 M NaCl, and for **10**, the addition of the stoichiometric amount of K^+ or Ba^{2+} to occupy the central cavity of the $\{As_4W_{40}O_{140}\}$ group.   In aqueous solution, NMR data demonstrate that the dimeric and polymeric structures undergo partial or complete dissociation into the W_{40} components.[5(b)]

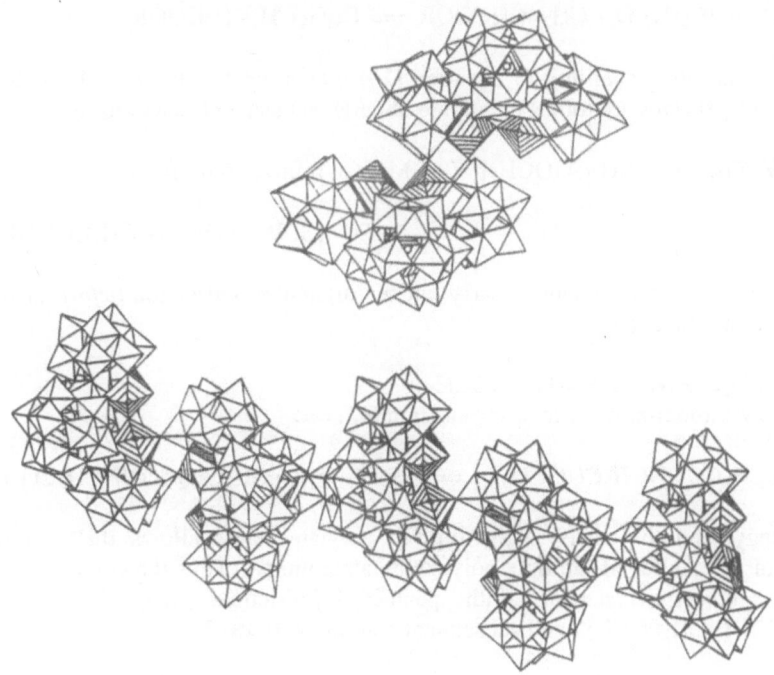

9 (upper) **10** (lower)

ORGANIC LINKERS

The interaction of potential bridging ligands such as pyrazine and 4,4N-bipyridine with transition-metal-substituted Keggin anions, e.g. $[SiW_{11}O_{39}Co(H_2O)]^{8-}$ has led to the formation of linked Keggin species (so-called "dumbbell" complexes).[18] These have for the most part been identified in solution by NMR spectroscopy, and are of course subject to dissociation via ligand exchange processes. The robustness of the dumbbell species will depend upon the nature and oxidation state of the substituent transition metal cation, but little has been done to explore further possibilities in this area. Some surprises may be in store: for example cobalt(III) proves to be quite labile regarding terminal ligand exchange when embedded in a polytungstate matrix.[18(c)]

We have chosen to examine organogermanium and organotin derivatives of polytungstates since the Ge-C and Sn-C bonds are extremely robust and hydrolysis-resistant. Moreover we have shown earlier that RMCl$_3$ react cleanly with lacunary anions to give the desired substituted species in high yield.[19] All of the following tungstate derivatives are stable in neutral or acidic aqueous solution.

$$C_6H_5GeCl_3 + [SiW_{11}O_{39}]^{8-} ==> [SiW_{11}O_{39}GeC_6H_5]^{5-} + 3\ Cl^- \qquad (10)$$

$$3\ n\text{-}C_4H_9SnCl_3 + [P_2W_{15}O_{56}]^{12-} + 3\ H_2O ==>$$

$$[P_2W_{15}O_{59}\{SnC_4H_9\}_3]^{9-} + 6\ H^+ + 9\ Cl^- \qquad (11)$$

Functionalization of the organic groups is readily achieved by standard procedures using the appropriate terminal alkene and trichlorogermane, or -stannane, e.g.

$$HGeCl_3 \cdot 2(C_2H_5)_2O + CH_2=CHCOOR ==> Cl_3GeCH_2CH_2COOR \qquad (12)$$

Subsequent reaction according to equation (10) yields the functionalized polytungstate.[20] Although such species are robust enough to undergo subsequent reactions, e.g.

$$[SiW_{11}O_{39}SnCH_2CH_2COOCH_3]^{5-} \text{ (0.5 M HCl, reflux, 40 min)} ==>$$

$$[SiW_{11}O_{39}SnCH_2CH_2COOH]^{5-}$$

it is generally more convenient to carry out the organic modification *before* incorporation into the polytungstate, e.g.

$Cl_3GeCH_2CH_2COOH + NH_2CH_2COOCH_3$
(dicyclohexylcarbodiimide/N-hydroxysuccinimide) ==>

$Cl_3GeCH_2CH_2CONHCH_2COOCH_3 \cdots ==> [SiW_{11}O_{39}GeCH_2CH_2CONHCH_2COOCH_3]^{5-}$

This strategy, and the ready availability of acrylate esters allows the straightforward synthesis of assemblies of multiple polyoxometalate units. Thus the composite star anion **11** was synthesized starting with pentaerythritol-tetraacrylate, the formation of $C(CH_2OCOCH_2CH_2GeCl_3)_4$, and subsequent reaction with $[SiW_{11}O_{39}]^{8-}$.

11

CONCLUSIONS

The directed synthesis of large inorganic polyoxometalates with specific structural and stereochemical features remains an elusive but not unattainable goal. We have shown that opportunities exist for straightforward assembly of stable or metastable polyoxometalate subunits using principles of both coordination chemistry and organic synthesis. Among the many advantages offered by such synthetic approaches is the generation of chiral structures for molecular recognition and catalytic selectivity that we are currently pursuing.

ACKNOWLEDGMENT

We thank the National Science Foundation (CHE-9727417), the Department of Energy (DE-FG07-96ER14695), and Mitsubishi Chemical Co. for recent and current research support.

REFERENCES

1. (a) A. Müller, E. Krickemeyer, J. Meyer, H. Bögge, F. Peters, W. Plass, E. Diemann, S. Dillinger, F. Nonnenbruch, M. Randerath, and C. Menke $[Mo_{154}(NO)_{14}O_{420}(OH)_{28}(H_2O)_{70}]^{(25\pm5)-}$: a water-soluble big wheel with more than 700 atoms and a relative molecular mass of about 24 000, *Angew. Chem. Int.Ed. Engl.* 34:2122 (1995); (b) K. Wassermann, M.H. Dickman, and M.T. Pope, Self-assembly of supramolecular polyoxometalates. The compact, water-soluble heteropolytungstate anion $[As^{III}_{12}Ce^{III}_{16}(H_2O)_{36}W_{148}O_{524}]^{76-}$, *Angew. Chem. Int. Ed. Engl.* 36:1445 (1997); (c) A. Müller, E. Krickemeyer, H. Bögge, M. Schmidtmann, and F. Peters, Organizational forms of matter: an inorganic superfullerene and keplerate based on molybdenum oxide, *Angew. Chem., Int. Ed. Engl.* 37:3360 (1998).
2. (a) A. Müller, F. Peters, M.T. Pope, and D. Gatteschi, Polyoxometalates: very large clusters-nanoscale magnets, *Chem. Rev.* 98:239 (1998); (b) J. M Clemente-Juan and E. Coronado, Magnetic clusters from polyoxometalate complexes, *Coord. Chem. Rev.* 193-195:361 (1999)
3. C. M. Flynn, Jr. and G. D. Stuckey, Crystal structure of sodium 12-niobomanganate(IV), $Na_{12}[MnNb_{12}O_{38}].50H_2O$, *Inorg. Chem.* 8:335 (1969)
4. C. M. Tourné, G. Tourné, and M. C. Brianso, Cesium uranium germanium tungstate, $Cs_{12}[U(GeW_{11}O_{39})_2]\cdot13-14 H_2O$, *Acta Crystallogr., Sect B* 36:2012 (1980)
5. (a) F. Robert, M. Leyrie, A Tézé, G. Hervé, and Y. Jeannin, Crystal structure of ammonium dicobalto(II)-40-tungstotetraarsenate(III). Allosteric effects in the ligand, *Inorg. Chem.* 19:1746 (1980); (b) K. Wassermann and M.T. Pope, Large cluster formation through multiple substitution with lanthanide cations (La, Ce, Sm, Eu and Gd) of the polyoxoanion $[B-\alpha-(AsO_3W_9O_{30})_4(WO_2)_4]^{28-}$. Synthesis and structural characterization, *Inorg. Chem.* 40:2763 (2001)
6. A. Tézé, M. Michelon, and G. Hervé, Syntheses and structures of the tungstoborate anions, *Inorg. Chem.* 36:505 (1997)
7. (a) J. Fischer, L. Ricard, and R. Weiss, The structure of the heteropolytungstate $(NH_4)_{17}Na[NaSb_9W_{21}O_{86}]$ $\cdot14 H_2O$. An inorganic cryptate, *J. Am. Chem. Soc.* 98:3050 (1976); (b) M. Michelon, G. Hervé, M. Leyrie, Synthesis and chemical behavior of the inorganic cryptates. Ammonium, alkali, and alkaline earth antimony tungstates $[MSb_9W_{21}O_{86}]^{(19-n)-}$ $Mn^+ = Na^+, K^+, NH_4^+, Ca^{2+}, Sr^{2+}$, *Inorg. Nucl. Chem.* 42:1583 (1980)
8. R.D. Peacock and T.J.R. Weakley, Heteropolytungstate complexes of the lanthanide elements. Part 1. Preparation and reactions, *J. Chem. Soc. (A)* 1836 (1971)
9. V.N. Molchanov, L.P Kazanskii, E.A. Torchenkova, and V.I. Simonov, Crystal structure of $K_{16}[Ce(P_2W_{17}O_{61})_2]\cdot n H_2O$ (n = 50), *Sov. Phys. Crystallog. (Engl. Transl.)* 24:96 (1979)
10. (a) Q.H. Luo, R.C. Howell, M. Dankova, J. Bartis, C.W. Williams, W. DeW. Horrocks, Jr., V.G. Young, Jr., A.L. Rheingold, L.C. Francesconi, and M.R. Antonio, Coordination of rare-earth elements in complexes with monovacant Wells-Dawson polyoxoanions, *Inorg. Chem.* 40:1894 (2001); (b) A. Ostuni, R.E. Bachman, A.K. Jameson, and M.T. Pope, Diastereomers of the Peacock-Weakley heteropolytungstates, $[Mn^+(\alpha_m-P_2W_{17}O_{61})_2]^{(20-n)-}$ $(M = An^{IV}, Ln^{III}; m=1,2)$. Syn- and anticonformations of the polytungstate ligands. Structures of the $\alpha_1\alpha_1$, $\alpha_1\alpha_2$ and $\alpha_2\alpha_2$ complexes, In preparation
11. (a) A.B. Yusov and A.M. Fedoseev, Effect of water molecules on photoluminescence of curium(III) and rare-earth metals(III) in complexes with polytungstate ligands, *Zh. Prikl. Spektrosk.* 49:929 (1988); *Chem. Abstr.* 111:67023p (1989); (b) J. Bartis, M. Dankova, J. J. Lessmann, Q.H. Luo, W. DeW. Horrocks, Jr., and L.C. Francesconi, Lanthanide complexes of the α-1 isomer of the $[P_2W_{17}O_{61}]^{10-}$ heteropolytungstate: Preparation, stoichiometry, and structural characterization by ^{183}W and ^{31}P NMR spectroscopy and europium(III) luminescence spectroscopy, *Inorg. Chem.* 38:1042 (1999)
12. (a) M. Sadakane, M.H. Dickman, and M.T. Pope, Controlled assembly of polyoxometalate chains from lacunary building blocks and lanthanide-cation linkers, *Angew. Chem. Int. Ed. Engl.* 39:2914 (2000); (b) M. Sadakane, M.H. Dickman, and M.T. Pope, Chiral polyoxotungstates. 1. Stereoselective interaction of amino acids with enantiomers of $[Ce^{III}(\alpha_1-P_2W_{17}O_{61})(H_2O)_x]^{7-}$. The structure of DL-$[Ce_2(H_2O)_8(P_2W_{17}O_{61})_2]^{14-}$, *Inorg. Chem.* 40:2715 (2001)

13. J.P. Ciabrini, and R. Contant, Mixed heteropolyanions. Synthesis and formation constants of cerium(III) and cerium(IV) complexes with lacunary tungstophosphates,, *J. Chem. Res. (S)*, 391 (1993); *(M)*, 2720 (1993)

14. A. Ostuni, Lanthanide and actinide complexes of the lacunary Wells-Dawson anions: synthesis, structure and spectroscopy, M.S. Thesis, Georgetown University, 1998

15. (a) J. Iball, J.N. Low, and T.J.R. Weakley, Heteropolytungstate complexes of the lanthanoid elements. III. Crystal structure of sodium decatungstocerate(IV) tricontahydrate, , *J. Chem. Soc., Dalton Trans.* 2021 (1974);   (b) T. Ozeki and T. Yamase, Effect of lanthanide contraction on the structures of the decatungstolanthanoate anions in $K_3Na_4H_2[LnW_{10}O_{36}]\cdot nH_2O$ (Ln = Pr, Nd, Sm, Gd, Tb, Dy) crystals, *Acta Crystallogr.,Sect. B* 50:128 (1994)

16. (a) T. Yamase, H. Naruke, and Y. Sasaki, Crystallographic characterization of the polyoxotungstate $[Eu_3(H_2O)_3(SbW_9O_{33})(W_5O_{18})_3]^{18-}$ and energy transfer in its crystalline lattices,   *J. Chem. Soc., Dalton Trans.* 1687 (1990);   (b) H. Naruke, and T. Yamase, A novel-type mixed-ligand polyoxotungstolanthanoate, $[Ln(BW_{11}O_{39})(W_5O_{18})]^{12-}$ (Ln = Ce^{3+} and Eu^{3+}),   *Bull. Chem. Soc. Jpn.* 73:375 (2000)

17. N. Belai, M. Sadakane, and M.T. Pope, Formation of unsymmetrical polyoxotungstates via transfer of polyoxometalate building blocks.   NMR evidence supports the kinetic stability of the pentatungstate anion, $[W_5O_{18}]^{6-}$, in aqueous solution,   *J. Am. Chem. Soc.* 123:2087 (2001)

18. (a) L.C.W. Baker and J.S. Figgis, A new fundamental type of inorganic complex: hybrid between heteropoly and conventional coordination complexes.   Possibilities for geometrical isomerisms in 11-, 12-, 17-, and 18-heteropoly derivatives, *J. Am. Chem. Soc.* 92:3794 (1970);   (b) J. Park, M. Ko, and H. So, NMR spectra of 4,4'-bipyridyl, pyrazine, and ethylenediamine coordinated to undecatungstocobalto(III)silicate and -borate anions. Identification of 1:1 and dumbbell-shaped 1:2 complexes, *Bull. Korean Chem. Soc.* 14:759 (1993);   (c) J.L. Samonte and M.T. Pope, Derivatization of polyoxotungstates in aqueous solution. Exploration of the kinetic stability of cobalt(II)- and cobalt(III) derivatives of lacunary anions with pyridine and pyridine-type ligands,   *Can. J. Chem..* In press

19. (a) G.S. Chorghade and M.T. Pope, Heteropolyanions as nucleophiles. I. Synthesis, characterization and and reactions of Keggin- and Dawson-Type tungstostannates(II),   *J. Am. Chem. Soc.* 109:5234 (1987);   (b) F.   Xin and M.T. Pope, Polyoxometalate derivatives with multiple organic groups. 1. Synthesis and structures of tris(organotin) β-Keggin and α-Dawson tungstophosphates, *Organometallics*, 13:4881 (1994);   (c) F. Xin, M.T. Pope, G.J. Long, and U. Russo, Polyoxometalate derivatives with multiple organic groups. 2. Synthesis and structures of tris(organotin) α,β-Keggin tungstosilicates,   *Inorg. Chem.*, 35:1207 (1996);   (d) F. Xin and M.T. Pope, Polyoxometalate derivatives with multiple organic groups. 3. Synthesis and structure of bis(phenyltin)bis(decatungstosilicate), $[(PhSnOH_2)_2(\gamma-SiW_{10}O_{36})_2]^{10-}$,   *Inorg.Chem.*, 35:5693 (1996);   (e) G. Sazani, M.H. Dickman, and M.T. Pope, Organotin derivatives of α-$[X^{III}W_9O_{33}]^{9-}$ (X = As, Sb) heteropolytungstates.   Solution and solid state characterization of $[\{(C_6H_5Sn)_2O\}_2H(AsW_9O_{33})_2]^{9-}$ and $[(C_6H_5Sn)_3Na_3(SbW_9O_{33})_2]^{6-}$, *Inorg. Chem.* 39:939 (2000)

20. G. Sazani, Synthesis and Characterization of Novel Functionalized Organotin and Organogermanium Derivatives of Heteropolyanions, Ph.D. Thesis, Georgetown University, 1999

COMPOSITE MATERIALS DERIVED
FROM OXOVANADIUM SULFATES

M. Ishaque Khan,*[,1] Sabri Cevik,[1] and Robert J. Doedens[2]

[1] Department of Biological, Chemical, and Physical Sciences
Illinois Institute of Technology Chicago, IL 60616
[2] Department of Chemistry, University of California, Irvine, CA 92697

INTRODUCTION

Vanadium-oxide compounds, including molecular systems and solids, are of current interest. Consequently, the vanadium oxide chemistry, in general, and of vanadium oxide clusters (polyoxovanadates) in particular, has seen exciting development in recent years[1]. A large number of clusters of different shapes and sizes and solids containing vanadium atoms in a variety of oxidation states and geometries have been prepared and characterized. Oxovanadate clusters containing up to hundreds of atoms, with dimensions up to several nanometers are known[2]. In principle, they offer suitable building units for the fabrication of nanostructured solids whose properties could possibly be correlated to the constituent clusters. The potential of this method has recently been demonstrated by us[3-4].

Polyoxometalates and their derivatives can be amalgamated with a variety of ligands. An impressive array of vanadium-oxide-phosphate based materials with new electronic and structural properties has been prepared by combining the tetrahedral {PO_4} ligand with oxovanadate moieties[5-8]. Some of these materials have open-framework containing very large cavities and channels similar to those observed in conventional aluminosilicate-based zeolites[9].

As compared to the $V/O/PO_4$, however, the chemistry of $V/O/SO_4$ system remains unexplored. The studies in this area have either been confined to few molecular compounds or concerned with the possible catalytic roles of VO/SO_4 compounds in the industrial production of sulfuric acid[10-15]. This is despite the recent molecular modeling studies which predicts the potential of the tetrahedral {SO_4} group as suitable building block for the construction of new metal-sulfate based materials[16]. Since vanadium exhibits rich coordination chemistry bonding to a variety of organic ligands[17], their incorporation in the

Polyoxometalate Chemistry for Nano-Composite Design
Edited by Yamase and Pope, Kluwer Academic/Plenum Publishers, 2002

V/O/SO$_4$ system offers opportunity for making new inorganic-organic hybrid (composite) materials. The suitable combination of oxovanadate sulfate fragments and organic ligands may, in principle, pave the way for new composite-, including nanocomposite materials. Such composite materials may exhibit new and interesting properties not found in purely organic or inorganic phases.

We have been exploring the synthesis of composite materials by employing vanadylsulfate (VO/SO4)-based motifs in combination with simple organic ligands. This article describes some of the progress we made in this direction. The discussion is limited to few representative examples from our lab. A more detailed review of the work will appear in a future publication.

SYNTHESIS

Molecular Precursors

The first step in our effort to prepare the desired materials was to identify suitable starting materials. Simple vanadium compounds and appropriate sulfate ion sources may be considered as the precursors for providing building blocks suitable for constructing new V/O/SO$_4^{2-}$ based solids. There are, however, relatively few vanadium(IV) molecular precursors, such as VCl$_4$, VO(acac)$_2$ and VOSO$_4$·xH$_2$O, available for this purpose. The preparation of the commonly available VOSO$_4$·xH$_2$O results in the formation of at least three distinct phases containing varying numbers of water molecule(s)[15d,e]. In an attempt to provide reproducible and convenient molecular precursors for this work, we have prepared two useful complexes- [V^{IV}OSO$_4$(H$_2$O)$_4$]·SO$_4$·[H$_2$N(C$_2$H$_4$)$_2$NH$_2$][19] and [V^{IV}OSO$_4$(H$_2$O)$_4$]·SO$_4$·[HN(C$_2$H$_4$)$_3$NH]. These molecular compounds, which we have fully characterized by single crystal x-ray structure analysis, TGA, manganometric titration and spectroscopic studies, are reproducible and readily prepared in decent yields by the following hydrothermal synthetic methods.

Synthesis of [V^{IV}OSO$_4$(H$_2$O)$_4$]·SO$_4$·[H$_2$N(C$_2$H$_4$)$_2$NH$_2$]. Method 1: A mixture of VO(acac)$_2$, Na$_2$SO$_4$, piperazine and dilute H$_2$SO$_4$ (2 M, 5 ml) in the molar ratio 3:10:6:10 was placed in a 23 ml Parr Teflon-lined autoclave which was subsequently heated for 48 hours in a programmable electric furnace maintained at 160°C. After cooling the autoclave at room temperature for 4 hours, the resultant solution was filtered and blue filtrate was kept in a closed glass tube that was allowed to stay at room temperature. The rectangular plate-shaped blue crystals, which separated over a period of 10-14 days, were filtered from mother liquor and dried in the air.

Method 2: A mixture of V$_2$O$_5$, V metal (-325 mesh), Na$_2$SO$_4$, piperazine, (C$_2$H$_5$)$_3$NHCl, CH$_3$NH$_3$Cl and dilute H$_2$SO$_4$ (1 M, 5 ml) in the molar ratio 5:3:5:5:3:3:5 was placed in a 23 ml Parr Teflon-lined autoclave which was subsequently heated for 72 hours in a programmable electric furnace maintained at 180°C. After cooling the autoclave at room temperature for 4 hours, the resultant solution was filtered and blue filtrate was kept in a closed glass tube that was allowed to stay at room at temperature. The rectangular plate-shaped blue crystals, which separated over a period of 10-14 days, were filtered from the mother liquor and dried in the air.

Synthesis of [V^{IV}OSO$_4$(H$_2$O)$_4$]·SO$_4$·[HN(C$_2$H$_4$)$_3$NH]: Method 1: A mixture of VOSO$_4$·xH$_2$O, NaSO$_4$, DABCO, H$_2$SO$_4$ (1 M 5 ml) and CH$_3$SO$_3$H in the molar ratio

4:5:6:5:10 was placed in a 23 ml Parr Teflon-lined autoclave which was subsequently heated for 48 hours in a programmable electric furnace maintained at 160°C. After cooling the autoclave at room temperature for 4 hours, the resultant solution was filtered and blue filtrate was kept in a closed glass tube that was allowed to stay at room temperature. The rectangular plate-shaped blue crystals, which separated over a period of 30 days, were filtered from the mother liquor and dried in the air.

Method 2: A mixture of $VOSO_4 \cdot xH_2O$, $NaSO_4$, DABCO, $(C_2H_5)_3NHCl$, CH_3NH_3Cl and dilute H_2SO_4 (1 M 5 ml) the molar ratio 4:5:5:3:3:5 was placed in a 23 ml Parr Teflon-lined autoclave which was subsequently heated for 48 hours in a programmable electric furnace maintained at 180°C. After cooling the autoclave at room temperature for 4 hours, the resultant solution was filtered and blue filtrate was kept in a closed glass tube that was allowed to stay at room temperature. The rectangular plate-shaped blue crystals, which separated over a period of 20 days, were filtered from the mother liquor and dried in air.

Both of these compounds are reproducible and easily prepared in decent yields (70% for $[V^{IV}OSO_4(H_2O)_4] \cdot SO_4 \cdot [H_2N(C_2H_4)_2NH_2]$[19], and 20% for $[V^{IV}OSO_4 (H_2O)_4] \cdot SO_4 \cdot [HN(C_2H_4)_3NH])$ (based on Vanadium) by the hydrothermal reactions described above. Although CH_3SO_3H, $(C_2H_5)_3NHCl$ or CH_3NH_3Cl do not appear in the characterized final products, their presence is needed in the reaction mixtures probably to maintain the appropriate pH and/or ionic strength of the reaction medium to obtain better single crystalline forms of the complexes. The compounds could also be prepared by employing slightly different stoichiometries and reactants. Crystals of $[V^{IV}OSO_4(H_2O)_4] \cdot SO_4 \cdot [H_2N(C_2H_4)_2 NH_2]$ and $[V^{IV}OSO_4(H_2O)_4] \cdot SO_4 \cdot [HN(C_2H_4)_3NH]$ are stable in air and soluble hot water.

Materials With Extended Structures

Although, there are some fragmentary reports on the oxovanadium sulfate solids containing reduced, oxidized, and mixed-valance vanadium centers[12-15], composite materials containing oxovanadium sulfate moieties incorporating organic ligands have been practically unknown. We have prepared some interesting type of inorganic-organic hybrid materials. These extended structure solids constitute some of the first examples of fully reduced oxovanadium sulfate ($V/O/SO_4^{2-}$) based composites. We will describe them with the help of two examples, $[V^{IV}_2O_2(OH)_2(SO_4)(2,2'-bipyridine)_2]$[18] and $[V^{IV}_2O_2(SO_4)_2(2,2'-bipyridine)_2]$. These compounds were prepared in crystalline form and in high yield by employing the following hydrothermal methods and characterized by a number of physicochemical techniques including infrared spectroscopy, thermogravimetry, elemental analysis, manganometric titration, bond valance sum calculations, and complete single crystal x-ray structure analysis. The molecular compounds $[V^{IV}OSO_4(H_2O)_4] \cdot SO_4 \cdot [H_2N(C_2H_4)_2NH_2]$ and $[V^{IV}OSO_4(H_2O)_4] \cdot SO_4 \cdot [HN(C_2H_4)_3NH]$. described above, are useful precursors for synthesizing the new materials.

Synthesis of $[V^{IV}_2O_2(OH)_2(SO_4)(2,2'-bipyridine)_2]$. A mixture of $[V^{IV}O(H_2O)_4SO_4] \cdot SO_4 \cdot [H_2N(C_2H_4)_2NH_2]$ [76], Na_2SO_4, 2,2'-bipyridine and water in the molar ratio 1.25:5:2:278 were placed in a 23 ml Parr Teflon-lined autoclave which was subsequently heated for 48 hours in a programmable electric furnace maintained at 180°C. After cooling the autoclave at room temperature for 4 hours, brown crystals of $[V^{IV}_2O_2(OH)_2(SO_4)(2,2'-bipyridine)_2]$ were filtered from mother liquor, carefully washed with water, and dried in the air at room temperature. The yield of the product was 55% (based on vanadium).

The compound can also be prepared by the following alternative method: A mixture of VO(acac)$_2$, Na$_2$SO$_4$, 2,2'-bipyridine and water in the molar ratio 3:10:10:277 was placed in a 23 ml Parr Teflon-lined autoclave which was subsequently heated for 48 hours in a programmable electric furnace maintained at 160°C. After cooling the autoclave at room temperature for 4 hours, brown crystals of [V$^{IV}_2$O$_2$(OH)$_2$(SO$_4$)(2,2'-bipyridine)$_2$] [86] were filtered from mother liquor, carefully washed with water, and dried in the air at room temperature. The yield of the product was 70% (based on vanadium).

Synthesis of [V$^{IV}_2$O$_2$(SO$_4$)$_2$(2,2'-bipyridine)$_2$]. A mixture of [VO(H$_2$O)$_4$SO$_4$]·SO$_4$(H$_2$N(C$_2$H$_4$)$_2$NH$_2$] [76], 2,2'-bipyridine, Na$_2$SO$_4$ and dilute sulfuric acid (2 M H$_2$SO$_4$) in the molar ratio 1.25:5:2:10 was placed in a 23 ml Parr Teflon-lined autoclave which was subsequently heated for 48 hours in a programmable electric furnace maintained at 180°C. After cooling the autoclave at room temperature for 4 hours, mono-phasic needle like green crystals of [V$^{IV}_2$O$_2$ (SO$_4$)$_2$(2,2'-bipyridine)$_2$] was filtered from the mother liquor, carefully washed with water, and dried in the air at room temperature. The yield of the product was 50% (based on vanadium).

Although during the course of our on-going work we have discovered some alternative methods, not described here, for preparing [V$^{IV}_2$O$_2$(OH)$_2$(SO$_4$)(2,2'-bipyridine)$_2$] and [V$^{IV}_2$O$_2$(SO$_4$)$_2$(2,2'-bipyridine)$_2$], the synthetic procedures given above provide convenient ways of synthesizing pure compounds in good yields. Crystals of [V$^{IV}_2$O$_2$(OH)$_2$(SO$_4$)(2,2'-bipyridine)$_2$] and [V$^{IV}_2$O$_2$(SO$_4$)$_2$(2,2'-bipyridine)$_2$] are stable in the air and insoluble in common solvents

SPECTROSCOPIC PROPERTIES

The FT-IR spectra of [V^{IV}OSO$_4$(H$_2$O)$_4$]·SO$_4$·[H$_2$N(C$_2$H$_4$)$_2$NH$_2$] and [V^{IV}OSO$_4$(H$_2$O)$_4$]·SO$_4$· [HN(C$_2$H$_4$)$_3$NH] exhibit, in addition to piperazine and DABCO peaks, very strong peaks at 977 cm^{-1} and 969 cm^{-1}, respectively, which is characteristic of v(V=O). There are multiple strong features in the 1000-1250 cm^{-1} region which are attributable to SO$_4^{2-}$ groups. Low energy peaks for v(V-O-S) are present in their characteristic region (below 900 cm^{-1}). The infrared spectra of [V$^{IV}_2$O$_2$(OH)$_2$(SO$_4$)(2,2'-bipyridine)$_2$] and [V$^{IV}_2$O$_2$(SO$_4$)$_2$(2,2'-bipyridine)$_2$] also exhibit intense absorption bands between 950 cm^{-1} and 1000 cm^{-1} due to v(V=O). The v(VO) bands appear at 976 cm^{-1} and 970 cm^{-1} for [V$^{IV}_2$O$_2$(OH)$_2$(SO$_4$)(2,2'-bipyridine)$_2$], and at 979 cm^{-1} for [V$^{IV}_2$O$_2$(SO$_4$)$_2$(2,2'-bipyridine)$_2$]. v(SO$_4$) modes appear at 1007 cm^{-1} and 1022 cm^{-1} for [V$^{IV}_2$O$_2$(OH)$_2$(SO$_4$)(2,2'-bipyridine)$_2$] and [V$^{IV}_2$O$_2$(SO$_4$)$_2$(2,2'-bipyridine)$_2$], respectively.

The multiple strong bands in 1000-1250 cm^{-1} region are attributable to the bridging SO$_4^{2-}$ ligands in the compounds. The strong splitting of the v(SO$_4^{2-}$) modes (1000-1250 cm^{-1} region) in the spectra signals the significant deformations of the tetrahedral geometry of SO$_4^{2-}$, which has been proven in each case by the single crystal structure analyses of [V$^{IV}_2$O$_2$(OH)$_2$(SO$_4$)(2,2'-bipyridine)$_2$][18] and [V$^{IV}_2$O$_2$(SO$_4$)$_2$(2,2'-bipyridine)$_2$]. Peaks at 1241 cm^{-1}, 1214 cm^{-1}, 1176 cm^{-1}, 1159 cm^{-1}, 1102 cm^{-1} and 1058 cm^{-1}, assigned to v_{as}(S-O), and 1017 cm^{-1} attributable to v_s(S-O) are due to the doubly bridging sulfate group in [V$^{IV}_2$O$_2$(OH)$_2$(SO$_4$)(2,2'-bipyridine)$_2$]. Peaks at 1250 cm^{-1}, 1234 cm^{-1}, 1158 cm^{-1}, 1124 cm^{-1}, 1112 cm^{-1}, 1097 cm^{-1}, 1056 cm^{-1}, 1044 cm^{-1} and 1030 cm^{-1} assigned to v_{as}(S-O) and 1022 cm^{-1} attributable to v_s(S-O) are due to the triply bridging sulfate groups in [V$^{IV}_2$O$_2$(SO$_4$)$_2$(2,2'-bipyridine)$_2$]. The intensities and positions of these peaks correspond to the observed values in the IR spectra of K$_6$[(VO)$_4$(SO$_4$)$_8$][12], K[VO$_2$(SO$_4$)(H$_2$O)$_2$]·H$_2$O[11], and related compounds containing doubly and triply bridging sulfate ligands[13-15]. The spectra of $[V^{IV}_2O_2(OH)_2(SO_4)(2,2'\text{-bipyridine})_2]$ and $[V^{IV}_2O_2(SO_4)_2(2,2'\text{-bipyridine})_2]$ also contain the characteristic features of 2,2'-bipyridine ligands.

STRUCTURE

The two compounds $[V^{IV}OSO_4(H_2O)_4]\cdot SO_4\cdot[H_2N(C_2H_4)_2NH_2]$[19] and $[V^{IV}OSO_4(H_2O)_4]\cdot SO_4\cdot[HN(C_2H_4)_3NH]$ have analogous structures. Their structures consist of layers of the coordination compound $[VO(H_2O)_4 SO_4]$ (Figure 1) which are separated by SO_4^{2-} anions and organic (piperazinium- $[H_2N(C_2H_4)_2 NH_2]^{2+}$ in $[V^{IV}OSO_4(H_2O)_4]\cdot SO_4\cdot[H_2N(C_2H_4)_2NH_2]$ and doubly protonated DABCO- $[HN(C_2H_4)_3NH]^{2+}$ in $[V^{IV}OSO_4(H_2O)_4]\cdot SO_4\cdot[HN(C_2H_4)_3 NH])$ cations. Figure 2 shows the unit cell content of $[V^{IV}OSO_4(H_2O)_4]\cdot SO_4\cdot[H_2N(C_2H_4)_2NH_2]$. Within a layer $[V^{IV}OSO_4(H_2O)_4]$ molecules are held through intermolecular hydrogen bonds involving sulfate oxygens and water ligands in $[V^{IV}OSO_4(H_2O)_4]\cdot SO_4\cdot[H_2N(C_2H_4)_2NH_2]$. The cations interact with $[V^{IV}OSO_4(H_2O)_4]$ layers through multiple hydrogen bonds involving –NH centers in $[V^{IV}OSO_4(H_2O)_4]\cdot SO_4\cdot[HN(C_2H_4)_3NH]$ and –NH₂ centers in $[V^{IV}OSO_4(H_2O)_4]\cdot SO_4\cdot[H_2N(C_2H_4)_2 NH_2]$. The oxygen atoms of the free sulfate anions interact with hydrogen atoms of water ligands in $[V^{IV}OSO_4(H_2O)_4]$ layers through multiple hydrogen bonds in both complexes.

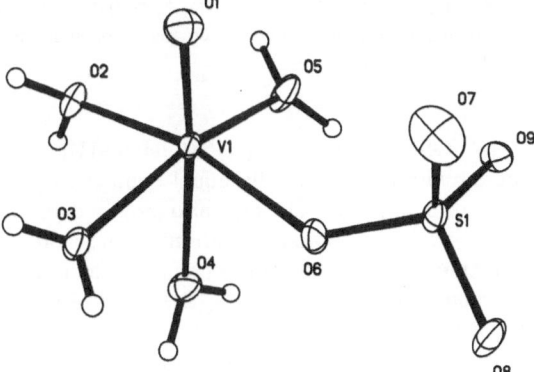

Figure 1. The structure of $[VO(SO_4)(H_2O)_4]$ in the crystals of $[V^{IV}OSO_4(H_2O)_4]\cdot SO_4\cdot[H_2N(C_2H_4)_2NH_2]$.

The sulfate group is directly bonded to the vanadium center of $[VO(H_2O)_4SO_4]$ molecule (Figure 1) in these complexes. The octahedral geometries around vanadium(IV) centers are defined by four oxygen atoms from water ligands, a terminal oxo ligand (O1) and another oxygen atom from tetrahedral sulfate ligand. The V-O1 bond length, 1.595 Å, is comparable with the corresponding value observed in $VOSO_4.5H_2O$ (1.591 Å)[15d] but shorter than those found in anhydrous $VOSO_4$ (1.630 Å)[15e] and $[V_2O_2(OH)_2(SO_4)(2,2'\text{-bipy})_2]$ (1.618 Å)[18]. The V-O(H₂) bond trans- to V=O(1) group is much longer (2.247 Å) than the remaining three long V-O(H₂) bonds (2.0038 - 2.0375 Å) and the V-center is slightly above the plane defined by four oxygens (O2, O3, O5, O6), a pattern observed earlier also[15d]. The S-O distances in the SO_4^{2-} group lie in 1.453 - 1.488 Å range, the longest one involving the oxygen atom coordinated to the vanadium(IV) center. These distances are comparable to the four S-O bond lengths (1.465 - 1.492 Å) found in the non-coordinated SO_4^{2-} group that is not covalently bonded to any other atom. Unlike near regular tetrahedral structure of the non-coordinated SO_4^{2-}, the overall geometry around sulfur in the coordinated sulfate group is a distorted tetrahedron.

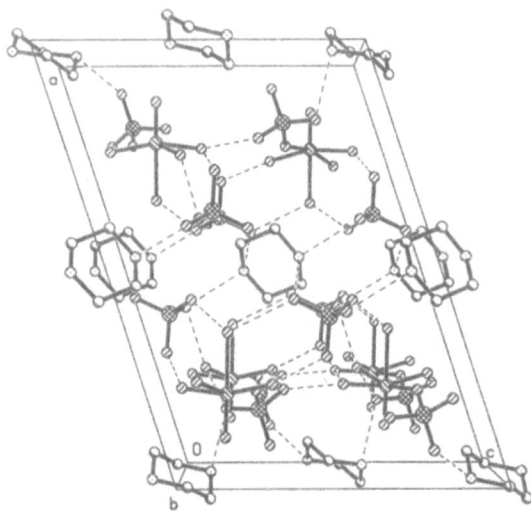

Figure 2. The unit cell content in the crystals of $[V^{IV}OSO_4(H_2O)_4]\cdot SO_4\cdot[H_2N(C_2H_4)_2NH_2]$
Dashed lines show hydrogen bonding scheme involving oxygen and nitrogen atoms. Hydrogen atoms are not shown.

The crystals of $[V^{IV}OSO_4(H_2O)_4]\cdot SO_4\cdot[H_2N(C_2H_4)_2NH_2]$ contain two types of piperazines (one disordered with two essentially equal components and one situated about the crystallographic center of symmetry). The crystallographic data, the results of the bond valence sum calculations[20] and manganometric titrations of the reduced vanadium(IV) sites, and charge balance consideration indicate that all piperazines are doubly protonated.

The extended structure of $[V^{IV}_2O_2 (OH)_2(SO_4)(2,2'\text{-bipyridine})_2]^{18}$ is shown in Figure 3. It is composed of the building block units given in Figure 4. The structure may be considered to contain ribbons constructed from infinite zigzag inorganic chains, $[-\{V_2O_2(OH)_2\}-\mu_2-SO_4-\{V_2O_2(OH)_2\}SO_4-]_n$, shown in Figure 5, incorporating 2,2'-bipyridine ligands. The neutral inorganic chains are composed of pairs of vanadium(IV) octahedra, $\{V(O_2N_2)-\mu_2-(OH)_2- V(O_2N_2)\}$, fused through a common edge, which are linked by sulfate tetrahedra through octahedral-tetrahedral corner sharing (Figure 5). The organic backbone of 2,2'-bipyridine ligands extend sideways from either face of the inorganic ribbons penetrating in the interchain regions separating the adjacent parallel inorganic chains from each another. Each organic stack is, in turn, bound by two consecutive stacks of inorganic chains.

The octahedral geometry around each vanadium atom in the structure is defined by $\{VO_2(OH)_2N_2\}$ with each V^{IV} center coordinated to a terminal oxo group, two μ_2-OH groups, two nitrogen donor atoms from a chelating 2,2'-bipyridine ligand, and an oxygen donor atom from a $\mu_2-SO_4^{2-}$ ligand. This arrangement generates four-memebered, $-V^{IV}-O(H)-V^{IV}-O(H)-$, rhombus-shaped metallacycles. The $\{O=V^{IV}-(\mu_2-OH)_2V^{IV}=O\}$ core represents a unique structural feature of the $V^{IV}/O/SO_4/2,2'$-bpy system. The two terminal oxo groups adopt *anti-* configuration with one $V-O_{terminal}$ vector pointing above and the other below the plane of the $V^{IV}-(\mu_2-OH)_2V^{IV}$ rhombus. The 2,2'-bipyridine is present in usual chelating bidentate mode forming five-membered -V-N-C-C-N- ring.

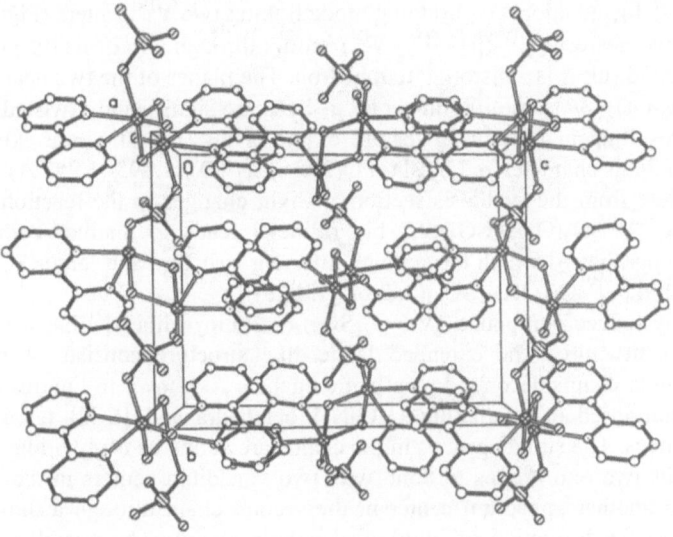

Figure 3. A view of the extended structure of $[V_2O_2(OH)_2(SO_4)(2, 2'\text{-bpy})_2]$ showing ribbons of $[\{V_2(OH)_2\}\text{-}\mu_2\text{-}SO_4\text{-}\{V_2(OH)_2\}SO_4]$ with 2,2'-bipyridine in interchain region.

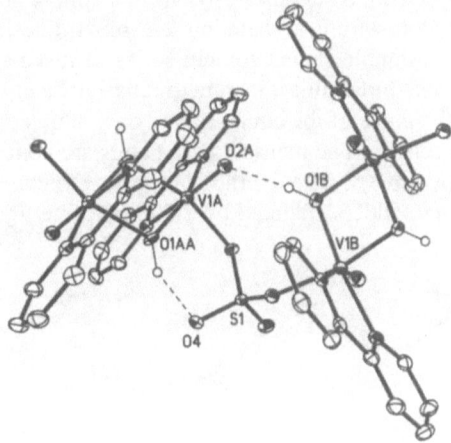

Figure 4. $\mu_2\text{-}SO_4$-Bridged binuclear units in the structure of $[V_2O_2(OH)_2(SO_4)(2,2'\text{-bpy})_2]$. Displacement ellipsoids are drawn at the 50% probability level.

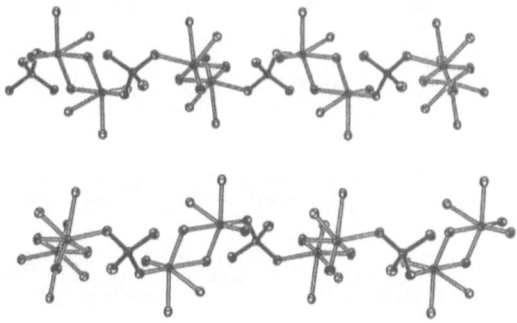

Figure 5. Ball and stick representation of the infinite inorganic chains, $[\text{-}\{V_2O_2(OH)_2\}\text{-}\mu_2\text{-}SO_4\text{-}\{V_2O_2(OH)_2\}SO_4\text{-}]_n$, present in the crystal structure of $[V_2O_2(OH)_2(SO_4)(2,2'\text{-bpy})_2]$.

The SO_4^{2-} ligand adopts μ_2-bridging mode linking two V^{IV} centers (Figure 4), one from each of the two nearest $\{V^{IV}-(\mu_2-OH)_2-V^{IV}\}$ motifs through two of its oxygens. The overall geometry around sulfur is a distorted tetrahedron. The planes of the two nearest $\{V^{IV}-(\mu_2-OH)_2-V^{IV}\}$ motifs, linked by μ_2-SO_4, are significantly twisted with respect to each other resulting in the zigzag feature of the chain. The structure exhibits intrachain hydrogen bondings characterized by short $O(2A)\ldots\ldots H-(O)$ (1.895- 1.953 Å).

As is clear from the synthesis section, a slight changes in the reaction condition used for preparing $[V^{IV}_2O_2(OH)_2(SO_4)(2,2'-bipyridine)_2]$ leads to another compound $[V^{IV}_2O_2(SO_4)_2(2,2'-bipyridine)_2]$ which has a much different structure and larger SO_4^{2-}:V ratio than that observed in $[V^{IV}_2O_2(OH)_2(SO_4)(2,2'-bipyridine)_2]$.

The fully reduce compound $[V^{IV}_2O_2(SO_4)_2(2,2'-bipyridine)_2]^{21}$ has a two-dimensional double chain structure. The extended ladder-like structure consists of parallel running inorganic double-chains decorated by organic ligands. As shown in Figures 6 the individual chains are composed of alternating $\{VO_4N_2\}$ octahedra and $\{SO_4\}$ tetrahedra joined by common vertices. The sulfate groups in the chains are acting as triply bridging groups. Each $\{SO_4\}$ uses its two oxo groups to bond with two vanadium centers in the same chain and coordinate to another vanadium center in the second chain through a third oxygen atom. This results in the formation of interlinked double chains. The overall geometry around sulfur is a slightly distorted tetrahedron. The octahedral geometry around each vanadium centers, is defined by a terminal V-O group (V-O = 1.582 Å), two μ_2-O from the adjacent sulfate groups in the same chain, an another μ_2-O from a sulfate group in the adjacent chain and two nitrogen donor atoms from a chelating 2,2'-bipyridine ligand (V-N = 2.133 and 2.259 Å). The cross-linked double chains contain series of fused eight-membered $\{-V^{IV}-O-S-O-V^{IV}-O-S-O-\}$ rings. 2,2'-bipyridines, symmetrically chelated to vanadium centers, are almost perpendicular to the plane of the equatorial oxygen atoms present in the coordination spheres of the vanadium centers. The metrical parameters are comparable to those observed in $[V^{IV}_2O_2(SO_4)_2(2,2'-bipyridine)_2]$. The overall composite structure may be viewed as made up of a series of 'inorganic ladders' flanked by 2,2'-bipyridine ligands which also separate the adjacent ladders.

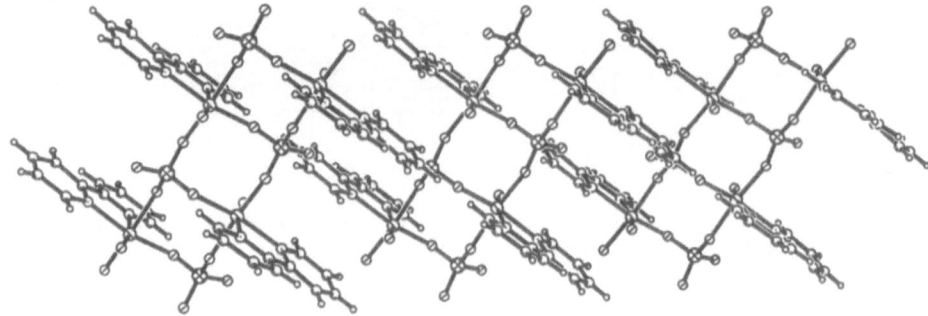

Figure 6. A view of the extended structure of $[V^{IV}_2O_2(SO_4)_2(2,2'-bpy)_2]$.

Bond lengths in $[V^{IV}_2O_2(OH)_2(SO_4)(2,2'-bipyridine)_2]$ and $[V^{IV}_2O_2(SO_4)_2(2,2'-bipyridine)_2]$ compare well to each other as well as to the corresponding distances involving V^{IV} centers in the reduced compounds $[(VO)_{1.5}SO_4 (1,10'-phenantroline)]^{21}$ and $[Bu^n_4N]_2[V^{IV}_4O_4(H_2O)_2(SO_4)_2\{(OCH_2)_3CCH_2CH_3\}_2]^{15c}$, in the mixed-valence species $K_6[(VO)_4(SO_4)_8]^{12}$, in general, slightly longer than the corresponding values involving V^V

sites (e.g., in $M[VO(SO_4)_2]$ (M = K, NH_4)[13], $[V_2O_3(SO_4)_2]$[13], $K[VO_2(SO_4)(H_2O)]$[11]. $K[VO_2(SO_4)(H_2O)_2] \cdot H_2O$[11] (sulfate as μ_2-SO_4) and in $K_6[(VO)_4(SO_4)_8]$[12]. $[(VO)_{1.5}SO_4(1,10'\text{-phenantroline})]$[21] (sulfate as μ_3-SO_4) in $Cs_4[(VO)_2O(SO_4)_4]$[14] (sulfate as a bidentate chelating ligand)).

MAGNETIC PROPERTIES

Figure 7(a) shows the temperature dependent magnetic susceptibility for $[V_2O_2(OH)_2(SO_4)(2, 2'\text{-bpy})_2]$[18]. The compound exhibits Curie Weiss paramagnetism at high temperature (T>140K) with the fitted parameters C = 0.807emu.K/mole, θ = -80.97K. g = 2.07 and TIP = 0.0087. The inverse magnetic susceptibility data are plotted in Figure 7(b) with the fitted curve. At lower temperature, the magnetic susceptibility exhibits a broad maximum which is consistent with the presence of short-range antiferromagnetic coupling between vanadium (IV) centers.

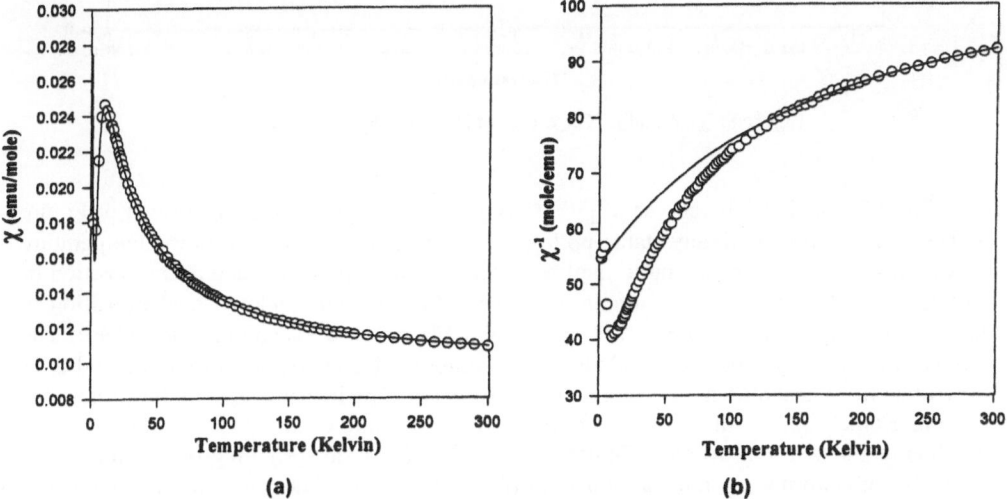

Figure 7. (a) The magnetic susceptibility of $[V_2O_2(OH)_2(SO_4)(2,2'\text{-bpy})_2]$ plotted as a function of temperature over the 1.7-300K temperature region. (b) The inverse magnetic susceptibility of $[V_2O_2(OH)_2(SO_4)(2,2'\text{-bpy})_2]$ plotted as a function of temperature (right).

THERMAL PROPERTIES

The compounds described in the foregoing paragraphs are fairly stable. The thermogravimetric analysis of $[V^{IV}OSO_4(H_2O)_4] \cdot SO_4 \cdot [HN(C_2H_4)_3NH]$ indicates no noticeable weight loss up to 120 °C. The coordinated water molecules are lost in two steps in the temperature range 126-260 °C. This is followed by the decomposition of organic cations and sulfate groups in the temperature range 300-390 °C leaving an incompletely characterized residue.

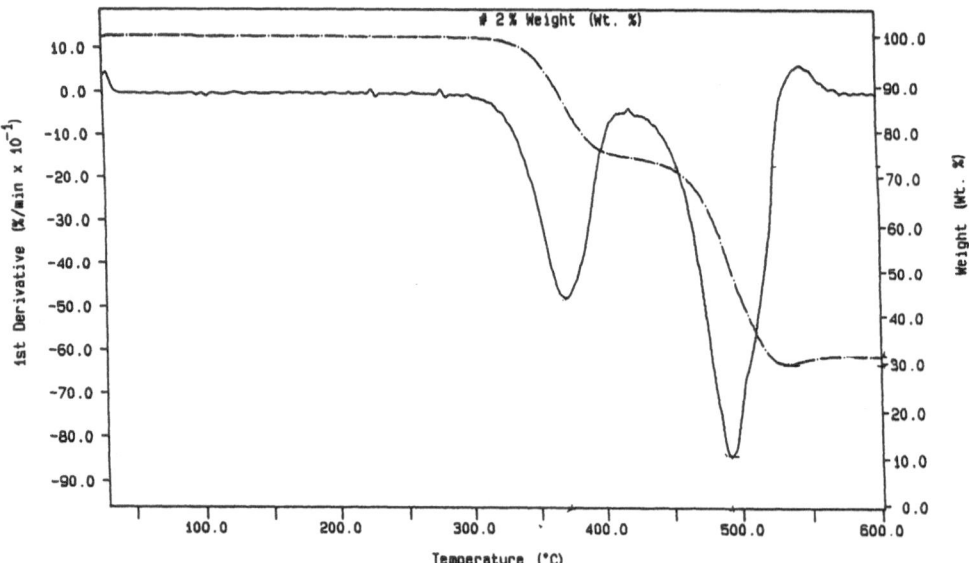

Figure 8. TGA and DTGA curves of $[V^{IV}_2O_2(OH)_2(SO_4)(2,2'\text{-bpy})_2]$

The TGA and DTGA curves of $[V^{IV}_2O_2(OH)_2(SO_4)(2,2'\text{-bipyridine})_2]$ shown in Figure 9. The compound is thermally stable up to about 300 °C. It loses weight in the temperature range 300-390 °C corresponding to the liberation of a water molecule and decomposition of sulfate into sulfur oxides. The organic (2,2'-bipyridine) ligands undergo decomposition at fairly high temperature (450-525 °C) leaving a stable reduced vanadium oxide phase. The observed total weight loss corresponds to the complete decomposition of sulfate and 2,2'-bipyridine ligands and liberation of a molecule of water. The IR bands due to the sulfate and 2,2'-bipyridine ligands disappeared from the IR spectrum of the black residue, which exhibits features at 998(m) cm^{-1}, 743(vs) cm^{-1} and 537(m) cm^{-1}. Heating the residue up to 1000 °C transforms it into a bluish-purple material that has not been completely characterized. The infrared spectrum of the bluish-purple material exhibit peaks at 998(m) cm^{-1}, 729(m) cm^{-1} and 540(m) cm^{-1}. The differential thermogram reveals that, like the removal of water molecules and sulfur oxides, the decompositions of the two organic ligands also proceed almost simultaneously.

The thermogravimetric analysis of $[V^{IV}_2O_2(SO_4)_2(2,2'\text{-bipyridine})_2]$ indicates that the compound is thermally stable up to 400 °C. The compound undergoes weight loss in the temperature range 400°C - 485°C, corresponding to the decomposition of 2,2'-bipyridine and sulfate ligands into gaseous products.

CONCLUSION

The molecular and solid systems described here represent an interesting class of compounds. $[V^{IV}_2O_2(OH)_2(SO_4)(2,2'\text{-bipyridine})_2]$ and $[V^{IV}_2O_2(SO_4)_2(2,2'\text{-bipyridine})_2]$ constitute the examples of fully reduced oxovanadium sulfate based hybrid materials with novel extended structures. Their preparations appear to be a significant step towards the development of V/O/SO$_4$/organic ligand systems and its potential in providing new materials. These results point out that a large number of oxovanadium sulfate phases incorporating organic ligands with novel structures remain to be discovered. The results also suggest that use of robust rod like organic ligands may lead to the formation of $V/O/SO_4^{2-}$/organic based composites with three-dimensional structures. Some of these possibilities are being explored in our laboratories and the outcome of this effort will be published elsewhere.

ACKNOWLEDGEMENTS

This work was partly supported by a grant to MIK from the American Chemical Society's Petroleum Research Fund (ACS-PRF# 35591-AC5). MIK expresses his sincere appreciation to his coworkers and collaborators whose names appear in the references.

REFERENCES

1. (a) M. T. Pope and A. Müller , Polyoxometalate chemistry: an old field with new dimensions in several disciplines, *Angew. Chem. Int. Ed. Eng.* 30:34 (1991). (b) C. L. Hill, Ed. Polyoxometalates, *Chem. Rev.* 98:359 (1998). (c) Jean-Pierre Jolivet, Metal Oxide Chemistry and Synthesis, John Wiley, New York (2000).

2. A. Müller, F. Peters,M. T. Pope, and D. Gatteschi, Polyoxometalates: very large clusters-nanoscale magnets, *Chem. Rev.*98:239 (1998).

3. M. I. Khan, Novel extended solids composed of transition metal oxide clusters, *J. Solid State Chemistry.* 152:105 (2000).

4. M.I. Khan, M. I. E.Yohannes, and D. Powell, Synthesis and characterization of a new mixed-metal-oxide framework material composed of vanadium oxide clusters: x-ray crystal structure of $(N_2H_5)_2[Zn_3V^{IV}_{12}V^V_6O_{42}(SO_4)(H_2O)_{12}]\cdot24H_2O$ *J. Chem. Soc., Chem. Com.*,23 (1999). (b) M. I. Khan. E.Yohannes, D. Powell, Metal oxide clusters as building blocks for the synthesis of metal oxide surfaces and framework materials: synthesis and x-ray crystal structure of $[Mn_3V_{19}H_6O_{46}(H_2O)_{12}]$ $\cdot30H_2O$. *Inorg. Chem.*, 38:212.(1999). (c) M. I. Khan, E. Yohannes, and R. J. Doedens, Metal oxide based framework materials composed of polyoxovanadate clusters: synthesis and x-ray crystal structures of $[M_3V_{18}O_{42}(H_2O)_{12}(XO_4)]\cdot24H_2O$ (M = Fe, Co; X = V, S), *Angew. Chem. Int. Ed. Engl.* 38:1292 (1999).

5. J. Zubieta, Clusters and solid phases of the oxovanadium phosphate and –organophosphonate systems. *Comments Inorg. Chem.* 16:153 (1994).

6. M. I. Khan, R. Haushalter, C. J. O'Connor, C. Tao, and Zubieta, An octahedral-tetrahedral vanadium phosphate one-dimensional chain incorporating 1-aminoethane-2-ammonium cations as ligands, *Chem. Mater.* 7: 593 (1995).

7. M. Roca, D. M. Marcos, P. Amorós, A. Beltrán-Porter, A. J. Edwards, and D. Beltrán-Porter, Synthetic strategies to obtain V-P-O open frameworks containing organic species as structural directing agents. Crystal structure of the V(IV)-Fe(III) bimetallic phosphate $[H_3N(CH_2)_2NH_3]_2[H_3N(CH_2)_2 NH_2]$ $[Fe^{III}(H_2O)_2(V^{IV}O)_8(OH)_4(HPO_4)_4(PO_4)_4]\cdot4H_2O$, *Inorg. Chem.* 35:5613 (1996).

8. J. R. DeBord, W. M. Reiff, R. Haushalter, and J. Zubieta, J. 3-D Organically templated mixed valence (Fe^{2+}/Fe^{3+}) iron phosphate with oxide-centered $Fe_4O(PO_4)_4$ cubes: Hydrothermal synthesis. crystal structure, magnetic susceptibility, and mössbauer spectroscopy of $[H_3NCH_2CH_2NH_3]_2[Fe_4O(PO_4)_4]\cdot H_2O$ *Chemistry of Materials,* 9:1994(1997).

9. M. I. Khan, L. Meyer,R. Haushalter, A. Schweitzer, J. Zubieta, and J. Dye, Giant voids in the hydrothermally synthesized microporous square pyramidal-tetrahedral framework vanadium phosphates $[HN(CH_2CH_2)_3NH]K_{1.35}[V_5O_9(PO_4)_2]\cdot xH_2O$ and $Cs_3[V_5O_9(PO_4)_2]\cdot xH_2O$, *Chem. Mater.* 8: 43 (1996).

10. S. Fehrmann, G. N. Boghosian, G. N. Papatheodorou, K. Nielsen, R. W. Berg, and N. J. Bjerrum, Crystal structure and infrared and Raman spectra of $K_4(VO)_3(SO_4)_5$" *Inorg. Chem.* 28:1847 (1989).

11. K. Richter and R. Mattes, Hydrated Ppases in the V_2O_5-K_2O-SO_3-H_2O system. preparation and structures of $K[VO_2(SO_4)(H_2O)]$ and $K[VO_2(SO_4)(H_2O)_2].H_2O$, *Inorg. Chem.*30: 4367(1991).

12. K. M. Eriksen, K. Nielsen, and R. Fehrmann, Crystal structure and spectroscopic characterization of $K_6(VO)_4(SO_4)_8$ containing mixed-valent vanadium(IV)-vanadium(V), *Inorg. Chem.* 35:480 (1996).

13. K. Richter and ˙ R. Mattes, Darstellung, Ramanspektren und kristallstrukturen von $V_2O_3(SO_4)_2$. $K[VO(SO_4)_2]$ und $NH_4[VO(SO_4)]$, *Z. Anorg. Allg. Chem.* 611:158 (1992).

14. K. Nielsen, R. Fehrmann, and K.M. Ericksen, Crystal structure of $Cs_4(VO)_2O(SO_4)_4$, *Inorg. Chem.* 32:4825 (1993).

15. (a) A. Müller, E. Krickmeyer, S. Dillinger, H. Bögge, and A. Stammler, $[As_4Mo_6V_7O_{39}(SO_4)]_4$: A species with an unusual structure and a model for the different host-guest properties of poly- vanadates and –molybdates, *J. Chem. Soc. Chem. Commun.* 2539 (1994). (b) L. M. Daniels, C. A. Murillo, and K. G. Rodriguez, Preparation of anhydrous vanadium(II) sulfate compounds from aqueous solutions: The synthesis and characterization of $[V(en)_3]SO_4$, $V(bipyridine)_2SO_4$ and $V(py)_4SO_4$ (en = eythylenediamine and py = pyridine), *Inorg. Chim. Acta* 229:27 (1995). (c) Y. D. Chang, Q. Chen, J. Salta, M. I. Khan, and J. Zubieta, Coordination chemistry of the tetrametalate core, $\{M_4O_{16}\}$: Synthesis from $[V_2O_2Cl_2\{(OCH_2)_2C(R)(CH_2OH)\}_2]$ and structures of the mixed metal cluster $[V_2Mo_2O_8(OMe)_2\{(OCH_2)_3CR\}_2]^{2-}$ and the reduced cluster $[V_4O_4(H_2O)^2(SO_4)_2\{(OCH_2)_3CR\}_2]^{2-}$, *Chem. Commun. J. Chem. Soc.* 1872 (1993). (d) C. J. Ballhausen, B. F. Djurinskij, and K. J. Watson, The polarized absorption spectra of three crystalline polymorphs of $VOSO_4 \cdot 5H_2O$, *J. Am. Chem. Soc.* 90:3305 (1968). (e) J. M. Longo and R. J. Arnott, Crystal structure of anhydrous $VOSO_4$, *J. Solid State Chem.* 1: 394 (1970).

16. B. I. Lazoryak, *Russ. Chem. Rev.* 65:287 (1996).

17. (a) M. I. Khan and J. Zubieta, In *Early Transition Metal Clusters With - Π Donor Ligands* , M. H. Chisholm, Ed., VCH: New York (1995). (b) M. I. Khan and J. Zubieta, In *Prog. Inorg. Chem.* K. D. Karlin , ed., John Wiley & Sons: New York (19996).

18. M. I. Khan, S. Cevik, D. Powell, S. Li, and C. J. O'Connor, Synthesis and characterization of novel $V/O/SO_4$ chains incorporating 2,2'-bipyridine ligands: Crystal structure of $[V_2O_2(OH)_2(SO_4)(2,2'-bpy)_2]$. *Inorg. Chem.* 37:81(1998).

19. M. I. Khan, S. Cevik, and R. J. Doedens, Hydrothermal synthesis and characterization of a new vanadyl sulfate compound: Crystal structure of $[V(IV)OSO_4(H_2O)_4] \cdot SO_4 \cdot [H_2N(C_2H_4)_2NH_2]$, *Inorg. Chim. Acta.* 292:112 (1999).

20. I. D. Brown In *Structure and Bonding in Crystals*, Vol. II, M O'Keefe and A. Navrotdky, Eds., Academic Press: New York, (1981).

21. M. I. Khan, et al. unpublished results.

SOLID STATE COORDINATION CHEMISTRY:
BIMETALLIC ORGANOPHOSPHONATE OXIDE
PHASES OF THE M/Cu/O/RPO$_3$$^{2-}$ FAMILY
(M = V, Mo).

Robert C. Finn and Jon Zubieta[*]

Department of Chemistry, Syracuse University, Syracuse, NY 13244

INTRODUCTION

Oxygen is not only the most abundant terrestrial element but is also highly reactive; consequently, oxides exist for most of the elements with the exceptions of radon and the lighter noble gases,[1] and inorganic oxides are ubiquitous in the geosphere, the biosphere,[2] and the noosphere of this group's collective imagination. In addition to the hydrogen oxide that forms the basis of life, oxygen is found in combination with silicon and aluminum in the form of complex aluminosilicates which make up the vast proportion of igneous rocks.[3] Not only are most ores and gems examples of solid state oxides, but bone, shells, teeth, spicules and wood represent structurally complex oxides fashioned through biomineralization.[4,5]

The intense contemporary interest in solid state oxides reflects their properties, which endow these materials with applications from heavy construction to microelectronic circuitry. The huge range of solid state properties is a result of the diversity of chemical compositions and structure types exhibited by the inorganic oxides.[6-30]

However, while many naturally occurring oxides and minerals and even synthetic oxides possess complex crystal structures, the majority are of relatively simple composition and exhibit highly symmetrical structures with rather small unit cells. Although such simple oxides possess undeniably useful properties, there exists, in general, a correlation between the complexity of the structure of a material and the functionality which it displays.[31] Thus, many of the remarkable materials fashioned by Nature contain mixtures of inorganic oxides coexisting with organic molecules, and it is now apparent that organic components can dramatically influence the microstructures of inorganic oxides, hence providing a method for the design of novel materials.

The control of inorganic structure by an organic component reveals an interactive structural hierarchy in materials. In such materials, the inorganic oxide contributes to the increased complexity, and hence functionality, through incorporation as one component in a multilevel structured material where there is a synergistic interaction between organic and inorganic components.[31] Since this interaction within these hybrid organic-inorganic materials derives from the nature of the interface between the organic component and the inorganic oxide, synthetic and structural studies of materials exhibiting such an interface

Polyoxometalate Chemistry for Nano-Composite Design
Edited by Yamase and Pope, Kluwer Academic/Plenum Publishers, 2002

will contribute to the development of structure-function relationships for these materials.

There are now four major classes of materials in which organic components exert a significant structural role in controlling the inorganic oxide microstructure:[32] zeolites,[33] mesoporous oxides of the MCM-41 class,[34] biomineralized materials,[2, 4, 35] and microporous octahedral-tetrahedral or square pyramidal-tetrahedral transition metal phosphate frameworks (**TMPO**) with entrained organic cations.[36,37]

The common feature distilled from the four classes of materials listed above is the influence of the organic component in controlling the nucleation and growth of the inorganic oxide. Organic components alter the inorganic oxide microstructure as a consequence of the synergism between the organic and inorganic components which allows imprinting of structural information from the organic molecule onto the inorganic framework. Implicit in the evolution of such organic-inorganic hybrid materials is a shift from the thermodynamic to the kinetic domain, such that equilibrium phases are replaced by structurally more complex metastable phases.[38] However, traditional solid-state syntheses produce thermodynamic products, often by solid-solid interactions at temperatures of $ca.$ 1000 ^{0}C, conditions which will not retain the structural features of the organic component. Necessarily, low temperature techniques, in an approach resembling small molecule chemistry, must be adopted. Hydrothermal synthesis[39-41] has been found to provide a powerful technique for the preparation of organic-inorganic hybrid materials with retention of the structural elements of the reactants in the final products.

Metal organophosphonate phases are representative hybrid materials which have been extensively studied with respect to the structural consequences of steric demands of the organic subunit, spacer length modifications, and additional functionality.[42, 43] Metal organophosphonate oxide phases, $[M_xO_y(RPO_3)_z]$, constitute a subclass of these hybrid materials which conflates the structural versatility of metal oxides with the complexity of organosphonate architectures.

In the specific case of oxovanadium-organophosphonate phases, the structural chemistry is profoundly influenced not only by the organic spacer, but also by the introduction of organic cations as charge compensating and space filling entities with multipoint hydrogen bonding interactions to the oxide component. [44] However, the structural consequences of introducing metal-organic subunits as potential structure-directing agents have not been explored. Furthermore, while metal organophosphonate chemistry in general has witnessed considerable development in recent years, the oxomolybdenum phosphonate family is relatively undeveloped. Curiously, while numerous soluble oxide clusters of the type $[Mo_xO_y(O_3PR)_3]^{n-}$ have been described, the only structurally characterized examples of solids are provided by the isomophous two-dimensional phases $[M_2(MoO_3)_3(O_3PR)]$.[45] In this contribution, we describe preliminary studies of the modification of oxometal-organodiphosphonate phases by secondary metal-organic subunits, thus exploiting both the spatial transmission of structural information by the diphosphonate ligand and the surface modification of the oxide by a multidentate organonitrogen ligand. Several prototypical structures of the M/O/Cu/RPO$_3$$^{2-}$ family (M = V and Mo) are described.

Results and Discussion

Significant advances in materials chemistry are driven by the discovery and design of novel solid phases. One strategy for the development of rational design for solid materials focuses on the introduction of organic components as molecular building blocks which imprint structural information on the solid or allow transmission of architectural motifs in a spatially extended construction. This approach offers the advantage that constituents of varying size, shape and functionality are available through the power of organic synthesis.

Furthermore, exploitation of fundamental coordination chemistry for solid state synthesis affords a variety of metal ions with preferred coordination geometries and numerous ligand architectures for the construction of metal-organic composite materials. Such hybrid inorganic-organic materials exhibit useful physical properties with applications to optics,[46] magnetism,[47] transport,[48] thermochromism,[49] and conductance.[50]

The coordination chemistry approach provides an effective strategy for the modification of the oxide microstructure. However, at the temperatures of most solid state reactions, there is little hope of accomplishing the assembly of a compound of predictable structure from two or more species, while retaining the bond relations between most of the constituent atoms. Furthermore, as consistently demonstrated with mineralogical examples, as well as in the laboratory, it is not possible to prepare open framework solids at elevated temperatures or pressures and, therefore, low temperature synthetic routes are obligatory for these metastable solids. However, it is difficult to achieve the intimate mixing and reaction of solid state starting materials at the low temperatures (usually < 250 °C) required to prevent collapse of the open framework oxides. Fortunately, the wealth of Nature's remarkable structurally complex mineral species indicates that hydrothermal synthesis provides a low-temperature pathway to crystalline inorganic-organic hybrid materials.

While organoamine constituents have been conventionally introduced as charge-balancing counterions in zeolite synthesis and in the preparations of V-P-O phases, in this application the organic component serves as a ligand to a secondary metal site, a first row transition or post-transition metal cation. Consequently, a coordination complex cation is assembled which serves to provide charge-compensation, space-filling and structure-directing roles.

The structure of the organoamine-secondary metal complex cation is derived, of course, from the geometrical requirements of the ligand as well as the coordination preferences of the metal. The ligand set may include chelating agents which coordinate to a single metal center or bridging ligands of various extensions which may provide a polymeric cationic scaffolding for the entraining of the V-P-O substructure. One strategy adopts appropriate stoichiometric control to form mononuclear metal-organoamine ligand chelate complex cations $\{M(N\text{---}N)_2\}^{n+}$ which are coordinatively unsaturated and hence capable of bonding to the oxo-groups of the oxide substructure, so as to provide linkage between the oxide subunits.

The properties of this cationic component may be tuned by exploiting the preferred coordination modes of various transition and post-transition metal cations. For example, while a Cu(II)-organoamine fragment will likely exhibit 4+2 or 4+1 coordination geometries, the Cu(I) counterpart will result in low coordination numbers with tetrahedral or trigonal planar geometries predominating. Similarly, a Ni(II)-based cation will adopt more regular octahedral coordination, while Zn(II) species may adopt various coordination modes.

The contemporary interest in metal organophosphonate coordination chemistry has received considerable impetus from the applications of such materials as sorbents, catalysts and catalyst supports.[51, 52] The structural chemistry of the metal organophosphonate system is extremely rich[53, 54] and is represented by mononuclear coordination complexes,[55] molecular clusters,[56-60] one-dimensional materials,[61] and layered phases.[62] Extensive investigations have been reported on layered phosphonates of divalent,[63-66] trivalent,[67, 68] tetravalent,[69-73] and recently hexavalent elements.[76] Several examples of three dimensional metal organophosphonate frameworks have also been described since the first report by Bujoli.[75-78]

Oxovanadium-organophosphonate chemistry[79] includes solid materials with one-, two-, and three-dimensional structures, incorporating either single valence V(IV) or V(V) sites or mixed valence V(IV)/V(V) sites,[80-85] as well as molecular species exhibiting a remarkable structural diversity.[56-59, 86-93] A noteworthy and common feature of these clusters is the organization of the V-P-O framework about a template, generally present as a guest or captive species. The range of templates is quite diverse, including neutral molecules, anions and even cations. Template effects are, of course, also a persistent theme of the solid state chemistry of the V-O-RPO$_3^{2-}$ system.

As noted above, oxometal organophosphonate phases are a well-documented family of materials for which structural modification through incorporation of an organic subunit is effected through variations in the steric demands of the organic, the tether length, and/or the presence of additional functional groups. The structural demands of the organophosphonate may be combined with the influences of secondary metal-organoimine components to exploit both the spatial transmission of structural information by the disphosphonate ligand and the surface modification of the oxide material by a multidentate organonitrogen ligand.

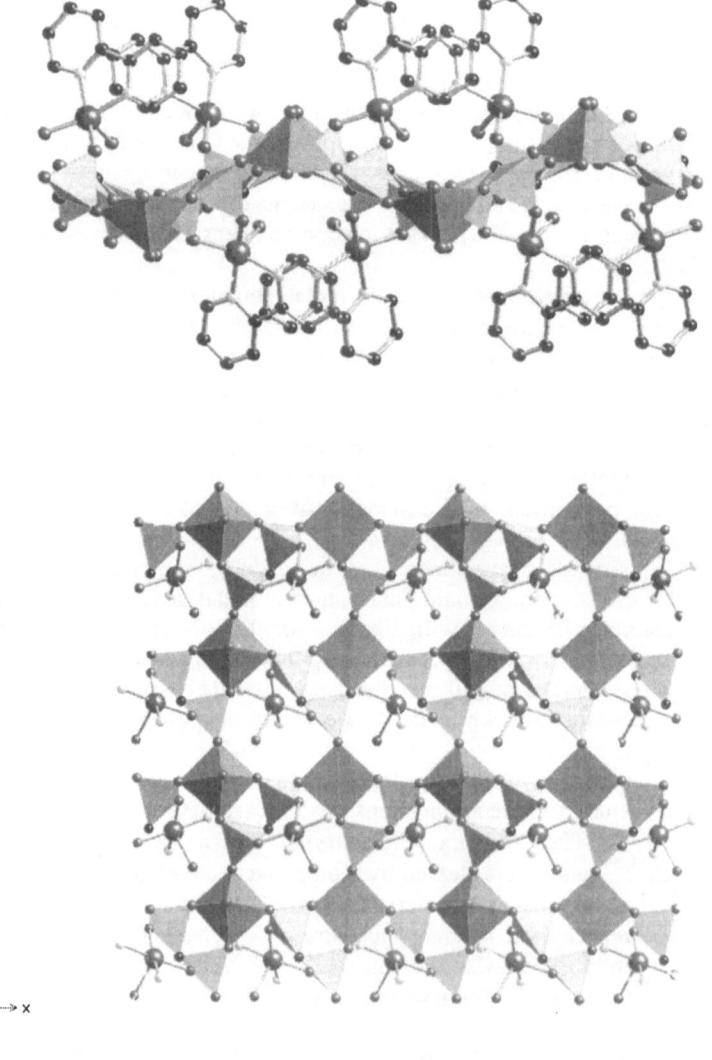

Figure 1. Top: A view of the layer structure of [{Cu(bpy)(H$_2$O)}(VO)(O$_3$PCH$_2$PO$_3$)]. Bottom: The {(VO)(O$_3$PCH$_2$PO$_3$)}$^{2-}$ substructure.

Since structural modifications are consequences of both the tether length of the disphosphonate ligand $\{O_3P(CH_2)_xPO_3\}^{4-}$ and of the nature of the M(II)-diimine subunit, one approach to the development of the chemistry is to vary the spatial extension of the organodiphosphonate for a variety of M(II)-organodiimine subunits. Two series have been studied to date, for the oxovanadium family $Cu(II)(LL)-O_3P(CH_2)_xPO_3$ with LL = 2, 2′ bipyridine, x = 1, 2 and 3 and with LL = 1, 10-phenanthroline, x = 1, 2 and 3.

The structure of $[\{Cu(2,2'\text{ -bpy})(H_2O)\}(VO)(O_3PCH_2PO_3)]$, shown in **Figure 1**, consists of two dimensional layers constructed from conrner-sharing Cu(II) square pyramids, V(V) square pyramids and phosphorus tetrahedra. The coordination sphere of the Cu(II) site is defined by the nitrogen donors of a bipyridine ligand and two oxygen donors of a chelating methylenediphosphonate ligand in the basal plane, and an aqua ligand in the apical position. The basal plane of the vanadium center consists of two oxygen donors from a chelating methylenediphosphonate ligand and two oxygen donors from each of two monodentate methylenediphosphonate groups; the apical position is occupied by the terminal oxo-group. Each methylenediphosphonate ligand links one copper and three vanadium sites.

The layer structure of $[\{Cu(bpy)(H_2O)\}(VO)(O_3PCH_2PO_3)]$ may be described as undulating $\{(VO)(O_3PCH_2PO_3)\}^{2-}$ networks with a period of 12.0Å and an amplitude of 4.5Å, decorated with $\{Cu(bpy)H_2O\}^{2+}$ groups which project into the interlamellar regions. The $\{(VO)(O_3PCH_2PO_3)\}^{2-}$ network, shown in **Figure 1b**, is constructed from seven polyhedral connect rings $\{V_3P_4O_6C\}$ linked through corner-sharing interactions into a two-dimensional covalently linked layer. The $\{Cu(bpy)(H_2O)\}^{2+}$ groups are sited above and below the cyclic cavities.

As shown in **Figure 2**, the structure of $[Cu(2,2'\text{ bpy})(VO)(O_3PCH_2CH_2PO_3)]$ consists of a two-dimensional network, again constructed from corner-sharing vanadium $\{VO_5\}$ square pyramids, phosphate tetrahedra and copper $\{CuO_3N_2\}$ square pyramids. The vanadium(IV) site is defined by four oxygen donors from each of four phosphonate groups in the basal plane and an apical oxo-group which bridges to the Cu(II) center. The copper geometry consists of the nitrogen donors of the bpy ligand and two oxygen donors from two phosphonate ligands in the basal plane and the bridging oxo-group in the apical position. The oxo-bridge is asymmetric with a V-O bond distance of 1.611(2)Å and a Cu-O distance of 2.394(2)Å. Each $\{PO_3\}$ terminus of the diphosphonate ligand bridges one copper and two vanadium sites. Consequently, the structure may be described as $\{Cu(bpy)(VO)(O_3PCH_2\text{-})_2\}$ chains, tethered through the ethylene bridges of the diphosphonate ligands into a two-dimensional covalently linked network. Within the one-dimensional oxide substructure, the common cyclic motif $\{V_2P_2O_4\}$ of corner-sharing $\{VO_5\}$ square pyramids and $\{O_3PR\}$ tetrahedra is observed. These rings are linked through the vanadium sites into a chain, which is decorated by the $\{Cu(bpy)\}^{2+}$ moieties, linked to the vanadium oxo-group and the oxygen donors of the diphosphonate ligands not involved in the $\{V_2P_2O_4\}$ ring formation. This results in the additional ring motifs $\{VCuP_2O_4\}$ and $\{VCuPO_3\}$. The connectivity gives rise to a distinctive partitioning of the layer structure into inorganic chain substructures and tethering organic domains. The $\{Cu(bpy)\}^{+2}$ subunits project into the interlamellar regions above and below the plane.

The structural consequences of tether lengthening in $[Cu(2,2'\text{ bpy})(VO)(O_3PCH_2CH_2PO_3)]$ are significant. While structure $[Cu(2,2'\text{ bpy})(VO)(O_3PCH_2PO_3)]$ contains vanadium centers with terminal oxo-groups, the oxo-groups in the ethylene analogue bridge to the copper sites. Consequently, although the copper geometries in both the methylene and the ethylene analogues are square pyramidal $\{CuN_2O_3\}$, one oxygen donor in the methylene analogue is provided by an aqua ligand, in contrast to the bridging oxo group in the ethylene case. The $[(VO)\{O_3P(CH_2)_nPO_3\}]^{2-}$ networks, shown in **Figures 1b** and **2b**, are quite distinct, with the former displaying fourteen-membered $[V_3P_4O_6C]$ rings and the latter fourteen-membered $[V_2P_4O_4C_4]$ rings.

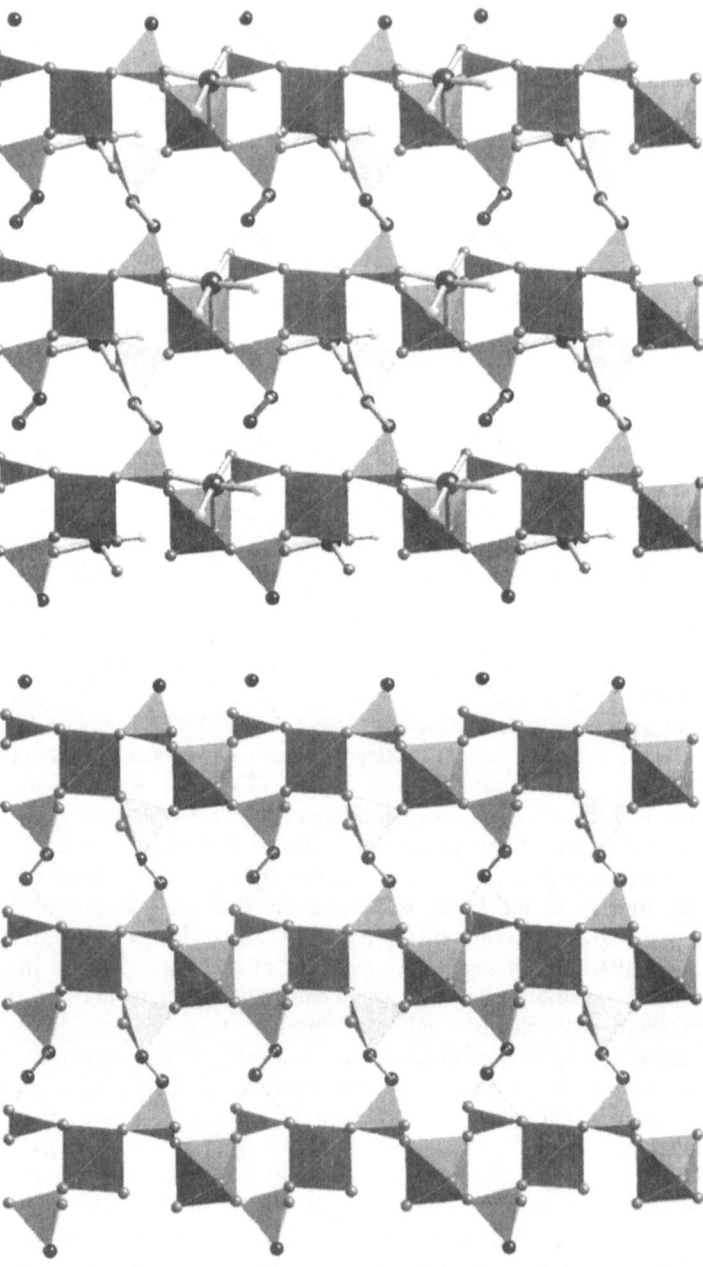

Figure 2. Top: A view of the structure of [Cu(2,2′ bpy)(VO)(O₃PCH₂CH₂PO₃)]. Bottom: The {(VO)(O₃PCH₂CH₂PO₃)}²⁻ substructure.

While the structure of the {(VO)(O₃PCH₂PO₃)}²⁻ network of [Cu(2,2′ bpy)(VO)(O₃PCH₂PO₃)] is quite dintinct from other examples of oxovanadium-methylenediphosphonate layered materials, the {(VO)(O₃PCH₂CH₂PO₃)}²⁻ network of the ethylene species is reminiscent of the oxovanadium-phosphonate layer of

[H₃NCH₂CH₂NH₃][(VO)(O₃PCH₂CH₂PO₃)]. The covalent connectivities of the oxovanadium-phosphonate network structures of [Cu(2,2' bpy)(VO)(O₃PCH₂PO₃)] and this latter material are identical, and the terminal oxo-group and pendant P-O groups observed in [H₃NCH₂CH₂NH₃][(VO)(O₃PCH₂CH₂PO₃)] serve to anchor the {Cu(bpy)}$^{2+}$ moiety to the layer structure in [Cu(2,2' bpy)(VO)(O₃PCH₂PO₃)].

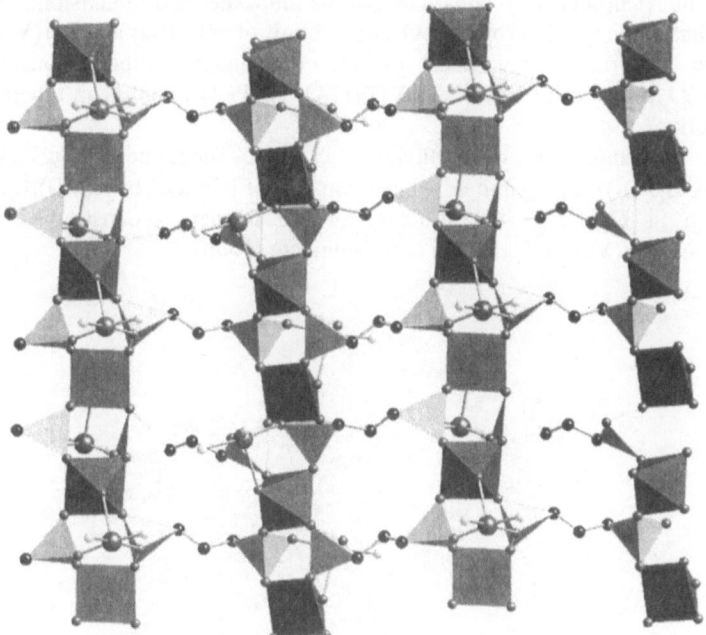

Figure 3.   The structure of [Cu(2,2' bpy)(VO)(O₃PCH₂CH₂CH₂PO₃)]•2H₂O.

The structure of the propylenediphosphonate derivative [Cu(2,2' bpy)(VO)(O₃PCH₂CH₂CH₂PO₃)]•H₂O is analogous to that of ethylene-bridged analogue, as shown in **Figure 3.** The structures share the common structural motifs of {Cu(bpy)(VO)(O₃PCH₂.)₂} chains and cyclic {V₂P₂O₄} rings within the chains. The propylene groups bridge chains into a two-dimensional network. However, the consequences of tether extension are apparent in the relative orientations of the {Cu(bpy)}$^{2+}$ moieties. In contrast to [Cu(2,2' bpy)(VO)(O₃PCH₂CH₂PO₃)] in which the {Cu(bpy)}$^{2+}$ units link to phosphonate oxygen donors on the same edge of the oxovanadium phosphonate chain and consequently orient the bpy plane approximately parallel to the V-V vectors of the chain, the {Cu(bpy)}$^{2+}$ groups of [Cu(2,2' bpy)(VO)(O₃PCH₂CH₂CH₂PO₃)] coordinate oxygen donors from phosphonate groups on opposite edges of the chain, with a concomitant orientation of the bpy plane perpendicular to the V-V vectors of the chain. It would appear that chain lengthening in the propylene analogue relieves the steric strain which would result from the perpendicular orientation of bpy units in a shorter tether system.

It is noteworthy that substitution of 1,10-phenanthroline for 2,2' bipyridine results in a structurally unique series of materials. For example, the network structure of [Cu(1,10-phen)(VO)(H₂O)(O₃PCH₂PO₃)], illustrated in **Figure 4,** contrasts with that of [{Cu(2,2' bpy)(H₂O)}(VO)(O₃PCH₂PO₃)] in several important features. The aqua ligand in the latter material resides on the Cu(II) site. Consequently, the copper links to the V-P-O network only through two phosphonate oxygen donors. The vanadium center coordinates to four phosphonate oxygen donors. However, the aqua ligand of [Cu(phen)(VO)(H₂O)(O₃PCH₂PO₃)] is situated on the V(IV) center which consequently links to only three phosphate oxygen donors. Whereas tracing the covalent connectivity of the oxovanadium-diphosphonate substructure of [{Cu(bpy)(H₂O)}(VO)O₃PCH₂PO₃]] reveals a network substructure, the V-O-diphosphonate substructure of [Cu(phen)(VO)(H₂O)(O₃PCH₂PO₃)] consists of one-dimensional chains. These chains are connected through the {Cu(phen)O₃} square pyramids into the two-dimensional structure. It is evident that while the {Cu(bpy)(H₂O)O₂} subunit of [{Cu(bpy)(H₂O)}(VO)(O₃PCH₂PO₃)] serves to decorate the surface of the oxovanadium-diphosphonate network, the {Cu(phen)O₃} subunit of [Cu(phen)(VO)(H₂O)(O₃PCH₂PO₃)] is an integral component of the network.

The structure of the ethylene bridged analogue [{Cu(1,10-phen)}₂(V₂O₅)(O₃PCH₂CH₂PO₃)] is also quite distinct from that of [Cu(2,2′ bpy)(VO)(O₃PCH₂CH₂PO₃)]. As shown in **Figure 5,** the structure of the former is constructed from chains of corner-sharing V(V) tetrahedra and diphosphonate groups,

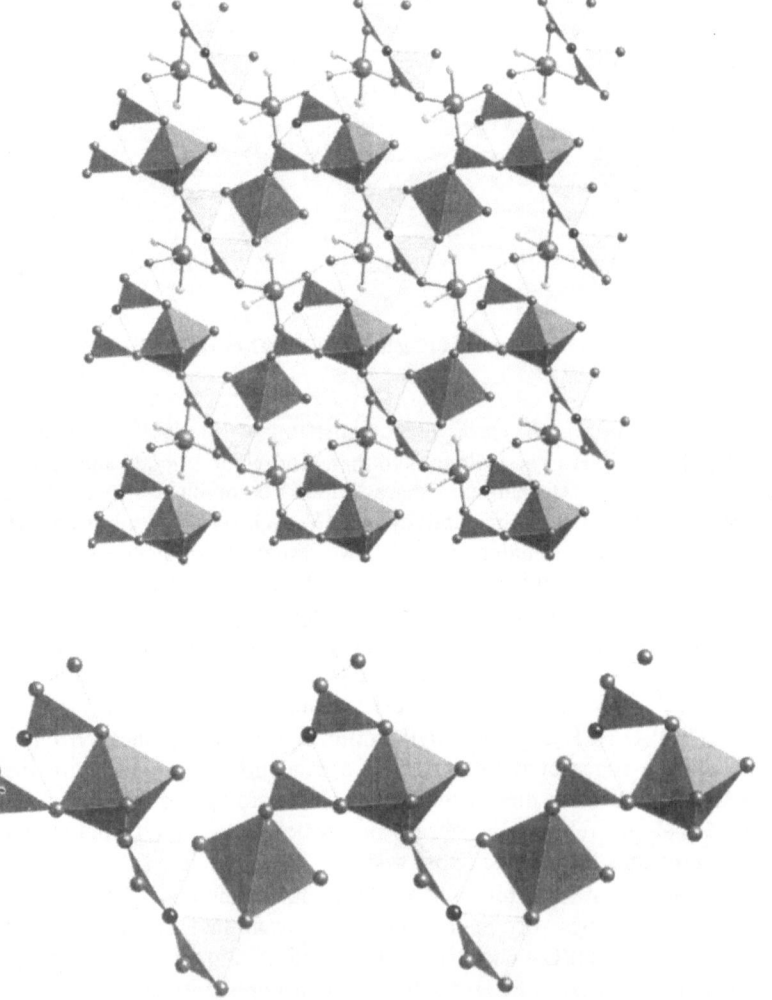

Figure 4. Top: The two-dimensional structure of [Cu(1,10-phen)(VO)(H₂O)(O₃PCH₂PO₃)]. Bottom: The one-dimensional V-O-diphosphonate substructure.

linked through copper square pyramids. Several features of the structure are note worthy. The presence of binuclear units of corner-sharing V(V) tetrahedra is unique, and the +5 oxidation state is unusual for oxovanadium-organophosphonate structures in general.

The structural complexity of this family of materials is also manifested in the structure of [Cu(1,10-phen)(V$_2$O$_3$)(O$_3$PCH$_2$CH$_2$CH$_2$PO$_3$)]. As shown in **Figure 6,** the network structure of this propylene diphosphonate analogue, like that of the previously described [Cu(2,2′-bpy)(VO)(O$_3$PCH$_2$CH$_2$CH$_2$PO$_3$)], is constructed from V-P-O-Cu chains

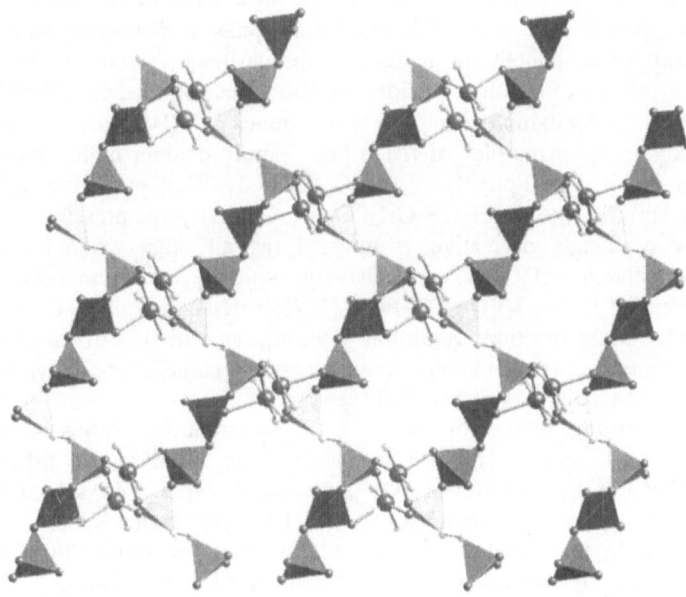

Figure 5. A view of the structure of [{Cu(1,10-phen)}$_2$(V$_2$O$_5$)(O$_3$PCH$_2$CH$_2$PO$_3$)].

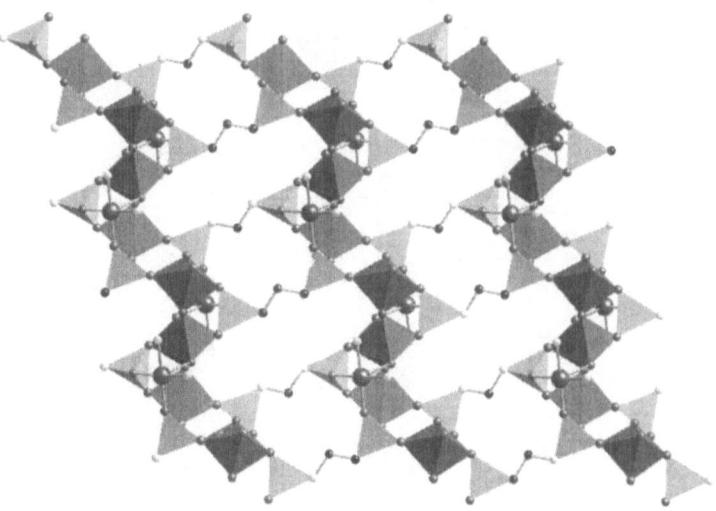

Figure 6. A view of the structure of [Cu(1,10-phen)(V$_2$O$_3$)(O$_3$PCH$_2$CH$_2$CH$_2$PO$_3$)].

linked through the propylene bridges of the disphosphonate into a two-dimensional network. However, the detailed connectivity within the inorganic chains of [Cu(1,10-phen)(V$_2$O$_3$)(O$_3$PCH$_2$CH$_2$CH$_2$PO$_3$)] is quite distinct from that of [Cu(2,2' bpy)(VO) (O$_3$PCH$_2$CH$_2$CH$_2$PO$_3$)]. The oxide chain of the o-phen derivative contains a trinuclear submotif of two V(IV) square pyramids corner-sharing with a central V(IV) octahedron. The trinuclear units are bridged by {O$_3$PC} tetrahedra to form the puckered one-dimensional chain. In addition, the central vanadium site of the trinuclear unit engages in a corner-sharing interaction through a bridging oxo-group with the {Cu(phen)}$^{2+}$ subunit, which serves to decorate the exterior of the chain.

In comparing the three 2,2' bipyridine phases discussed above to the three 1,10-phenanthroline containing materials, the striking feature is the contrast in the nuclearities of the vanadium oxide substructures of these materials. While the 2,2' bipyrinde family exhibits exclusively mononuclear V(IV) sites as submotifs, that is, no V-O-V bridging, the examples of the o-phen family contain higher oligomers, binuclear V(V) sites in [{Cu(1,10-phen)}$_2$(V$_2$O$_5$)(O$_3$PCH$_2$CH$_2$PO$_3$)] and trinuclear V(IV) sites in [Cu(1,10-phen)(V$_2$O$_3$)(O$_3$PCH$_2$CH$_2$CH$_2$PO$_3$)]. Similarly, the presence of V(V) sites in the ethylenediphosphonate derivative is unusual, as all other examples of this class of compounds contain V(IV) sites, exclusively, with the exception of the mixed valence [{Cu(bpydicarbH$_2$)}$_2$(V$_3$O$_5$)(O$_3$PCH$_2$PO$_3$)$_2$(H$_2$O)]•2H$_2$O, described below. We can only conclude that under reaction conditions favoring the isolation of metastable species, the complexities inherent in the variable oxidation states, coordination polyhedra and degree of aggregation of vanadium oxides are fully realized.

Another curious feature of the Cu-1,10 phenanthroline series is the isolation of a three-dimensional phase with (O$_3$PCH$_2$PO$_3$)$^{4-}$, in addition to the two-dimensional [Cu(1,10-phen)(VO)(H$_2$O)(O$_3$PCH$_2$PO$_3$)], discussed above. As shown in **Figure 7**, the structure of [Cu(1,10-phen)(V$_2$O$_3$)(O$_3$PCH$_2$PO$_3$)] consists of {Cu(phen)(VO)$_2$ (O$_3$PCH$_2$PO$_3$)} layers linked through vanadium square pyramids into a covalently linked framework. The layers are constructed from chains of corner-sharing V(V) and (O$_3$PC) tetrahedra, linked through Cu(II) square pyramids. Each diphosphonate

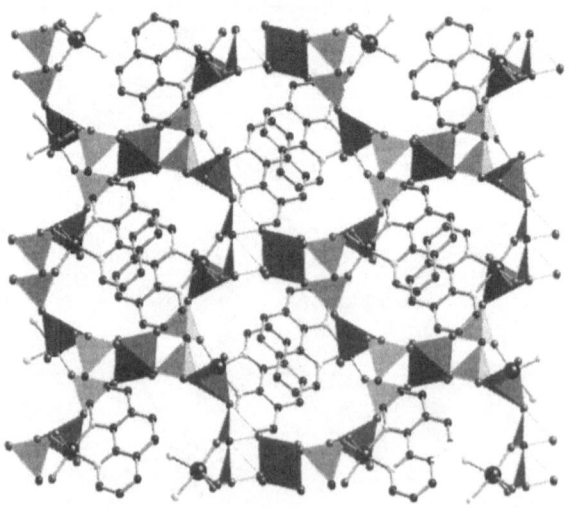

Figure 7. The structure of [Cu(1,10-phen)(V$_2$O$_3$)(O$_3$PCH$_2$PO$_3$)]

ligand projects two oxygen vertices into the interlamellar region, which bond to the square pyramidal V(IV) sites which serve to link the V-P-O networks.

Attempts to modify the structures by introducing steric bulk and/or additional functionality on the organoimine ligand are represented by [{Cu(2,2'-bpydicarbH$_2$)}$_2$ (V$_3$O$_5$)(O$_3$PCH$_2$PO$_3$)$_2$(H$_2$O)]•2H$_2$O. As shown in **Figure 8**, the V-P-O network is constructed from corner-sharing V(IV) and V(V) square pyramids and {O$_3$PC} tetrahedra. The square pyramidal {Cu(bpydicarbH$_2$)O$_3$} subunits decorate the surfaces of the layer. The {V$_3$O$_5$} motif consists of a trinuclear unit of corner-sharing polyhedra. The central vanadium site exhibits a terminal oxo-group and bonds to the two peripheral vanadium centers through oxo-groups and to two phosphonate oxygen donors. Each peripheral vanadium site is additionally coordinated to three phosphonate oxygen atoms and to a copper site through a V-O-Cu bridging oxo-group. Each methylenediphosphonate ligand acts as a bidentate chelator of a Cu(II) site and bridges four vanadium centers from three trinuclear units. The unusual connectivity pattern generates 16 membered rings {V$_4$P$_4$O$_8$},

Figure 8. A view of the structure of [{Cu(2,2'-bpydicarbH$_2$)}$_2$(V$_3$O$_5$)(O$_3$PCH$_2$PO$_3$)$_2$(H$_2$O)]· H$_2$O.

resulting in considerable void volume which is occupied by water molecules of crystallization.

The oxomolybdenum core based materials are represented by [Cu(o-phen)(Mo$_2$O$_5$) (O$_3$PCH$_2$PO$_3$)(H$_2$O)], [{Cu(o-phen)}$_2$(Mo$_4$O$_{12}$)(H$_2$O)$_2$(O$_3$PCH$_2$CH$_2$PO)]•2H$_2$O and the propylene-diphosphonate materials [{Cu(H$_2$O)$_2$(o-phen)}{Cu(o-phen)$_2$}(Mo$_5$O$_{15}$) (O$_3$PCH$_2$CH$_2$CH$_2$PO$_3$)] and [Cu(bpy)(Mo$_2$O$_5$)(O$_3$PCH$_2$CH$_2$CH$_2$PO$_3$)].

The structure of [Cu(o-phen)(Mo$_2$O$_5$)(O$_3$PCH$_2$PO$_3$)(H$_2$O)], shown in **Figure 9**, consists of one-dimensional chains constructed from corner-sharing Cu(II) square pyramids and phosphorus tetrahedra, and corner-and edge-sharing Mo(VI) octahedra. The coordination sphere of the Cu(II) site is defined by the nitrogen donors of the chelating o-phen ligand, an oxygen donor from each {PO$_3$} group of the chelating methylenediphosphonate ligand and an oxo-group bridging to a molybdenum site.

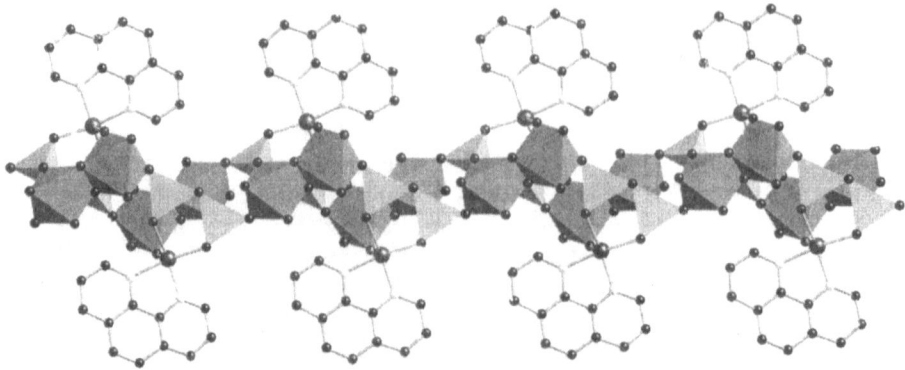

Figure 9.   The one-dimensional structure of [Cu(o-phen)(Mo$_2$O$_5$)(O$_3$PCH$_2$PO$_3$)(H$_2$O)].

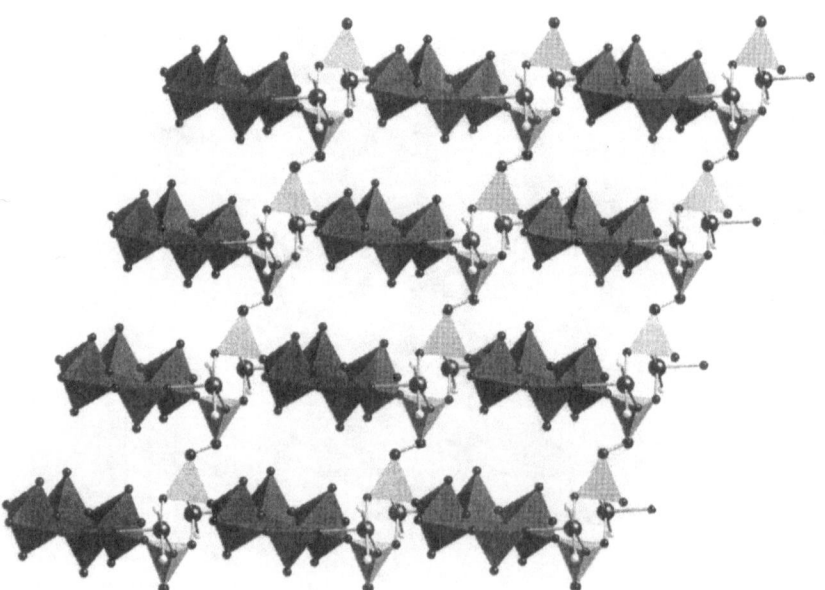

Figure 10.   A view of the two-dimensional structure of [{Cu(o-phen)}$_2$(Mo$_4$O$_{12}$)(H$_2$O)$_2$
(H$_2$O)$_2$(O$_3$PCH$_2$CH$_2$PO$_3$)]· H$_2$O.

Binuclear Mo sites constructed from edge-sharing octahedra are embedded in the chain. The geometry about the Mo1 site is defined by two cisoid terminal oxo-groups, three oxygen donors from the diphosphonate ligands, one of which bridges to the Mo2 site, and a bridging oxo-group.   The second molybdenum center exhibits a terminal oxo-group, an oxo-group bridging to the Cu site, an oxo-group bridging to the Mo1 center, two oxygen donors from the phosphonate ligand and an aqua ligand.   Each molybdate binuclear unit bonds to three diphosphonate ligands.   The diphosphonate units, in turn, chelate a Cu site and employ the remaining four oxygen donors to link three adjacent binuclear molybdenum sites.

The profound structural consequences of increasing the diphosphonate tether length by one methylene group are evident in the two-dimensional structure of $[\{Cu(o\text{-phen})\}_2(Mo_4O_{12})(H_2O)_2(O_3PCH_2CH_2PO_3)]\cdot 2H_2O$, shown in **Figure 10**. The structure may be described as bimetallic oxide phosphate chains linked through ethylene bridges of the diphosphonate ligands into a layer structure. The $\{Cu(o\text{-phen})(Mo_2O_6)(H_2O)(O_3P\text{-})\}_n$ chains are constructed from noncylic $\{Mo_4O_{16}\}$ clusters linked through diphosphonate ligands and square pyramidal Cu(II) sites. The copper centers bond to two nitrogen donors from a chelating phenanthroline ligand, two oxygen donors from two different diphosphonate ligands and an oxo-group bridging to a molybdenum site. Each $\{-PO_3\}$ unit of the diphosphonate provides two oxygen donors to adjacent copper sites in an μ^2-O, O' type bridge while the third oxygen bonds to an octahedral molybdenum center. The ethylene groups of the diphosphonate ligands serve to bridge adjacent inorganic chains. The tetranuclear molybdate cluster embedded in the chain structure exhibits several unusual features. The molybdenum centers are arranged in a non-linear chain of edge-sharing polyhedra with octahedra at the termini and square pyramids in the interior locations. The octahedral sites are defined by two cisoid terminal oxo-groups, an oxo-group bridging to a square pyramidal site, a triply-bridging oxo-group, an oxygen donor from the diphosphonate ligand and an aqua ligand. The five-coordination at the second site results from a terminal oxo-group, an oxo-group bridging to the copper site, two triply-bridging oxo-groups and a doubly-bridging oxo-group.

The contrasting dimensionalities of $[Cu(o\text{-phen})(Mo_2O_5)(O_3PCH_2PO_3)(H_2O)]$ and $[\{Cu(o\text{-phen})\}_2(Mo_4O_{12})(H_2O)_2(O_3PCH_2CH_2PO_3)]$ are not totally unprecedented. In the oxovanadium diphosphonate family, $[H_2N(CH_2CH_2)_2NH_2](VO)(O_3PCH_2PO_3)]$ is one-dimensional, while $[H_3N(CH_2)_2NH_3](VO)(O_3PCH_2CH_2PO_3)]$ is two-dimensional, reflecting the propensity of the methylene-bridged diphosphonate to form six membered chelate rings $\{M\text{-}O\text{-}P\text{-}C\text{-}P\text{-}O\}$ rather than provide extension and tethering of neighboring structural units, the modality adopted by the ethylene-bridged ligand.

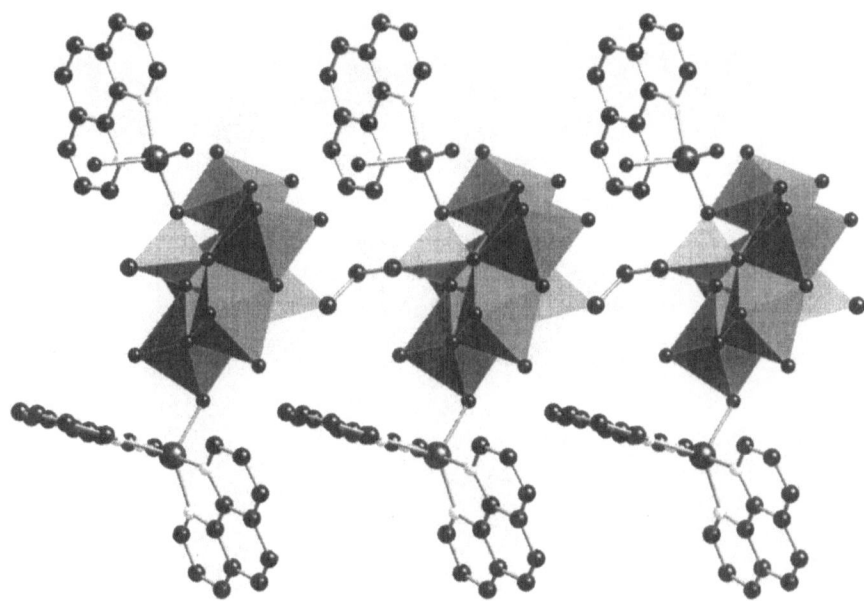

Figure 11. The one-dimensional structure of $[\{Cu(H_2O)_2(o\text{-phen})\}\{Cu(o\text{-phen})_2\}(Mo_5O_{15})(O_3PCH_2CH_2CH_2PO_3)]$.

Somewhat surprisingly, the structure of the propylene derivative, [{Cu(H$_2$O)$_2$(o-phen)}{Cu(o-phen)$_2$}(Mo$_5$O$_{15}$)(O$_3$PCH$_2$CH$_2$CH$_2$PO$_3$)], shown in **Figure 11**, is one-dimensional. The structure may be described as {Mo$_5$O$_{21}$} cyclic clusters, linked through diphosphonate ligands into a chain which is decorated by {Cu(H$_2$O)$_2$(o-phen)}$^{2+}$ and {Cu(phen)$_2$}$^{2+}$ subunits. There are, thus, two distinct Cu(II) environments: a square pyramidal site defined by four nitrogen donors from two o-phen ligands and a bridging oxo-group to a molybdate cluster; and an octahedral site linked to a single o-phen chelate, two aqua ligands, a phosphonate oxygen donor and a bridging oxo-group from the cluster. The diphosphonate ligands serve to bridge adjacent clusters, coordinating exclusively to molydenum sites at one {-PO$_3$} terminus and to Mo and Cu centers at the other. The molybdate cluster is structurally analogous to the well known [Mo$_5$O$_{15}$(O$_3$PR)$_2$]$^{4-}$ molecular cluster,[94,95] consisting of a ring of edge-sharing octahedral. The cyclic structure of the cluster of [{Cu(H$_2$O)$_2$(o-phen)}{Cu(o-phen)$_2$}(Mo$_5$O$_{15}$)(O$_3$PCH$_2$CH$_2$CH$_2$PO$_3$)] is distorted from that of the symmetrical molecular analogue by the linking of the {Cu(o-phen)(H$_2$O)$_2$}$^{2+}$ subunit which expands the Mo-O(phosphonate) bond length at this doubly-bridging site.

The isolation of this one-dimensional material suggested that a building block approach to such chain structures is conceivable. However, attempts to link preformed clusters through diphosphonate ligands of various tether lengths under non-hydrothermal conditions proved futile. Furthermore, adopting pH and stoichiometry conditions favoring such cluster formation under hydrothermal conditions did not yield the analogous [(Mo$_5$O$_{15}$){O$_3$P(CH$_2$)$_x$PO$_3$}]$^{4-}$ chains (n = 1, 2 or 4) but rather required the presence of Cu(II) and phenanthroline to yield [Cu(o-phen)(Mo$_2$O$_5$)(O$_3$PCH$_2$PO$_3$)(H$_2$O)] and [{Cu(o-phen)}$_2$(Mo$_4$O$_{12}$)(H$_2$O)$_2$(O$_3$PCH$_2$CH$_2$PO$_3$)] and an unidentified material, respectively. While cluster preorganization and linkage is an attractive strategy for structural design, it appears that such synthetic building blocks persist only under narrowly defined conditions, whereas in general dissociation is evident and the system "selects" the appropriate structural unit from the species present in the hydrothermal mix.

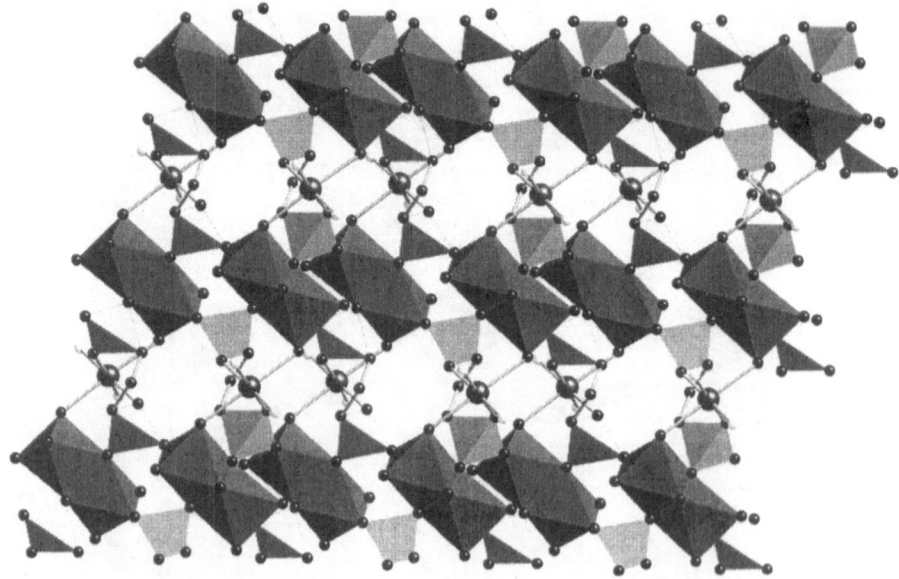

Figure 12.   A view of the structure of [Cu(bpy)(Mo$_2$O$_5$)(O$_3$PCH$_2$CH$_2$CH$_2$PO$_3$)].

These arguments are reinforced by the isolation of $[Cu(bpy)(Mo_2O_5)$ $(O_3PCH_2CH_2CH_2PO_3)]$ under the same conditions as those employed for the synthesis of $[\{Cu(H_2O)_2(o\text{-phen})\}\{Cu(o\text{-phen})_2\}(Mo_5\ O_{15})\ (O_3PCH_2CH_2CH_2PO_3)]$. As shown in **Figure 12,** the structural consequences of replacing o-phen by bpy are profound. The structure of the bpy-containing material consists of two-dimensional networks constructed from binuclear molybdate subunits, phosphonate tetrahedra and '4+2' Cu(II) octahedra. A most unusual feature of the structure is the presence of binuclear units of *face-sharing* $\{MoO_6\}$ octahedra. Each Mo· site is defined by a terminal oxo-group, an oxo-group bridging to the second Mo of the unit, an oxo-group bridging to a Cu site, and two phosphonate oxygen donors. The copper site is bound in the equatorial plane to the nitrogen donors of the bpy ligand, and two oxygen donors from the diphosphonate ligand chelating through an oxygen of each (-PO_3) terminals, in addition the Cu is weakly bound by axial interactions to the oxo-groups of Mo sites from two adjacant binuclear units. As shown in **Figure 12,** the structure may be described as oxomolybdate phosphate chains, linked by propylene bridges into a two-dimensional Mo/O/P/C network. The $\{Cu(bpy)\}^{2+}$ groups decorate both surfaces of this layer.

The isolation of these hybrid phases illustrates the synergistic influence of diphosphonate tether lengths and the metal-organic component on the structures of hybrid oxide materials. The organic tether serves to expand the distance between molybdate polyhedra or clusters and may consequently influence the dimensionality of the material. However, the structural influences of the variabilities of polyhedral type and polyhedral connectivities available to the oxometalate substructures are also determinants which may produce unanticipated structural motifs. For example, the common tendency of the V/P/O and Mo/P/O families to form two-dimensional substructures, based on the $MO(PO_4)$ prototype, would suggest that an oxomolybdate material incorporating a diphosphonate of sufficiently extended tether length to buttress such layers would allow adoption of a three dimensional connectivity. The steric constraints of the $\{Cu(o\text{-phen})_2\}^{2+}$ or $\{Cu(bpy)\}^{2+}$ subunit can, of course, passivate the surface of the oxide and result in lower dimensional materials, as noted for $[Cu(2,2'\text{-bpy})(VO)(O_3PCH_2CH_2CH_2PO_3)]\bullet H_2O$. The one-dimensional structure of $[\{Cu(H_2O)_2(o\text{-phen})\}\{Cu(o\text{-phen})_2\}(Mo_5O_{15})$ $(O_3PCH_2CH_2CH_2PO_3)]$ may reflect several influences: the persistence of the $\{Mo_5O_{15}(O_3P\text{-})_2\}^{4\text{-}}$ core and the incorporation of significant numbers of passivating subunits, in this case both $\{Cu(H_2O)_2(o\text{-phen})\}^{2+}$ and $\{Cu(o\text{-phen})_2\}^{2+}$.

Acknowledgment

This work was supported by a grant from the Natural Science Foundation (CHE9987471).

References

1. Greenwood, N.N.; Earnshaw, A., *Chemistry of the Elements,* Pergamon Press, New York, 1984.
2. Hench, L.L., *Inorganic Biomaterial,* in *Materials Chemistry, an Emerging Discipline,* L.V. Interrante, L.A. Casper, A.B. Ellis, eds. ACS series 245, chapter 21, pp. 523, 1995.
3. Mason, B., *Principles of Geochemistry,* 3rd ed., Wiley, New York. 1966.
4. Mann, S., Biomineralization - The Hard Part of Bioinorganic Chemistry *Chem. Soc., Dalton Trans.* **1993,** *1* 1; Mann, S., Biomineralization - The Form(Id)Able Part of Bioinorganic Chemistry *Chem. Soc., Dalton Trans.* **1997,** *21* 3953.
5. Zaremba, C.M.; Belcher, A.M.; Fritz, M.; Le, Y.; Mann, S.; Hansma, P.K.; Morse, D.E.; Speck, J.S.; Stucky, G.D., Critical Transitions in the Biofabrication of Abalone Shells and Flat Pearls *Chem. Mater.* **1996,** *8,* 679.

6. Well, A.F., *Structural Inorganic Chemistry*, 4th ed., Oxford University Press, Oxford, 1975.
7. Reynolds, T.G.; Buchanan, R.C., *Ceramic Materials for Electronics* 2nd ed., R.C. Buchanan, ed., Dekker, New York, 1991, p. 207.
8. Büchner, W.; Schliebs, R.; Winter, G., Büchel, K.H., *Industrial Inorganic Chemistry*, VCH, New York, 1989.
9. McCarroll, W.H., *Encyclopedia of Inorganic Chemistry*, R.B. King, ed., John Wiley and Sons, New York, 1994, vol. 6, p. 2903.
10. Haertling, G.H., *Ceramic Materials for Electronics*, 2nd. Ed., R.C. Buchanan, ed., Dekker, New York, 1991, p. 129.
11. Leverenz, H. W., *Luminescence of Solids*, New York, 1980.
12. Einzinger, R., Metal Oxide Varistors *Ann. Rev. Mater. Sci.* **1987**, *17*, 299.
13. Tarascon, J.M.; Barboux, P.; Miceli, P.F.; Greene, L.H.; Hull, G.H.; Eibschutz, Perovskite, M.; Sunshine, S.A., Structural and Physical Properties of the Metal (*M*) substituted $YBa_2Cu_{3-x}M_xO_{7-x}$, *Phys. Rev.* **1988**, *B37*, 7458.
14. Matkins, D.I., D.W. Jones eds., Academic Press, New York, 1972, p. 235.
15. Rao, C.N.R.; Rao, K.J., *Ferroics*, in *Solid State Compounds*, A.K. Cheetham and P. Day eds., Clarendon Press, Oxford, 1992, 281.
16. Bierlein, J.D.; Arweiler, C.B., Electro-optic and dielectric properties of KTiOPO *Appl. Phys. Lett.* **1986**, *49*, 917.
17. Centi, G.; Trifuro, F.; Ebner, J.R.; Franchetti, V.M., Mechanistic Aspects of Maleic Anhydride Synthesis from C_4 Hydrocarbons over Phosphorous Vanadium Oxide, *Chem. Rev.* **1988**, *88*, 55.
18. Grasselli, R.K., Selectivity and Activity Factors in Bismuth-molybdate Oxidation Catalysts, *Appl. Catal.* **1985**, *15*, 127.
19. Gasior, M.; Gasior, I.; Grzybowska, B., o-Xylene Oxidation on the V_2O_5-TiO_2 Oxide System 1. Dependence of Catalytic Properties on the Modification of TiO_2, *Appl. Catal.* **1984**. *10*, 87.
20. Okuhara, T.; Misono, M., *Encyclopedia of Inorganic Chemistry*, R.R. King, ed., John Wiley and Sons, New York 1994, vol. 6, p. 2889.
21. Niiyama, H.; Echigoya, E., Hydrogen Transfer Reaction between Alcohols and Acetone, *Bull. Chem. Soc. Jpn.* **1972**, *45*, 938.
22. Yamaguchi, T., Recent Progress in Solid Superacid *Appl. Catal.* **1990**, *61*, 1.
23. Clearfield, A., Role of Ion Exchange in Solid State Chemistry, *Chem. Rev.* **1988**, *88*, 125.
24. Newsam, J.M., *Zeolites*, A.K. Cheetham and P. Day, eds., Clarendon Press, , Oxford, 1992, p. 234.
25. Ruthven, D. M., *Principles of Absorption and Adsorption Processes*, Wiley-Interscience, New York, 1984.
26. Szostak, R., *Molecular Sieves – Principles of Synthesis and Identification*, Van Nostrand Rheinhold, New York, 1988.
27. Raleo, J.A., *Zeolite Chemistry and Catalysis*, American Chemical Society, Washington , DC, 1997.
28. Y. Murakami, A. Iijima, J.W. Ward ed., *New Developments in Zeolite Science*, Elsevier, Amsterdam, 1986.
29. (a) Vaughan, D.E.W., *Properties and Applications of Zeolites*, Chem. Soc. Special Publ. No. 33, R.P. Townsend, ed.: The Chemical Society, London, 1979, p. 294. (b) Cheetham, A.K., Advanced Inorganic Materials - An Open Horizon *Science* **1994**, *264*, 794.
30. Venuto, P.B., A perspective on reaction paths within micropores *Microporous Materials* **1994**, *2*, 297.
31. Whitesides, G.M.; Ismagilov, R.F., Complexity in Chemistry *Science* **1999**, *284*, 89.
32. Stupp, S.I.; Braun, P.V., Molecular Manipulation of Microstructures - Biomaterials, Ceramics, and Semiconductors *Science* **1997**, *277*, 1242.
33. (a) Smith, V. J., Topochemistry of Zeolites and Related Materials. 1. Topology and Geometry, *Chem. Rev.* **1988**, 88, 149. (b) Davis, M.E.; Katz, Ahmad, W.R., Rational Catalyst Design via Imprinted Nanostructured Materials *Chem. Mater.* **1996**, *8*, 1920.
34. Kresge, C.T.; Leonowicz, M.E.; Roth, W.J.; Vartuli, J.C.; Beck, J.S., Ordered Mesoporous Molecular-Sieves Synthesized by a Liquid-Crystal Template Mechanism *Nature* **1992**, *359*, 710.
35. Mann, S., Molecular Tectonics in Biomineralization and Biomimetic Materials Chemistry *Nature* **1993**, *365*, 499.
36. Haushaulter, R.C.; Mundi, L.A., Reduced Molybdenum Phosphates - Octahedral Tetrahedral Framework Solids with Tunnels, Cages, and Micropores *Chem. Mater.* **1992**, *4*, 31.
37. Khan, M.I.; Meyer, L.M.; Haushalter, R.C.; Schweitzer, C.L.; Zubieta, J.; Dye, J.L., Giant Voids in the Hydrothermally Synthesized Microporous Square Pyramidal Tetrahedral Framework Vanadium Phosphates $(HN(CH_2CH_2)_3NH)K_{1.35}(V_5O_9(PO_4)_2)\cdot XH_2O$ and $Cs_3(V_5O_9(PO_4)_2)\cdot XH_2O$ *Chem. Mater.* **1996**, *8*, 43.
38. Stein, A.; Keller, S.W.; Mallouk, T.E., Turning Down the Heat - Design and Mechanism in Solid-State Synthesis *Science* **1993**, *259*, 1558.
39. Laudise, R.A., Hydrothermal Synthesis of Crystals, *Chem. Eng. News* **1987**, Sept. 28, p. 30.

40. Gopalakrishnan, J., Chimie Douce Approaches to the Synthesis of Metastable Oxide Materials *Chem. Mater.* **1995**, *7*, 1265.

41. The power of this technique is also illustrated in metal-halide systems where compositional space diagrams are yielding fundamental information on key reaction variables. Francis, R.J.; Halayamani, P.S.; Bee, J.S.; O'Hare, D., Variable Dimensionality in the Uranium Fluoride 2-Methyl-Piperazine System - Syntheses and Structures of UFO-5, UFO-6, and UFO-7 - Zero- Dimensional, One-Dimensional, and 2-Dimensional Materials with Unprecedented Topologies *J. Am. Chem. Soc.* **1999**, *121*, 1609 and references therein.

42. Alberti, G. in *Comprehensive Supramolecular Chemistry*, J.L. Atwood, J.E.D. Davis, F. Vogel, eds.; Pergamon Press, New York, 1996, vol. 9, G. Alberti, T. Bein, eds., p. 152.

43. Clearfield, A., in *Comprehensive Supramolecular Chemistry*, eds.; Pergamon Press, new York, vol. 9; g. Albert, T. Bein, eds., p. 107.

44. Soghomonian, V.; Chen, Q.; Haushalter, R.C.; Zubieta, J., Investigations into the Targeted Design of Solids - Hydrothermal Synthesis and Structures of One-Dimensional, 2-Dimensional, and 3-Dimensional Phases of the Oxovanadium-Organodiphosphonate System *Angew. Chem. Int. Ed. Engl.* **1995**, *34*, 223.

45. Harrison, W.T.A.; Dussach, L.L.; Jacobson, A.J., Syntheses, Crystal-Structures, and Properties of New Layered Alkali-Metal Molybdenum(VI) Methylphosphonates - $Cs_2(MoO_3)_3PO_3CH_3$ and $Rb_2(MoO_3)_3PO_3Ch_3$ *Inorg. Chem.* **1995**, *34*, 4774.

46. Mitzi, D.M., Synthesis, Crystal-Structure, and Optical and Thermal-Properties of (C4H9Nh3)(2)Mi(4) (M=ge, Sn, Pb) *Chem. Mater.* **1996**, *8*, 791.

47. Marsden, I.R.; Allan, M.L.; Friend, R.H.; Kurmoo, M.; Kanazawa, D.; Day, P.; Bravic, G.; Chasseau, D.; Ducasse, L.; Hayes, W., Crystal and Electronic-Structures and Electrical, Magnetic, and Optical-Properties of 2 Copper Tetrahalide Salts of bis(Ethylenedithio)- Tetrathiafulvalene *Phys. Rev. B: Condens. Matter* **1994**, *50*, 2118.

48. Mitzi, D.B.; Wang, S.; Feild, C.A.; Chess, C.A.; Guloy, A.M., Conducting Layered Organic-Inorganic Halides Containing (110)-Oriented Perovskite Sheets *Science* **1995**, *267*, 1473.

49. Mostafa, M.F.; Abdel-Kader, M.M.; Arafat, S.S.; Kandeel, E.M., Thermochromic Phase-Transitions in 2 Aromatic Tetrachlorocuprates *Phys. Sci.* **1991**, *43*, 627.

50. Wang, S.; Mitzi, D.B.; Landrum, G.A.; Genis, H.; Hoffmann, R., S., Synthesis and Solid-State Chemistry of CH_3BiI_2 - A Structure with an Extended One-Dimensional Organometallic Framework *J. Am. Chem. Soc.* **1997**, *119*, 724.

51. Burwell, D.A.; Valentine, K.G.; Timmermans, J.H.; Thompson, M.E., Structural Studies of Oriented Zirconium bis(Phosphonoacetic Acid) Using Solid-State P-31 and C-13 NMR *J. Am. Chem. Soc.* **1992**, *114*, 4144.

52. Clearfield, A., The Role of Ion Exchange in Solid State Chemistry *Chem. Rev.* **1988**, *88*, 125.

53. Zhang, Y. and Clearfield, Synthesis, Crystal-Structures, and Coordination Intercalation Behavior of 2 Copper Phosphonates A., *Inorg. Chem.* **1992**, *31*, 2821 and references therein.

54. Rong, D.; Hong, H.-G.; Kim, Y.I.; Kreuger, J.S.; Mayer, J.E.; Mallouk, T.E., Electrochemistry and Photoelectrochemistry of Transition-Metal Complexes in Well-Ordered Surface-Layers *Coord. Chem. Rev.* **1990**, *97*, 237.

55. Clark, E.T.; Rudolf, P.R.; Martell, A.E.; Clearfield, A. Structural Investigation of the Cu(II) Chelate of N-Phosphonomethylglycine - X-Ray Crystal-Structure of $Cu(II)(O_2CCH_2NHCH_2PO_3).Na(H_2O)_{3.5}$ *Inorg. Chim. Acta* **1989**, *164*, 59.

56. Huan, G.; Day, V.W.; Jacobson, A.J.; Goshorn, D.P. Synthesis and Crystal-Structure of a Spherical Polyoxovanadium Organophosphonate Anion - $[H_{12}(VO_2)_{12}(C_6H_5PO_3)_8]^{4-}$ *J. Am. Chem. Soc.* **1991**, *113*, 3188.

57. Huan, G.; Jacobson, A.J.; Day, V.W., An Unusual Polyoxovanadium Organophosphonate Anion, $[H_6(VO_2)_{16}(CH_3PO_3)_8]^{8-}$ *Angew. Chem. Int. Ed. Engl.* **1991**, *30*, 422.

58. Müller, A.; Hovemeier, K.; Rohlfing, R., A Novel Host Guest System with a Nanometer Large Cavity for Anions and Cations - $[2NH^{4+},2Cl^-\subset V_{14}O_{22}(OH)_4(H_2O)_2(C_6H_5PO_3)_8]^{6-}$ *Angew. Chem. Int. Ed. Engl.* **1992**, *31*, 1192.

59. Müller, A.; Hovemeier, K.; Krickemeyer, E.; Boegge, H. Modeling the Remote-Controlled Organization of Particles in a Nanodimensional Cavity - Synthesis and Properties of $(Et_3NH_3)('BuNH_3)_2Na_2[(H_2O)_2,N^-_3\subset V_{14}O_{22}(OH)_4(PhPO_3)_8]\cdot 6H_2O\cdot 2DMF$ *Angew. Chem. Int. Ed. Engl.* **1995**, *34*, 779.

60. (a) Chang, Y.-D. and Zubieta, J., Investigations into the Syntheses and Structures of Clusters of the Mo-O-REO$_3{}^{2-}$ Systems (E=P and As) *Inorg. Chim. Acta* **1996**, *245*, 177; (b) Khan, M.I.; Chen, Q.; Zubieta, J., Oxomolybdenum(V) Polyanion Clusters - Hydrothermal Syntheses and Structures of $(NH_4)_5Na_4[Na\{Mo_6O_{12}(OH)_3(O_3PC_6H_5)_4\}_2]\cdot 6H_2O$ and

$(C_6H_5CH_2NMe_3)_4K_4[K_2\{Mo_6O_{12}(OH)_3(O_3PC_6H_5)_4\}_2]\cdot10H_2O$ and Their Relationship to the Binuclear $(Et_4N)[Mo_2O_4Cl_3(H_2O)_3]\cdot5H_2O$ *Inorg. Chim. Acta* **1995**, *235*, 135.

61. Bujoli, B.; Palvadeau, P.; Rouxel, J., Synthesis of New Lamellar Iron(III) Phosphonates *Chem. Mater.* **1990**, *2*, 582.

62. Cao, G.; Lee, H.; Lynch, V.M.; Mallouk, T.E., Synthesis and Structural Characterization of Homologous Series of Divalent-Metal Phosphonates, $M^{II}(O_3PR)\cdot H_2O$ and $M^{II}(HO_3PR)_2$, *Inorg. Chem.* **1988**, *27*, 2781.

63. Martin, K.; Squattrito, P.J.; Clearfield, A., The Crystal and Molecular-Structure of Zinc Phenylphosphonate *Inorg. Chim. Acta* **1989**, *155*, 7.

64. Cao, G. and Mallouk, T.E., Shape-Selective Intercalation Reactions of Layered Zinc and Cobalt Phosphonates *Inorg. Chem.* **1991**, *30*, 1434.

65. LeBideau, J.; Bujoli, B.; Jouanneaux, A.; Payen, C.; Palvadeau, P.; Rouxel, J., Preparation and Structure of $Cu^{II}(C_2H_5PO_3)$ - Structural Transition Between Its Hydrated and Dehydrated Forms *Inorg. Chem.* **1993**, *32*, 4617.

66. Bhardwaj, C.; Hu, H.; Clearfield, A., Synthesis and Crystal-Structures of 3 Zinc (Chloromethyl)Phosphonates *Inorg. Chem.* **1993**, *32*, 4294.

67. Poojary, D.M. and Clearfield, A., Coordinative Intercalation of Alkylamines into Layered Zinc Phenylphosphonate - Crystal-Structures from X-Ray-Powder Diffraction Data *J. Am. Chem. Soc.* **1995**, *117*, 11278 and references therein.

68. Cao, G.; Lynch, V.M.; Swinnea, J.S.; Mallouk, T.E., Synthesis and Structural Characterization of Layered Calcium and Lanthanide Phosphonate Salts *Inorg. Chem.* **1990**, *29*, 2112.

69. Alberti, G.; Casciola, M.; Biswas, R.K., Preparation and Some Properties of Zr Phosphate Hypophosphite and Zr Phosphate Dimethylphosphinate with Gamma-Layered Structure *Inorg. Chim. Acta* **1992**, *201*, 207 and references therein.

70. Huan, G.; Jacobson, A.J.; Johnson, J.W.; Corcoran, Jr. E.W., Hydrothermal Synthesis and Single-Crystal Structural Characterization of the Layered Vanadium(IV) Phosphate $VOC_6H_5PO_3.H_2O$ *Chem. Mater.* **1990**, *89*, 220.

71. Huan, G.; Johnson, J.W.; Jacobson, A.J.; Merola, J.S., Hydrothermal Synthesis and Single-Crystal Structural Characterization of $(VO)_2[CH_2(PO_3)_2]\cdot4H_2O$ *J. Solid State Chem.* **1990**, *89*, 220.

72. Snover, J.L. and Thompson, M.E., Synthesis and Study of Zirconium Viologen Phosphonate Thin-Films Containing Colloidal Platinum *J. Am. Chem. Soc.* **1994**, *116*, 765.

73. Ortiz-Avila, C.Y.; Bhardwaj, C.; Clearfield, A., Zirconium Polyimine Phosphonates, a New Class of Remarkable Complexing Agents *Inorg. Chem.* **1994**, *33*, 2499.

74. Harrison, W.T.A.; Dussack, L.L.; Jacobson, A.J., Syntheses, Crystal-Structures, and Properties of New Layered Alkali-Metal Molybdenum(VI) Methylphosphonates - $Cs_2(MoO_3)_3PO_3CH_3$ and $Rb_2(MoO_3)_3PO_3CH_3$ *Inorg. Chem.* **1995**, *34*, 4774.

75. LeBideau, J.; Payen, C.; Polvadeau, P.; Bujoli, B., reparation, Structure, and Magnetic-Properties of Copper(II) Phosphonates - β-$Cu^{II}(CH_3PO_3)$, an Original 3-Dimensional Structure with a Channel-Type Arrangement *Inorg. Chem.* **1994**, *33*, 4885; Drumel, S.; Janvier, P.; Deniaud, D.; Bujoli, B., Synthesis and Crystal-Structure of $Zn(O_3PC_2H_4NH_2)$, the First Functionalized Zeolite-Like Phosphonate *J. Chem. Soc. Chem. Commun.* **1995**, 1051; Drumel, S.; Janvier, P.; Barboux, P.; Bujoli-Doeuff, M.; Bujoli, B., Synthesis, Structure, and Reactivity of Some Functionalized Zinc and Copper(II) Phosphonates *Inorg. Chem.* **1995**, *34*, 148.

76. Poojary, M.; Grohol, D.; Clearfield, A., Synthesis and X-Ray-Powder Structure of a Novel Porous Uranyl Phenylphosphonate Containing Unidimensional Channels Flanked by Hydrophobic Regions *Angew. Chem. Int. Ed. Engl.* **1995**, *34*, 1508; Poojary, M.; Cabeza, A.; Aranda, M.A.G.; Bruque, S.; Clearfield, A., Structure Determination of a Complex Tubular Uranyl Phenylphosphonate, $(UO_2)_3(HO_3PC_6H_5)_2(O_3PC_6H_5)_2\cdot H_2O$, from Conventional X-Ray- Powder Diffraction Data *Inorg. Chem.* **1996**, *35*, 1468.

77. Maeda, K.; Akimoto, J.; Kiyozumi, Y.; Mizukami, F., Almepo-Alpha - A Novel Open-Framework Aluminum Methylphosphonate with Organo-Lined Unidimensional Channels *Angew. Chem. Int. Ed. Engl.* **1995**, *34*, 1199; Maeda, K.; Akimoto, J.; Kiyozumi, Y.; Mizukami, F., Structure of Aluminum Methylphosphonate, Almepo-Beta, with Unidimensional Channels Formed from Ladder-Like Organic-Inorganic Polymer-Chains *J. Chem. Soc. Chem. Commun.* **1995**, 1033.

78. Harrison, W.T.A.; Dussack, L.L.; Jacobson, A.J., Syntheses and Properties of New Layered Alkali-Metal/Ammonium Vanadium(V) Methylphosphonates - $M(VO_2)_3(PO_3CH_3)_2$ (M=NH_4,K,Rb,Tl) - Single- Crystal Structures of $K(VO_2)_3(PO_3CH_3)_2$ and $NH_4(VO_2)_3(PO_3CH_3)_2$ *Inorg. Chem.* **1996**, 35, 1461.

79. Khan, M.I. and Zubieta, J., Oxovanadium and Oxomolybdenum Clusters and Solids Incorporating Oxygen- Donor Ligands *Prog. Inorg. Chem.* **1995**, *43*, 1.

80. Khan, M.I.; Lee, Y.-S.; O'Connor, C.J.; Haushalter, R.S.; Zubieta, J., Hydrothermal Synthesis and Structural Characterization of an Organically Templated Layered Oxovanadium(IV) Organophosphonate - $(C_2H_5NH_3)_2[V_3O_3(H_2O)(PhPO_3)_4]$ *Inorg. Chem.* **1994**, *33*, 3855.

81. Khan, M.I.; Lee, Y.-S.; O'Connor, C.J.; Haushalter, R.S. Zubieta, J., Organically Templated Layered Vanadium(IV)-Oxo-Organophosphonate Exhibiting Intercalated Organic Cations and Open-Framework V/P/O Layers - $[(C_2H_5)_2NH_2][(CH_3)_2NH_2][V_4O_4(OH)_2(C_6H_5PO_3)_4]$ *J. Am. Chem. Soc.* **1994**, *116*, 4525.

82. Khan, M.I.; Lee, Y.-S.; O'Connor, C.J.; Haushalter, R.S.; Zubieta, Hydrothermal Synthesis and Structure of $[(C_2H_5)_4N][(VO)_3(OH)(H_2O)(C_2H_5PO_3)_3].H_2O$ - An Inorganic-Organic Polymer Exhibiting Undulating V/P/O Layers and Interlamellar Organic Templates J., *Chem. Mater.* **1994**, *6*, 721.

83. Soghomonian, V.; Chen, Q.; Haushalter, R.C.; Zubieta, J., Investigations into the Targeted Design of Solids - Hydrothermal Synthesis and Structures of One-Dimensional, 2-Dimensional, and 3-Dimensional Phases of the Oxovanadium-Organodiphosphonate System *Angew. Chem. Int. Ed. Engl.* **1995**, *34*, 223.

84. Soghomonian, V.; Diaz, R.; Haushalter, R.C.; O'Connor, C.J.; Zubieta, J., Hydrothermal Syntheses and Crystal-Structures of 2 Oxovanadium- Organodiphosphonate Phases - $[H_2N(C_2H_4)_2NH_2][(VO)_2(O_3PCH_2CH_2CH_2PO_3H)_2]$, a Stair-Step Structure Incorporating an Organic Cationic Template, and $((VO)(H_2O)(O_3PCH_2NH(C_2H_4)_2NHCH_2PO_3))$, a Layered Structure Pillared by a Piperazinium-Tethered Diphosphonate *Inorg. Chem.* **1995**, *34*, 4460.

85. Soghomonian, V.; Haushalter, R.C.; Zubieta, J., Hydrothermal Synthesis and Structural Characterization of V(III)- Containing Phases of the Vanadium Organodiphosphonate System - Crystal- Structures of the V(III) Species $(H_3O)[V_3(O_3PCH_2Ch_2PO_3)(HO_3PCH_2CH_2PO_3H)_3]$ and of the Mixed-Valence V(III)/V(IV) Material $(H_3O)_2[(VO)V_2(OH)_2(O_3PCH_2CH_2PO_3)_2].H_2O$ *Chem. Mater.* **1995**, *7*, 1648.

86. Chen, Q.; Salta, J.; Zubieta, J., Synthesis and Structural Characterization of Binuclear Vanadium Oxo Organophosphonato Complexes - Building-Blocks for Oxovanadium Organophosphonate Solids *Inorg. Chem.* **1993**, *32*, 4485.

87. Chen, Q. and Zubieta, J., Control of Aggregation Through Anion Encapsulation in Clusters of the $V/O/RPO_3^{2-}$ System - Synthesis and Crystal and Molecular-Structures of the Tetranuclear Clusters $(^nBu_4N)[V_4(V)O_6F](PHPO_3)_4$ and $(^nBu_4N)_2(V(V)_3(V)(VI)O_6F(PHPO_3)_4)$ *J. Chem. Soc. Chem. Commun.* **1994**, 2663.

88. Chen, Q. and Zubieta, J., Vanadium Phosphate Framework Solid Constructed of Octahedra, Square Pyramids, and Tetrahedra with a Cavity Diameter of 18.4 Angstrom *Angew. Chem. Int. Ed. Engl.* **1993**, *32*, 610.

89. Salta, J.; Chen, Q.; Chang, Y.-D.; Zubieta, J., The Oxovanadium-Organophosphonate System - Complex Cluster Structures $[(VO)_6(^tBuPO_3)_8Cl]$, $[(VO)_4\{PhP(O)_2OP(O)_2Ph\}_4Cl]^-$, and $[V_{18}O_{25}(H_2O)_2(PhPO_3)_{20}Cl_4]^{4-}$ with Encapsulated Chloride Anions Prepared from Simple Precursors *Angew. Chem. Int. Ed. Engl.* **1994**, *33*, 757.

90. Chang, Y.-D.; Salta, J.; Zubieta, J., Synthesis and Structure of an Oxovanadium(V)Organophosphonate Cluster Encapsulating a Chloride-Ion - $[ClV_7O_{12}(O_3PC_6H_5)_6]^{2-}$ *Angew. Chem. Int. Ed. Engl.* **1994**, *33*, 325.

91. Chen, Q. and Zubieta, J., Investigations of the Vanadium-Oxo-Organophosphonato System - Preparation and Structural Characterization of a Mixed-Valence $V-^V-V-^{IV}$ Cluster Encapsulating Chloride Anions, $(Bu_4N)_2[(V_8O_{16})\{V_4O_4(H_2O)_{12}\}(PhPO_3)_8Cl_2]_2Et_2O\cdot2MeOH\cdot4H_2O$ *J. Chem. Soc. Chem. Commun.* **1994**, 1635.

92. Khan, M.I. and Zubieta, J. The Oxovanadium-Organoarsonate System - $(2CH_3OH\subset[V_{12}O_{14}(OH)_4(PhAsO_3)_{10}]^{4-}$ and $[2H_2O\subset V_{12}O_{12}(OH)_2(H_2O)_2(PhAsO_3)_{10}(PhAsO_3H)_4]^{2-}$, Clusters with Nanometer Dimensions and Cavities for Neutral Molecules,*Angew. Chem. Int. Ed. Engl.* **1994**, *33*, 760.

93. Salta, J. and Zubieta, J., *J. Cluster Sci.* **1996**, *7*, 531.

94. Pope, M.T., *Heteropoly and Isopoly Oxometalates,* Springer, New York, 1983.

95. Chang, Y.-D.; Zubieta, J., Investigations into the syntheses and structures of clusters of the $Mo-O-REO_3^{2-}$ systems (E=P and As) *Inorg. Chim. Acta* **1995**, *245*, 177.

POLYOXOTHIOMOLYBDATES DERIVED FROM THE $\{Mo_2O_2S_2(H_2O)_6\}^{2+}$ BUILDING UNIT

Francis Sécheresse, Emmanuel Cadot, Anne Dolbecq-Bastin, and Bernadette Salignac

Institut Lavoisier, UMR CNRS 8637, Université de Versailles St Quentin 78035, Versailles, France

INTRODUCTION

The substitution of oxygen by sulfur atoms in the framework of a polyoxoanion is expected to modify both the chemical and redox properties of those species. Unfortunately, this substitution is generally accompanied by the partial reduction of metal centers together with the degradation of the oxometallic structure. The first works reported by Holm[1] in this area evidenced the efficiency of the hexamethyldisilathiane as a good sulfurating agent of polyoxoanions but only sulfur containing compounds of limited nuclearity were obtained through this way.

A different strategy was imagined by Klemperer[2] consisting in introducing in the polyanion backbone a less reducible metal such as Ta^V or Nb^V. Based on this approach, α-$[PNbW_{11}EO_{39}]^{4-}$ E=$S^{3,4}$, Se^4, were isolated as solids and fully characterized after the oxo-parent was sulfurated by the Lawesson reagent[3] or another adequate sulfur donor[4]. We have developed a new strategy consisting in the stereospecific addition of a preformed thiofragment on a vacant polyoxoanion. The $[Mo_2O_2S_2(H_2O)_6]^{2+}$ precursor was obtained by oxidation in water of the $[Mo_2(O)_2(\mu-S)_2(\eta^2-S_2)(\eta^2-S_4)]^{2-}$ anion by iodine. The oxidation was performed in acidic medium in which the dithiocation is very stable and does not self-condense, or reacts with the parent complex.[5] The homologous $[W_2O_2S_2(DMF)_6]^{2+}$ was obtained in DMF, but conversely to $[Mo_2O_2S_2(H_2O)_6]^{2+}$, it revealed poorly stable in aqueous acid conditions to rapidly degrade with H_2S evolution. Thus, the addition of $[M^V_2O_2S_2]^{2+}$ on the divacant γ-$[XW_{10}O_{36}]^{n-}$ X=P^V, Si^{IV}, gave easily γ-$[XM_2S_2W_{10}O_{38}]^{n-}$ since a perfect adequacy in size and electronic behavior exist between the thiofragment and the cavity of the polyoxoanion.[6] This adequacy can be broken by varying the size of the vacant polyanion which can display a cavity too large or too small to contain the dithiocation. This is typically the case for the trivacant α-A- $[PW_9O_{34}]^{9-}$: the thio-precursor cannot fill the large vacancy but bridges two polyanion units to give $[P_2W_{18}Mo_6S_6O_{74}]^{12-}$, a sandwich-type polyanion.[7] The same type of structure was obtained with the monovacant $[PW_{11}O_{39}]^{7-}$ for which the cavity revealed too small to insert the dithiocation: two $[M^V_2O_2S_2]^{2+}$ units bridge two polyanions with the possibility of different isomers for the resulting

Polyoxometalate Chemistry for Nano-Composite Design
Edited by Yamase and Pope, Kluwer Academic/Plenum Publishers, 2002

59

$[P_2W_{22}Mo_4S_4H_4O_{84}]^{6-}$.[8] Considering the large number of available polyvacant structures, it is clear that a wide family of thio-polyanions can be obtained on these bases. However, these strategies do not permit to prepare sulfur rich polyanions.

By analogy with polyoxometalates,[9] we recently postulated that the condensation of a sulfur containing precursor remained the best way to prepare new species having both high sulfur stoichiometry and high nuclearity.[10]

THE FIRST MEMBERS OF A NEW FAMILY

A facile synthetic approach toward the formation of condensed species derived from the $[Mo_2O_2S_2(H_2O)_6]^{2+}$ precursor consists in the addition of a base to a solution of the hexaaquo-dication. Thus, on adding a concentrated solution of KOH to a solution of the dithiocation up to about pH=4, in the presence of iodide, a yellow solid is isolated and analyzed as $\{K_{0.3}(NMe_4)_{0.1}I_{0.4}[Mo_2O_2S_2(OH)_2]. 3H_2O\}_n$ (1).[11]
The dissolution of (1) in water leads to the crystallization of the neutral wheel $[Mo_{12}O_{12}S_{12}(OH)_{12}(H_2O)_6]$. $19H_2O$ (2) represented in Figure 1 and in Equations 1 and 2.[10]

(1) $n[(H_2O)_3Mo_2O_2S_2(H_2O)_3]^{2+} + 2nOH^- \xrightarrow{\text{[KI]=1 mol.L}^{-1}} \{I_{0.4}[Mo_2O_2S_2(OH)_2]^{0.4-}\}_n + 6n\,H_2O$
$$\textbf{(1)}$$

(2) $\{I_{0.4}[Mo_2O_2S_2(OH)_2]^{0.4-}\}_n \xrightarrow{\text{water}} [Mo_{12}O_{12}S_{12}(OH)_{12}(H_2O)_6] + 0.4nI^-$
$\quad\quad\quad\textbf{(1)} \quad\quad\quad\quad\quad\quad\quad\quad\quad\quad\quad\quad\quad\quad\quad\quad\textbf{(2)}$

The structure of (2) consists in a ring formed of six dinuclear $\{Mo_2O_2S_2\}$ units connected by 12 hydroxo-double bridges delimiting a central cavity of 11Å lined by six water molecules.

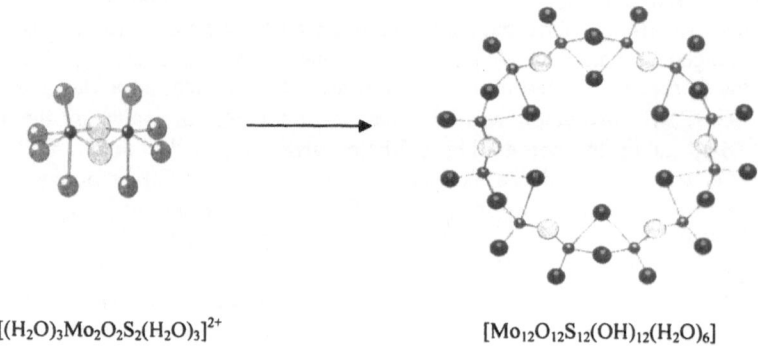

$[(H_2O)_3Mo_2O_2S_2(H_2O)_3]^{2+}$ $[Mo_{12}O_{12}S_{12}(OH)_{12}(H_2O)_6]$

Figure 1. Representation of the molecular structure of $[Mo_{12}O_{12}S_{12}(OH)_{12}(H_2O)_6]$ in ball and stick model (with permission of Wiley-VCH).

The molecular ring is composed of alternately edge- and face-sharing octahedra, this type of enchainments being rare in polyoxoanion chemistry.

In DMF containing halides X=I⁻, Cl⁻, the recrystallization of (1) gives the two dianions $[X_2Mo_{10}O_{10}S_{10}(OH)_{10}(H_2O)_5]^{2-}$ X=I⁻ (3),[12] Cl⁻ (4).[13] The molecular structure, common to (3) and

(4), is remarkable for displaying a cyclic arrangement of ten Mo^V atoms doubly bridged by hydroxo groups around a central cavity capped by two halide ions.

The reaction scheme illustrating the syntheses of these different compounds is given in Eq 1, 2, and 3.

$$(3) \quad \{I_{0.4}[Mo_2O_2S_2(OH)_2]^{0.4-}\}_n \xrightarrow{\text{DMF, NBu}_4X} [X_2Mo_{10}O_{10}S_{10}(OH)_{10}(H_2O)_5]^{2-} + 0.4nI^-$$
$$(2) \qquad\qquad\qquad\qquad\qquad\qquad\qquad\qquad X= I, \quad (3); \ X= Cl, (4)$$

The molecular structures of (3) and (4) have been determined by X-Ray diffraction methods. The two halides are located on each side of the plane containing the wheel, and penetrate deeply in the ring as illustrated in Figure 2 : the X-X distances are short, 4.78Å for iodides (sum of ionic radii, 4.40Å) and 3.82Å for chlorides (sum of ionic radii, 3.62Å).

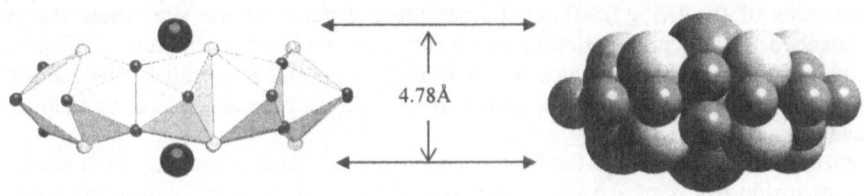

4.78Å

Figure 2. Polyhedra and space filling representations of the $[I_2Mo_{10}O_{10}S_{10}(OH)_{10}(H_2O)_5]^{2-}$ anion showing the interaction of the I-I group with the neutral $\{Mo_{10}O_{10}S_{10}(OH)_{10}(H_2O)_5\}$ ring (with permission of the RSC).

An aqueous solution of (1) has a pH of about 4.3 for a concentration of 6.10^{-3} mol. L^{-1} in $\{Mo_2O_2S_2\}$. By addition of 2.4 equivalents of KOH and excess of KCl, the pH of the solution raised to 5.9 allowing the crystallization of $K_3[ClMo_{10}O_{10}S_{10}(OH)_{12}(H_2O)_3]. \ 14.25 \ H_2O$ (5).

The cyclic backbone of (5) contains ten molybdenum atoms bonded through Mo-$(OH)_2$-Mo double bridges and is capped by a chloride ion. This situation is quite comparable to that observed for (3) and (4) except only one halide ion is capping the cyclic ring.[13] The (3-) negative charge of the anion is balanced in the solid state by three potassium cations which imposes the presence in the ring of three aquo and two hydroxo ligands instead of the five water molecules observed in (4). These two types of ligands were clearly identified by the length of the Mo-O bonds shorter in the case of OH-ligands (2.337(9)-2.3805(10)Å) than for aquo-ligands (2.487-2.517(13)Å). A view of the molecular structure of $[ClMo_{10}O_{10}S_{10}(OH)_{12}(H_2O)_3]^{3-}$ is given in Figure 3a.

In the solid state, all the cations are engaged in connections with the Mo_{10}-wheels generating the nice porous three-dimensional structure represented in Figure 3b. Two types of channels (16x8Å for the largest and 8x10Å for the smallest) are observed, resulting from the perpendicular arrangements of two $\{ClMo_{10}\}$ containing sheets.

The various channels do not contain any cation and are occupied only by water molecules which give the structure its microporous character. The thermal stability of the structure is quite poor because of the large number of water molecules completing the coordination sphere of the cations or filling the channels.

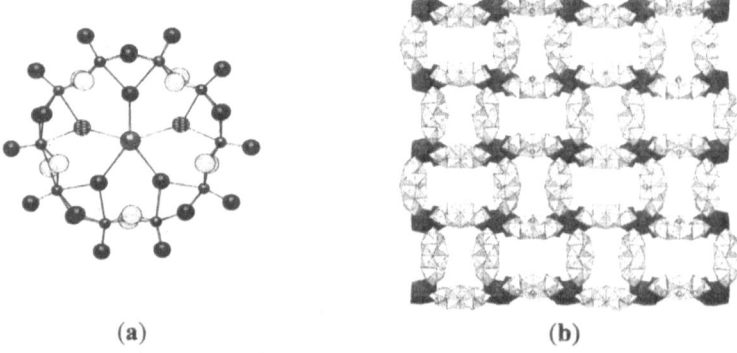

(a)   (b)

Figure 3. Representations of the structure of (**5**), (**a**) ball and stick view of the molecular anion, (**b**) view of the 3D-arrangement, potassium cations are black grey polyhedra.

The acid properties of the $[Mo_2O_2S_2(H_2O)_6]^{2+}$ precursor depend on the site where the water molecule is attached. The deprotonation of equatorial water molecules (the most acid) is at the origin of the polymerization process which finally leads to Mo_{10} and Mo_{12} rings. The acido-basic equilibria are reversible so that the dithiocation is restored by simple addition of an acid to any soluble condensed architectures.

The axial coordination sites, trans to the Mo=O bond, are less acid explaining the presence of water molecules inside the rings. As the condensation takes place at the equatorial positions, soluble inorganic chains are formed and are closed by a double olation-reaction. The formation of rings is probably due to an entropic factor since a cyclic conformation, rather than a linear chain, increases the entropy of the solvent, generates the elimination of two water molecules and reduces the interactions with the solvent by minimizing the surface.

The solubility of (**1**) is closely dependent on the halide concentration. When (**1**) is dissolved in pure water, the neutral Mo_{12} ring rapidly precipitates. Conversely, concentrated solutions of (**1**) can be obtained by maintaining a high halide concentration. This observation confirms that soluble species such as $\{X_2Mo_{10}\}^{2-}$, characterized in the solid state, unambiguously exist in solution. The specific interaction between the neutral ring and halide ions provokes a structural rearrangement of the metallic backbone, the nuclearity decreasing from 12 to 10 to induce the formation of the charged $\{X_2Mo_{12}\}^{2-}$ anion.

The negative charge of the neutral and poorly soluble $\{Mo_{12}\}$ ring to form $\{X_2Mo_{12}\}^{2-}$ permits the dissolution of the inorganic ring in aqueous medium. On this basis it can be assumed that the structure of (**1**), formed in solutions containing high concentrations of halide, corresponds to the formula $K_{1.5}(NMe_4)_{0.5}[I_2Mo_{10}O_{10}S_{10}(OH)_{10}(H_2O)_5]$. The different equilibria related to the $[Mo_2O_2S_2(H_2O)_6]^{2+}$ precursor are represented in Figure 4.

The molecules of water located in the cavity achieve the coordination of the Mo-centers. The length of these Mo-OH$_2$ bonds (2.56-2.75Å) suggests the water molecules are labile as would be molecules of solvent. This property, together with the electrophilic potentiality of the cavity, due to the presence of Mo^V cations, can favor the insertion in the ring of anionic groups of adequate size.

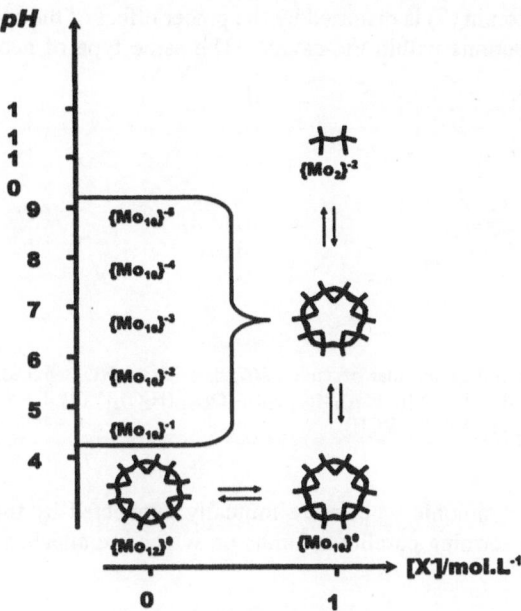

Figure 4. Diagrammatic representation of the equilibria involving the $[Mo_2O_2S_2(H_2O)_6]^{2+}$ precursor and related condensed species.

STRUCTURATING INFLUENCE OF XO₄ GROUPS, X=P, As

Owing to the sizes of the former ten- and twelve-membered rings, phosphate and arsenate groups were chosen to play the role of quite ideal guests for these cationic hosts since their charge can be easily tuned by the control of the pH of the solution. Two different types of compounds were obtained, depending on the initial concentration in phosphate or arsenate.

Weak XO₄ content

With weak concentrations in XO₄, in the range 0.02-0.08 mol.L^{-1} and for the ratios 1<Mo/X<3, the cyclic $[(HPO_4)Mo_{10}O_{10}S_{10}(OH)_{11}(H_2O)_2]^{2-}$ (**6**), noted {PMo₁₀}, and $[(HXO_4)_2Mo_{12}O_{12}S_{12}(OH)_{12}(H_2O)_2]^{4-}$, noted {XMo₁₂}, X=P, (**7**), X=As, (**8**), were crystallized.[14] The cavity of about 10Å in diameter contains a diprotonated phosphate group H₂PO₄$^-$. As the gobal charge of the anion is balanced in the solid by two tetramethylammonuim cations, it was necessary to introduce a hydroxo group in the cavity which was characterized by the short Mo-O distances of 2.354(10) and 2.422(10)Å.
The structure of $[(HPO_4)Mo_{10}O_{10}S_{10}(OH)_{11}(H_2O)_2]^{2-}$ (**6**) represented in Figure 5 is based on the cyclic enchainment of five {Mo₂O₂S₂} fragments bonded by double hydroxo bridges.

The 12-membered skeleton in (**7**) is distorted by the pincer effect of the two phosphate anions and hydrogen bonding interactions within the cavity. The same type of geometry is obtained with AsO_4 replacing PO_4.

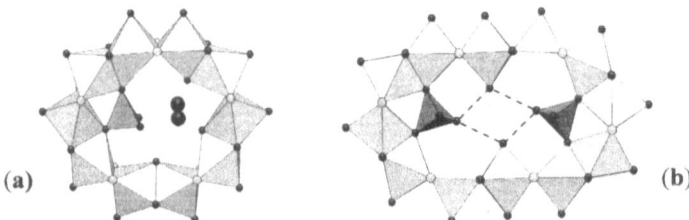

(a)   (b)

Figure 5. (**a**) Representation of the molecular structure of (**6**) showing the XO_4 tetrahedron and the two out of plane water molecules, (**b**) view of the elliptic $[(HPO_4)_2Mo_{12}O_{12}S_{12}(OH)_{12}(H_2O)_2]^{4-}$ (**7**) illustrating the cis positions of the two PO_4 groups (with permission of Wiley-VCH).

In the solid state, the polyanionic wheels are mutually connected by the sodium cations. The cations are edge bonded forming parallel columns on which are attached the different wheels as represented in Figure 6.[14]

Figure 6. 3D-arrangement of (**7**) showing the Mo_{12}-wheels attached to Na^+ columns (with permission of Wiley-VCH).

^{31}P NMR spectra of solutions containing $Na_4[(HPO_4)_2Mo_{12}O_{12}S_{12}(OH)_{12}(H_2O)_2]\cdot23H_2O$ at a concentration of 0.04 mol.L^{-1} were recorded at 298K and are represented in Figure 7. Three peaks are observed at $\delta_1 = 1.05$, $\delta_2 = -2.7$ and $\delta_3 = -3.4$ ppm, respectively. The first peak located at 1.05 ppm corresponds to uncoordinated phosphate while the two others are attributed to coordinated phosphates.

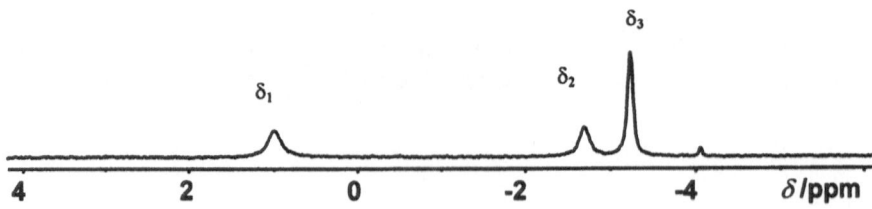

Figure 7. ^{31}P NMR of a solution of $\{P_2Mo_{12}\}$, T = 278K, $[\{P_2Mo_{12}\}] = 0.04$ mol. L^{-1} (with permission of Wiley-VCH).

The attribution of these two peaks was established through the study of P:As exchange from synthetic mixtures of $\{PMo_{12}\}$ and $\{AsMo_{12}\}$ and from spectra of solutions of $\{PMo_{12}\}$ at different concentrations.

Peak	δ_1: 1.05 ppm	δ_2: -2.7 ppm	δ_3: -3.4 ppm
Attribution	Free Phosphate	$\{P_2Mo_{12}\}$	$\{PMo_{10}\}$

These results show that both $\{P_2Mo_{12}\}$ and $\{PMo_{10}\}$ exist in equilibrium in solution. The structures of the two species have already been characterized in the solid state, (6), (7). That means that the dissolution of $\{P_2Mo_{12}\}$ in water is accompanied by a structural change, $\{P_2Mo_{12}\}$ being fluxional enough to generate $\{PMo_{10}\}$ and *vice versa*.[14]

High XO$_4$ content

When an excess of phosphate or arsenate XO_4 is added to a solution of $[Mo_2O_2S_2(H_2O)_6]^{2+}$, X/Mo >2, the structurating role of XO_4 groups leads to the formation of the hexanuclear wheel $[H_7X_4Mo_6S_6O_{25}]^{5-}$ X=P, (8) and X=As, (9), noted $\{X_4Mo_6\}^{5-}$.[15] These compounds are stable in solution only in the presence of excess of XO_4 but can be isolated quite quantitatively by precipitation or crystallization. The molecular structure of (8) represented in Figure 8 consists of the assembly of three dithiocations condensed around a central PO_4 group and connected to three outer phosphates. The inter-fragment connections are ensured by an oxo and a hydroxo groups.

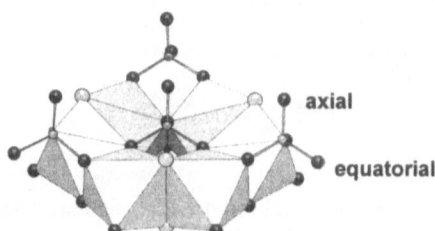

Figure 8. Polyhedral view of $[H_7X_4Mo_6O_{25}]^{5-}$ X = P, As, showing the six membered ring with the axial and equatorial positions of the X-OH bonds (with permission of Wiley-VCH).

Two types of connections between $\{Mo_2O_2S_2\}$ units have been encountered so far, face sharing in $\{Mo_{2n}\}$ and edge sharing in $\{X_4Mo_6\}$ structures. These two different bonding schemes are represented in Figure 9.

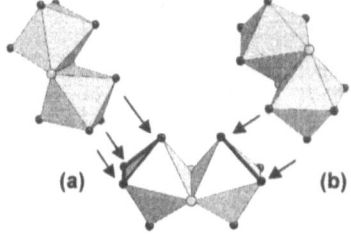

Figure 9. Two modes of connection of the $\{Mo_2O_2S_2\}^{2+}$ units, (a) face sharing, (b) edge sharing.

STRUCTURATING INFLUENCE OF CARBOXYLIC ACIDS

Acetic acid

The condensation of the $[Mo_2O_2S_2]^{2+}$ fragment leads to the formation of cyclic structures, the size and shape of which seem to be controlled by the template anion present in solution. With phosphate, the nuclearity of the metallic skeleton fits with the size of the anion which can be a single phosphate group like in $\{PMo_{10}\}$ or two phosphate groups like in $\{P_2Mo_{12}\}$. The size and reactivity of the acetate anion are comparable with that of phosphates explaining that acetates can be introduced in the inorganic ring which adapts its shape to the new template. The dissolution of (**1**) in acetate buffer gives after crystallization yellow needles analyzed as $(NMe_4)_3[Mo_{12}O_{12}S_{12}(OH)_{12}(CH_3COO_3)]$. The molecular structure of (**10**) is represented in Figure 10 showing the Mo_{12}-ring is fluxional enough to arrange around three acetate groups in a nice heart-shape.

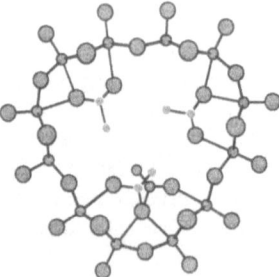

C$_{17}$Mo$_{12}$S$_{12}$O$_{12}$O$_{41}$N$_3$ (**10**)
Orthorhombic, *Pbam*
a = 33.4176(5)Å
b = 34.9007(3)Å
c = 7.5281(1)Å
Z = 6
R = 0.11

Figure 10. Crystal parameters and molecular structure of $[Mo_{12}O_{12}S_{12}(OH)_{12}(CH_3COO)_3]^{3-}$ (**10**), the three acetate groups are disordered over two positions.

Dicarboxylic acids

There is a double interest to use dicarboxylic acids as templates. The first one is they are easily available with various alkyl chain lengths which permits to tune the size of the molybdic ring by the length of the carbon chain. The second interest is they are good chelating agents through the two terminal carboxylate groups that limits the exchange phenomena observed with phosphates or acetates. Thus, the three following dicarboxylates were chosen, oxalate $[C_2O_4]^{2-}$, glutarate $[C_5H_6O_4]^{2-}$, and pimelate $[C_7H_{10}O_4]^{2-}$. Two protocoles of syntheses can indifferently be used, consisting in the addition of carboxylate anions to a solution of (**1**), or to a solution of $[Mo_2O_2S_2]^{2+}$. In both cases the pH of the solution is adjusted to about 5-6 by addition of a base and the reaction mixture cristallyzed.

Oxalic acid. $[Mo_8O_8S_8(OH)_8(C_2O_4)]^{2-}$ (**11**) was obtained as tetramethylammonium or lithium salt starting from (**1**). Four $\{Mo_2O_2S_2\}$ units are linked by hydroxo-groups and condensed around a central oxalate in a quite perfect circular arrangement as represented in Figure 11. All the Mo atoms in the cyclic skeleton adopt an octahedral geometry formed of a terminal M=O double bond, two hydroxo and two sulfido bridging groups and an oxygen of the oxalate.[19]

Glutaric acid. $Cs_2[Mo_{10}O_{10}S_{10}(OH)_{10}(C_5H_6O_4)]\cdot nH_2O$ (**12**) was obtained by precipitation by CsCl of a solution of (**1**) previously hydrolyzed about pH = 6-10. Crystals for X-ray diffraction determinations were obtained by recrystallysation of the precipitate in water. The molecular structure of (11) represented in Figure 13 shows a distorted cyclic arrangement of the inorganic backbone. The distorted inorganic ring contains ten molybdenum atoms, five of them being pentacoordinated, probably because of steric constraints due to the longer C_5-alkyl chain.[19]

Pimelic acid. $Cs_2[Mo_{12}O_{12}S_{12}(OH)_{12}(C_5H_6O_4)]\cdot nH_2O$ (13) was isolated through the same protocole of preparation.[19] The distorted ring contains twelve Mo-atoms, six of them being pentacoordinated like in (12), see Figure 11.

As represented in Figure 11, a direct connection exists between the nuclearity of the wheels and the size of the central carboxylate, C_2-Mo_8, C_5-Mo_{10}, C_7-Mo_{12}. The wheel adapts its size and shape to the size of the template during the condensation process.

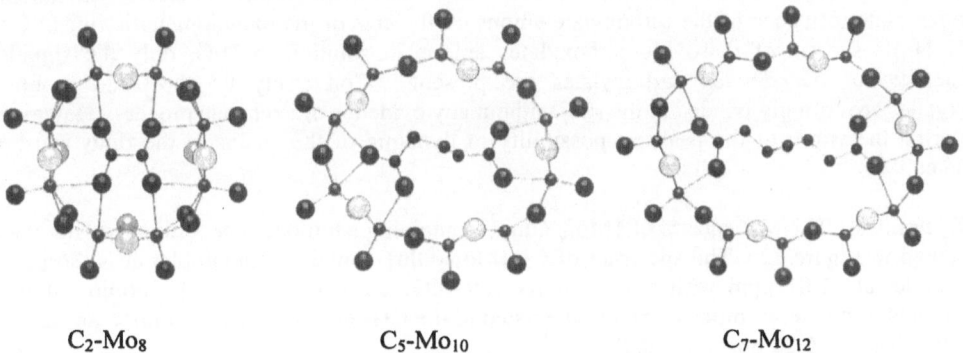

C_2-Mo_8 C_5-Mo_{10} C_7-Mo_{12}

Figure 11. View of $[Mo_nO_{2n}S_{2n}(OH)_{2n}(dicarbox)]^{2-}$ anions illustrating the adaptation of the nuclearity of the Mo-inorganic ring to the size of the templating carboxylate (with permission of the ACS).

Selectivity of the template effect. The selectivity of the method of synthesis is illustrated by the narrow distribution of the polyanions in solution. Dilute mixtures of $\{Mo_2O_2S_2\}^{2+}$ and carboxylic acid at pH = 4.5 have been examined by Electro Spray Mass Spectroscopy (ESMS). Each ESMS spectrum reveals that the parent peak of the expected anion is unambiguously observed in solution as predominant species. The experimental m/z values correspond exactly to that deduced from single-crystal X-ray diffraction, Table 1.[19]

Table 1. ESMS data for mixtures of $\{Mo_2O_2S_2\}^{2+}$ solutions containing oxalate, or glutarate or pimelate at pH = 4.5 (with permission of the ACS).

Polyanion	M/z exp. (I, %)	Formula	M/z th.
Mo_8-Ox	686(92) 664(2) 321(6)	$[Mo_8O_8S_8(OH)_8(C_2O_4)]^{2-}$	688
Mo_{10}-Glu	868(93) 638(4) 334(2) 321(1)	$[Mo_{10}O_{10}S_{10}(OH)_{10}(H_6C_5O_4)]^{2-}$	870

Table 1. (*Continued*)

Mo$_{12}$-Pim	1043(52)	[Mo$_{12}$O$_{12}$S$_{12}$(OH)$_{12}$(H$_{10}$C$_7$O$_4$)]$^{2-}$	1045
	785(5)		
	685(7)		
	638(26)		
	461(2)		
	334(5)		
	321(3)		

For {Mo$_{12}$-pim}, the additional peaks observed in the range 321<*m/z*<785 have been attributed to template free anionic species.

The selective preparation of {Mo$_8$-Ox}$^{2-}$, {Mo$_{10}$-Glu}$^{2-}$, and {Mo$_{12}$-Pim}$^{2-}$ is clearly dependent on the template influence of the carboxylate anions on the size of the oxothiometallic rings. On the ^{1}H NMR spectra of these Mo-carboxylates anions in solution in D$_2$O, only the signals characteristic of *coordinated* carboxylates are present. Conversely to phosphate groups, carboxylates are strongly bonded in the ring without any evidence of exchange process. However, considering the width of the peaks, a possibility of dynamic of the chains in the rings can be postulated.

Dynamics. ^{1}H NMR spectra of {Mo$_{10}$-Glu}$^{2-}$ containing additional uncoordinated Glu^{2-} are represented in Figure 12. The spectrum of Cs$_2${Mo$_{10}$-Glu} contains a quintuplet at –0.86 ppm and a triplet at –1.01 ppm with a 1:2 relative intensity, attributed to the CH$_2$ groups of the encapsulated glutarate. Similarly, the peaks related to the CH$_2$ groups of {Mo$_{12}$-Pim}$^{2-}$ appear as two broad lines at –0.20 ppm and –0.53 ppm.[19]

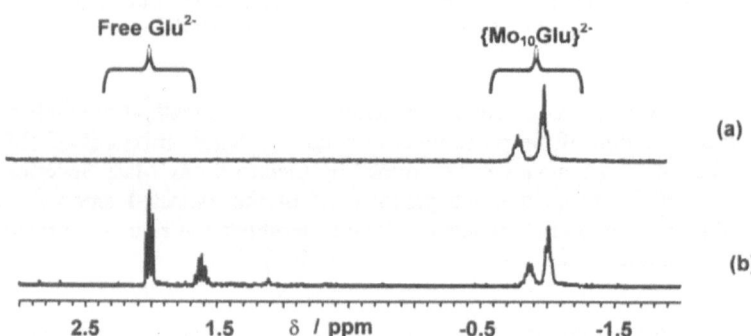

Figure 12. (**a**) ^{1}H NMR spectra of Cs$_2$ {Mo$_{10}$-Glu} in D$_2$O, (**b**) with addition of free Glu^{2-} (with permission of the ACS).

On the basis of its intensity, the –0.20 ppm line was attributed to two equivalent CH$_2$ groups, while the other at –0.53 ppm was attributed to the overlapped resonances of the two remaining CH$_2$ groups.

The ^{1}H NMR spectra of Li$_2${Mo$_8$-Ox} represented in Figure 13 contains two sharp lines at 2.31 and 8.03 ppm attributed to uncoordinated water and to eight equivalent hydroxo-bridges, but does not show any temperature dependence.

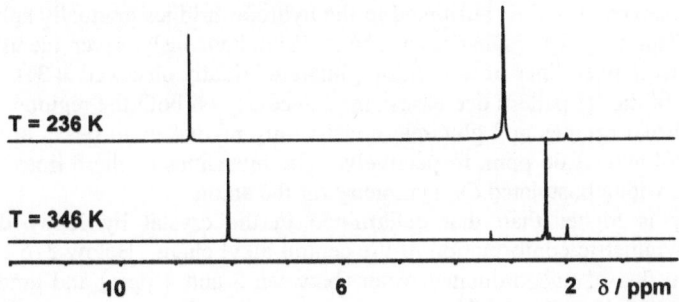

Figure 13. ^{1}H NMR spectra of Li$_2$[Mo$_8$O$_8$S$_8$(OH)$_8$(Ox)] in CD$_3$CN at 296K and 336K (with permission of the ACS).

That means that the eight-membered ring is strongly maintained in a rigid shape by the central oxalate.

In Figure 14 are given the ^{1}H NMR spectra of Li$_2$[Mo$_{10}$O$_{10}$S$_{10}$(OH)$_{10}$(C$_5$H$_6$O$_4$)] in CD$_3$CN. Three groups of lines are observed, as expected for the three types of protons present in the alkyl chain, in water and in hydroxo bridges.

The signals related to the CH$_2$ groups of the carboxylate are located at about 0.0 and -1 ppm, values close to that observed in D$_2$O. The water lines are located in the 4-2 ppm range, the remaining 7-10 ppm lines being attributed to the hydroxo-bridges. The relative intensity of the signals of the protons of the akyl chain and of the hydroxo bridges is 0.6, as expected for the six protons of the three CH$_2$'s and for the 10 protons of the hydroxo-bridges.

The decrease of the temperature provokes dramatic changes in the spectrum of {Mo$_{10}$-Glu}$^{2-}$, both in the region of the hydroxo-bridges and of the CH$_2$'s.

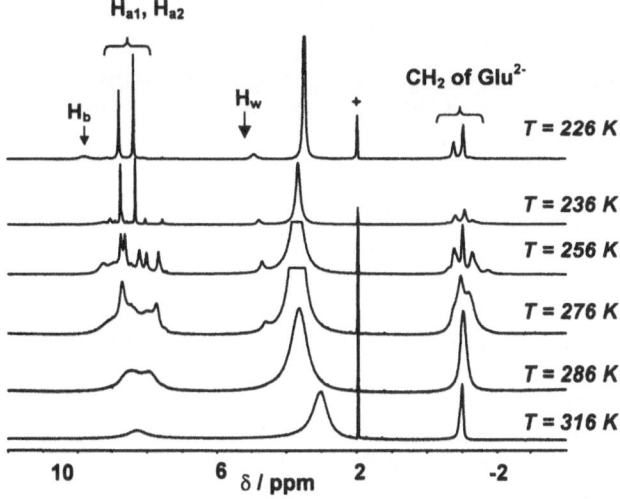

Figure 14. Variable temperature ^{1}H NMR spectra of Li$_2$[Mo$_{10}$O$_{10}$S$_{10}$(OH)$_{10}$(C$_5$H$_6$O$_4$)] in CD$_3$CN (with permission of the ACS).

The broad peak observed at 316 K, attributed to the hydroxo-bridges gradually splits into several sharp lines up to displaying ten resonances at 256 K. Simultaneously, seven identified lines arise from the two 1:2 overlapped lines of the central glutarate initially observed at 316 K. Below 256 K, the complexity of the ^1H pattern decreases simultaneously on both the regions. At 226 K, the resonances of hydroxo bridges and glutarate consist only of five main peaks at 9.78, 8.78 and 8.36 ppm and –0.78 and –1.06 ppm, respectively. The intensities of these lines (1:2:2 and 1:2, respectively) agree with a postulated C_{2v} symmetry for the anion.

Such a symmetry is higher than that determined in the crystal by X-ray diffraction and corresponds to a symmetric conformation of the central alkyl chain. Below 276 K an additional line separates from that of uncoordinated water (between 3 and 4 ppm) and gradually shifts to reach 4.91 ppm at 226 K. This additional peak was attributed to water exchange because the molybdenum atoms in a square pyramidal coordination can bind a single water molecule. A possible polyhedral representation of the C_{2v} conformer is given in Figure 15.

On the basis of its intensity, the broad signal at 9.80 ppm is assigned to the two equivalent protons labelled Hb while the two remaining lines at 8.80 and 8.40 ppm are attributed to the four equivalent Ha1 and Ha2 protons.

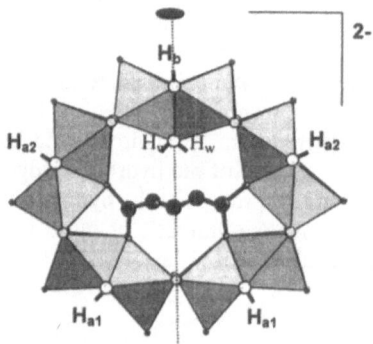

Figure 15. Postulated frozen conformation of $Li_2[Mo_{10}O_{10}S_{10}(OH)_{10}(C_5H_6O_4)]$ in CD_3CN at 226K. The four groups of protons, 4Ha1, 4Ha2, 2Hb, 2Hw, are deduced from the C_{2v} symmetry of the wheel (with permission of the ACS).

The different changes observed in the ^1H NMR spectra of $[Mo_{10}-glu]^{2-}$ recorded at variable temperature in CD_3CN are directly related to the fluxionality of the molecule. At room temperature, the central guest glutarate moves in the cavity as the hand of a watch owing to the concerted hopping of the two terminal carboxylate groups over the ten molybdenum atoms in the ring.

Such a dynamic is strongly supported by the versatility of the Mo^V atoms which can adopt both octahedral and pyramidal coordinations. At 226 K, the hopping of the dicarboxylate is totally blocked on the NMR time scale, which gives a single frozen conformation containing a water molecule (2Hw) coordinated to a Mo atom.

The same type of behavior is observed with $\{Mo_{12}-Pim\}^{2-}$ leading to the possibility of two frozen arrangements.[19] The ^1H NMR study allowed to evidence, between 226K and 336K, the fluxionality of $\{Mo_{10}-Glu\}^{2-}$ and $\{Mo_{12}-Pim\}^{2-}$ in contrast to the rigidity of $\{Mo_8-Ox\}^{2-}$. Replacing the alkyl carboxylic acid by an aromatic tri- or tetra-carboxylic acid blocks the motions in the wheel.

CONCLUSION AND PERSPECTIVES

$[Mo_2O_2S_2]^{2+}$ represents a powerful building unit which allows to generate new polyanionic architectures, unknown in polyoxo-chemistry, by controlling the pH of the solutions, and the type and number of templating anions involved in the condensation process. Through their flexibility, the wheels are relevant to molecular recognition for selective encapsulation of anions with various charges and shapes and are also relevant of solid state chemistry for giving infinite solids exhibiting porosity.

Such a chemistry can probably be extended to tungsten analogues by condensation of the isostructural and isoelectronic $[W_2O_2S_2]^{2+}$, but other thio-precursors are also available and represents good candidates for this chemistry, $[Mo_3S_4]^{4+}$ being a good example of such a possibility.[17]

REFERENCES

1. Y. Do, E. D. Simhon, and R. H. Holm, Oxo/sulfido ligand substitution in $[Mo_2O_7]^{2-}$: reaction sequence and characterization of the final product, $[MoS_3(OSiMe_3)]^-$, *Inorg. Chem.* 24: 1831 (1985)

2. W. G. Klemperer and C. Schwartz, Synthesis and characterization of the polyoxothioanions $TaW_5O_{18}S^{3-}$ and $NbW_5O_{18}S^{3-}$, *Inorg. Chem.* 24:4459 (1985).

3. E. Cadot, V. Béreau, and F. Sécheresse, Synthesis and characterization of the polyoxothioanion α-$[PW_{11}NbSO_{39}]^{4-}$ derived from the Keggin structure, *Inorg. Chim. Acta* 239:39 (1995).

4. E. Radkov, Y. J; Liu, and R. H. Beer, *Inorg. Chem.* 35:551 (1996).

5. D. Coucouvanis, A. Topadakis, and A. Hadjikyriakou, Synthesis of thiomolybdenyl complexes with $[Mo_2(S)_2(O)_2]^{2+}$ cores and substitutionally labile ligands. Crystal and molecular structure of the $[Mo_2O_2S_4(DMF)_3]$ complex, *Inorg. Chem.* 27:3273 (1988).

6. E. Cadot, V. Béreau, S. Halut, and F. Sécheresse, Synthesis and characterization of γ-$[SiW_{10}M_2S_2O_2O_{38}]^{6-}$, M = Mo, W, two Keggin oxothio-heteropolyanions with a metal-metal bond, *Inorg. Chem.* 95:551 (1996).

7. V. Béreau, E. Cadot, H. Bögge, A. Müller, and F. Sécheresse, Addition of $\{Mo_2O_2S_2\}^{2+}$, M = Mo, W, to A-α-$[PW_9O_{39}]^{9-}$. Synthesis and structural characterizations in the solid state and in solution, *Inorg. Chem.* 38:5803 (1999).

8. E. Cadot, B. Salignac, A. dolbecq, and F. Sécheresse, Polyoxometalates: Self Assembled Beautiful Structures, Adaptable Properties, Industrial Applications, A. Müller and M. T. Pope Eds, Kluwer, Dordrecht (2000).

9. M.T. Pope, Heteropoly and Isopoly Oxometalates, Springer, Berlin (1983).

10. E. Cadot, B. Salignac, S. Halut, and F. Sécheresse, $[Mo_{12}S_{12}O_{12}(OH)_{12}(H_2O)_6]$: a cyclic molecular cluster based on the $[Mo_2S_2O_2]^{2+}$ building block, *Angew. Chem. Int. Ed.* 37:611 (1998).

11. B. Salignac, Thesis, University of Versailles (2001).

12. E. Cadot, B. Salignac, J. Marrot, A. Dolbecq, and F. Sécheresse, $[Mo_{10}S_{10}O_{10}(OH)_{10}(H_2O)_5]$: a novel decameric molecular ring showing supramolecular properties, *Chem. Commun.* 261 (2000).

13. E. Cadot, A. Dolbecq, and F. Sécheresse, From Molecular Rings to the 3D-Solid: Ionization of the Neutral $[Mo_{10}S_{10}O_{10}(OH)_{10}(H_2O)_5]$ Molecular Ring For the Building Block Strategy, *J. Phys. Chem. Solids* 2000, submitted.

14. E. Cadot, B. Salignac, T. Loiseau, A. Dolbecq, and F. Sécheresse, Syntheses and ^{31}P NMR studies of cyclic oxothiomolybdate(V) molecular rings: exchange properties and crystal structures of the monophosphate decamer $[(H_2PO_4)Mo_{10}O_{10}S_{10}(OH)_{10}(H_2O)_2]^{2-}$ and the diphosphate dodecamer $[(HPO_4)_2Mo_{12}O_{12}S_{12}(OH)_{12}(H_2O)2]^{2-}$, *Chem. Eur. J.* 5: 3390 (1999).

15. E. Cadot, A. Dolbecq, B. Salignac, and F. Sécheresse, Self-condensation of $[Mo^VO_2S_2O_2]^{2+}$ with phosphate or arsenate ions by acid-base process in aqueous solutions: syntheses, crystal structures and reactivity of $[(HXO_4)Mo_6S_6O_6(OH)_3]^{5-}$, X = P, As, *Chem. Eur. J.* 5: 2396 (1999).

16. B. Salignac, S. Riedel, A. Dolbecq, F. Sécheresse, and E. Cadot, "Wheeling Templates" in molecular oxothiomolybdates rings: syntheses, structures, and dynamics, *J. Am. Chem. Soc.* 122: 10381 (2000).

17. A. Müller, V. Fedin, C. Kuhlmann, H. D. Fenske, G. Baum, H. Bögge, B. Hauptfleisch, Adding stable functional complementarity, nucleophilic and electophilic clusters: a synthetic route to $[\{(SiW_{11}O_{39})Mo_3S_4(H_2O)_3(\mu$-$OH)\}_2]^{10-}$ and $[\{(P_2W_{17}O_{61})Mo_3S_4(H_2O)_3(\mu$-$OH)\}_2]^{14-}$, *Chem. Commun.* 1189 (1999).

LANTHANIDE POLYOXOMETALATES:
BUILDING BLOCKS FOR NEW MATERIALS

Qunhui Luo, Robertha C. Howell, Lynn C. Francesconi*

Department of Chemistry, Hunter College of the City
University of New York, New York City, NY 10021

INTRODUCTION

Polyoxometalates can be versatile inorganic building blocks for construction of molecular-based materials. The use of polyoxometalates, in combination with inorganic or organic components has been realized in the construction of molecule-based materials. Prominent examples of inorganic materials can be found in the work by Muller on giant molybdates.[1-5] We have been interested in materials comprised of lanthanide polyoxoanion building blocks. We report herein our efforts in the studies of lanthanide complexes of $[\alpha\text{-}1\text{-}P_2W_{17}O_{61}]^{10-}$ and $[\alpha\text{-}2\text{-}P_2W_{17}O_{61}]^{10-}$, isomers of the monovacant Wells-Dawson anion, and $[\alpha\text{-}B\text{-}PW_9O_{34}]^{9-}$ as building blocks to new lanthanide polyoxometalate materials.

By means of their multiple coordination requirements and oxophilicity, lanthanide (Ln) cations are suitable to link polyoxometalates together to form new classes of materials with extended metal oxygen frameworks. For example, lanthanide and actinide ions can complex to surface oxygen atoms: surface supported actinide, Th(IV), and lanthanide, Er(III), complexed to a triad of surface oxygen atoms of the $UMo_{12}O_{48}^{8-}$ units anion form oligomeric species.[6] Recently, complexes of lanthanide ions bound to the surface of $Mo_8O_{26}^{4-}$ units have been reported.[7] In the case of tungstates, lacunary XW_9 ($SbW_9O_{33}^{7-}$ and $AsW_9O_{34}^{9-}$) units and $W_5O_{18}^{6-}$ units have been linked by lanthanide ions to form small , $Eu_3(H_2O)_3(W_5O_{18})_3(SbW_9O_{33})]^{18-}$ clusters[8] and recently, the largest tungstate recorded, a cyclic assembly $[Ln_{16}As_{12}W_{148}O_{524}(H_2O)_{36}]^{76-}$.[9] Extended polymeric structures in the solid state have been isolated from La(III) and Ce(III) complexes with the lacunary $SiW_{11}O_{39}^{4-}$ species.[10]

We are studying the solution and solid state properties of lanthanide complexes of the monovacant Wells-Dawson anions. A firm understanding of the solution and solid state chemistry is a prerequisite for the use of these units as building blocks in the preparation of new molecular based materials.

Polyoxometalate Chemistry for Nano-Composite Design
Edited by Yamase and Pope, Kluwer Academic/Plenum Publishers, 2002

MONOVACANT WELLS-DAWSON ANIONS

We have previously reported studies of the Ln complexes of both the $[\alpha\text{-}1\text{-}P_2W_{17}O_{61}]$ and $[\alpha\text{-}2\text{-}P_2W_{17}O_{61}]$ isomers, including complete multinuclear NMR data.[11,12] Under synthetic conditions of room temperature, lithium acetate buffer at pH=4.7, the 1:1 $[Ln(H_2O)_4(\alpha\text{-}1\text{-}P_2W_{17}O_{61})]$ species form. Under similar conditions, in the case of the $[\alpha\text{-}2\text{-}P_2W_{17}O_{61}]$ isomer, the 1:2 species forms. No complete single-crystal X-ray diffraction structures have been determined before now because of difficulties in obtaining suitable-quality crystals. The sole available structure determination, showing the tungsten framework only, was reported in 1979 for the 1:2 Ce(IV) complex $K_{16}[Ce(P_2W_{17}O_{61})_2]\cdot 50H_2O$.[13] No Ce-O distances were reported because the O atoms were not located in the low-grade (R = 19 %) structure determination. We have solved and refined the complete crystal and molecular structures of two lutetium complexes, $K_7[Lu(\alpha\text{-}1\text{-}P_2W_{17}O_{61})]$ (1) and $K_{17}[Lu(\alpha\text{-}2\text{-}P_2W_{17}O_{61})_2]$ (2). The former structure (1) is the first of its kind for any metal complex of $\alpha\text{-}1\text{-}[P_2W_{17}O_{61}]$. Both structures provide information on the Lu^{3+}-O coordination environments, which have been compared with metrical results from the corresponding XAFS (X-ray absorption fine structure) analyses of the solid-state complexes.[14]

$[Lu(H_2O)_4(\alpha\text{-}1\text{-}P_2W_{17}O_{61})]^{7-}$ (1) Elemental analysis was performed on the crystals: $K_7[Lu(\alpha\text{-}1\text{-}P_2W_{17}O_{61})]\cdot 18.7H_2O\cdot^1/2CH_3COOK$. Calc.: $K_8W_{17}P_2LuO_{80.7}H_{38.9}$: W, 62.34; Lu, 3.49; P, 1.23; K, 5.85; C, 0.24; H, 0.77. Found: W, 59.49; Lu, 3.01; P, 1.15; K, 5.80; C, 0.16; H, 0.40. TGA on the crystals shows 18.7 H_2O. Lithium was not found in the elemental analysis data.

The crystal structure of anion 1, **Figure 1**, shows that the Lu(III) ion is substituted for a $[WO]^{4+}$ unit in the "belt" region of the tungsten-oxygen framework of the parent Wells-Dawson ion, $[\alpha\text{-}P_2W_{18}O_{62}]^{6-}$. The bond lengths from Lu to the four oxygen atoms of the framework are 2.26(2), 2.34(2), 2.24(2), and 2.34(2) Å. Moreover, the crystal structure shows that four water molecules are bound to Lu(III), with Lu-O distances of 2.38(3), 2.44(2), 2.39(3) and 2.45(2) Å, so that Lu is fully coordinated with 8 O atoms in a square antiprism geometry. The presence of 4 H_2O molecules bound to Lu(III) is consistent with luminescence lifetime measurements of the analogous Eu complex, which has four water molecules bound to Eu(III).

Five of the seven potassium ions required for neutrality were found in the crystal structure in close proximity to the anion. The other potassium ions (at least 8 per unit cell, 2 per molecule) are likely disordered along with an unknown amount of water in the channels. The elemental analysis on the crystals is consistent with this assessment: two to three potassium ions, water and some acetate per molecule are disordered in the infinite channels that lie parallel to the b axis in the unit cell.

$[Lu(\alpha\text{-}2\text{-}P_2W_{17}O_{61})_2]^{17-}$. The crystal structure of anion 2 shown in **Figure 2** demonstrates that the Lu(III) ion substitutes for two $[WO]^{4+}$ units in the "cap" regions of two $\alpha\text{-}2\text{-}[P_2W_{17}O_{61}]^{10-}$. The Lu(III) ion is in a square antiprismatic coordination environment with 8 oxygen atoms—four from each of the two $\alpha\text{-}2\text{-}[P_2W_{17}O_{61}]^{10-}$ ligands. The Lu-O bond lengths exhibit a range of 2.249(18) to 2.388(18) Å. The two polyoxometalate "lobes" are disposed in a *syn* fashion. This structure is similar to the partial structure of $[Ce^{4+}(\alpha\text{-}2\text{-}P_2W_{17}O_{61})_2]^{16-}$. The molecule has C_2 point group symmetry, consistent with the ^{31}P and ^{183}W solution NMR spectroscopic results.[11]

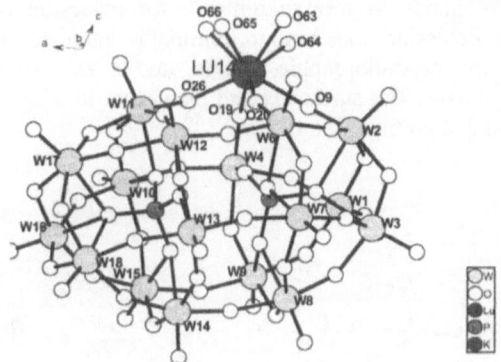

Figure 1. Crystal Structure of $[Lu(H_2O)_4(\alpha\text{-}1\text{-}P_2W_{17}O_{61})]^{7-}$ (**1**).

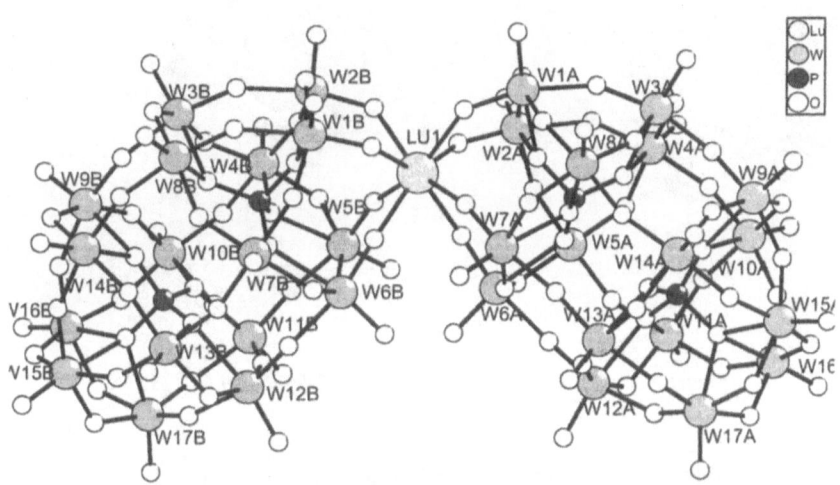

Figure 2. Crystal Structure of $[Lu(\alpha\text{-}2\text{-}P_2W_{17}O_{61})_2]^{17-}$ (**2**)

LACUNARY POLYOXOMETALATES AS BUILDING BLOCKS

1:1 Ln: (α-2-P$_2$W$_{17}$O$_{61}$) (3) Reaction of the lacunary K$_{10}$[α-2-P$_2$W$_{17}$O$_{61}$] with excess EuCl$_3$ and precipitation with KCl, followed by recrystallization from hot water, resulted in the 1:1 [Eu(α-2-P$_2$W$_{17}$O$_{61}$)]$^{7-}$ species. A polyoxometalate array is assembled through the building block, [Eu(α-2-P$_2$W$_{17}$O$_{61}$)]$^{7-}$. The crystal structure of the anion, **3**, **Figure 3**, shows clearly that two identical [Eu((H$_2$O)$_3$(α-2-P$_2$W$_{17}$O$_{61}$)]$^{7-}$ moieties are connected through two Eu–O–W bonds to form a dimer. An inversion center relates the two polyoxometalate units. Potassium ions bind to terminal oxygen atoms of two dimers giving rise to chains along the crystallographic *a* axis, and a mesh-like structure along the crystallographic *b* and *c* axes. The surface bound potassium ions results in the formation of discrete channels forming a porous 3D structure.[15,16]

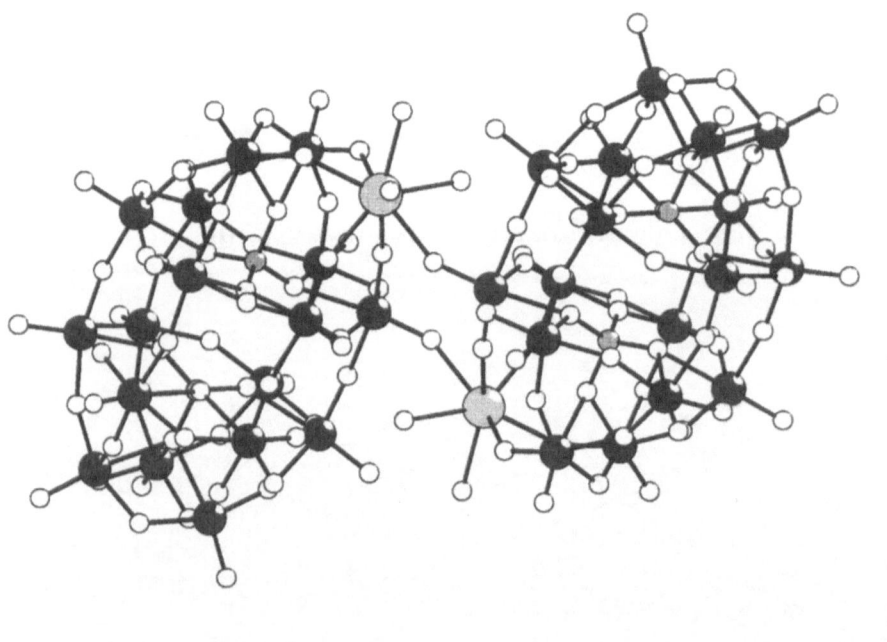

O Eu
● W
○ P
○ O

Figure 3. Crystal Structure of [Eu((H$_2$O)$_3$(α-2-P$_2$W$_{17}$O$_{61}$)]$^{7-}$ (**3**).

3 is less soluble in water than anions of similar size, probably due to the covalent bonding of the Eu(III) to a terminal oxygen of an adjacent polyoxoanion and the surface binding of the cations. The luminescence data in a dilute solution provide compelling data that the extended structure dissociates upon dissolution in water and allows one to follow the 1:1 $\Leftrightarrow$ 1:2 equilibrium.

Figure 4 shows the excitation spectrum resulting from dilute solutions (100 nM to 250 mM) of the 1:2 [Eu(α-2-P$_2$W$_{17}$O$_{61}$)$_2$]$^{17-}$ complex. Two peaks are observed: the peak at 580.44 nm corresponds to the [Eu(α-2-P$_2$W$_{17}$O$_{61}$)$_2$]$^{17-}$ complex; the peak at 579.81 nm corresponds to the 1:1 [Eu(α-2-P$_2$W$_{17}$O$_{61}$)]$^{7-}$ species. These concentrations bracket the equilibrium of the 1:1 and 1:2 Eu: α-2-[P$_2$W$_{17}$O$_{61}$]$^{10-}$ species. The formation constant for the equilibrium

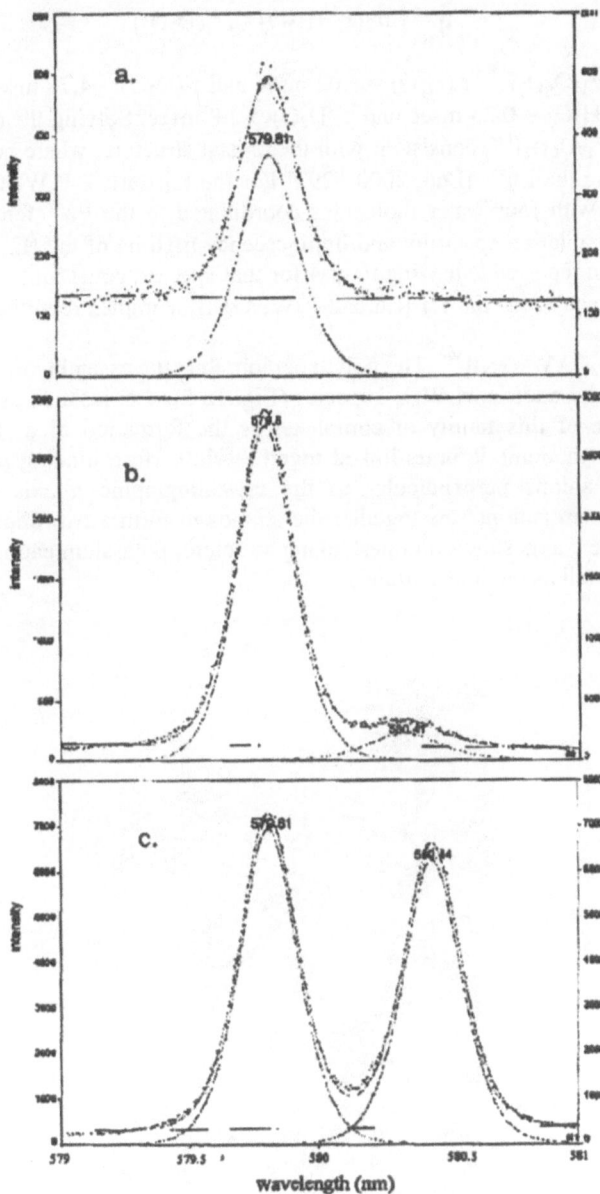

Figure 4. Excitation Spectra for $[Eu(\alpha\text{-}2\text{-}P_2W_{17}O_{61})_2]^{17-}$ taken in H_2O. Concentrations: a. 100 nM, b. 10 μM, c. 250 μM. Emission monitored at 614 nm.

$[Eu(\alpha\text{-}2\text{-}P_2W_{17}O_{61})]^{7-}$ + $\alpha\text{-}2\text{-}[P_2W_{17}O_{61}]^{10-}$ $\Leftrightarrow$ $[Eu(\alpha\text{-}2\text{-}P_2W_{17}O_{61})_2]^{17-}$ is in the micromolar range, consistent with the measurements of other workers for the Eu(III) and Ce(III) complexes.[17,18]

Luminescent lifetime measurements in H_2O and D_2O allow determination of the number of coordinated water molecules, q, according to the following equation.

$$q = 1.05[\tau^{-1}(H_2O) - \tau^{-1}(D_2O)]$$

For $[Eu(\alpha\text{-}2\text{-}P_2W_{17}O_{61})_2]^{17-}$, $\tau(H_2O) = 3.02$ msec and $\tau(D_2O) = 4.79$ msec, whereas for the 1:1 complex, $\tau(H_2O) = 0.23$ msec and $\tau(D_2O) = 2.8$ msec. Solving the equation, $q = 0.12$ for $[Eu(\alpha\text{-}2\text{-}P_2W_{17}O_{61})_2]^{17-}$, consistent with the crystal structure, where no water molecules are coordinated to the Ln^{3+}. [Luo, 2000 #292] For the $K_7[Eu(\alpha\text{-}2\text{-}P_2W_{17}O_{61})]$ complex, $q = 4.19$, consistent with four water molecules coordinated to the Eu^{3+} for the 1:1 complex. Previously, the excitation spectrum and luminescence lifetime of the $[Eu(\alpha\text{-}1\text{-}P_2W_{17}O_{61})]^{7-}$ species have been reported,[12] leading to $q=4$ for that species, consistent with the number of bound water molecules for the 1:1 $[Eu(\alpha\text{-}2\text{-}P_2W_{17}O_{61})]^{7-}$ complex reported here.

$[Y_8(PW_{10}O_{37})_4(W_3O_{18})]^{30-}$ The polyoxoanion, **4**, is the assembly of 4 $PW_{10}O_{37}^{9-}$ units, 8 Y^{3+} cations and 3 additional WO_6^{6-} species (**Figure 5**). **4** is isolated as a potassium salt. A unique feature of this family of complexes is the formation of a "tertiary" structure derived from the monomeric units linked together via surface binding potassium cations. Packing of the anions perpendicular to the crystallographic *a* axis shows a zig-zag geometry; potassium cations sew together the strands to form a two dimensional network. Packing along the *c* axis shows an interlocking structure; potassium cations tie together the dimeric units as well as the linear strands.

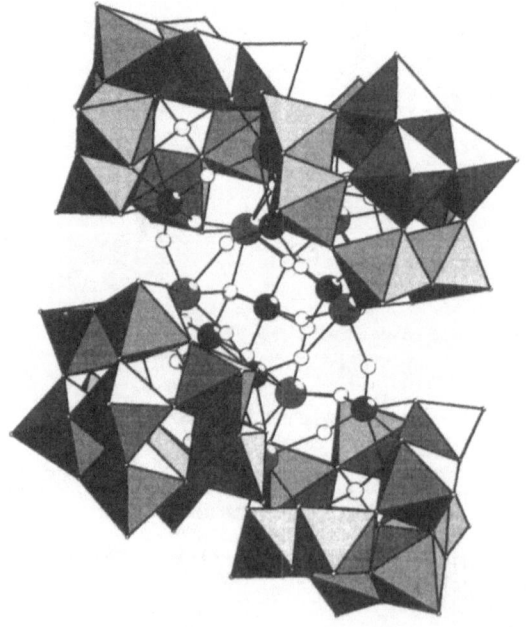

Figure 5. Crystal Structure of $[Y_8(PW_{10}O_{37})_4(W_3O_{18})]^{30-}$ (**4**).

Both ^{183}W and ^{31}P NMR provide good evidence to suggest that the cluster remains intact in solution. The PW_9 unit, itself, is not stable under aqueous conditions forming other polyoxoanions upon decomposition.[19] The Y, W network discussed above apparently holds the PW_9 units together tightly according to spectroscopic data. Two resonances are observed in the ^{31}P NMR spectrum consistent with the C_2 symmetry. The [183]W NMR spectrum of the Y analog shows 19 resonances with one of intensity 1 and three of about double intensity than the other fifteen, suggesting overlapping resonances. 22 resonances are expected as the complex consists of 22 non-equivalent tungsten atoms.

Reaction of Eu(III)and $(PW_9O_{34})^{9-}$ and Aluminum cations, 5

Isolation of the reaction product of Eu(III) and $(PW_9O_{34})^{9-}$ with aluminum cations results in the formation of a $[Eu((H_2O)_3(\alpha\text{-}2\text{-}P_2W_{17}O_{61})]_2^{14-}$ dimer, identical to 3, above; however in 5 the dimeric units are connected by aluminum (III) cations binding to terminal oxygen atoms of adjacent dimers forming a porous network, shown in **Figure 6**.

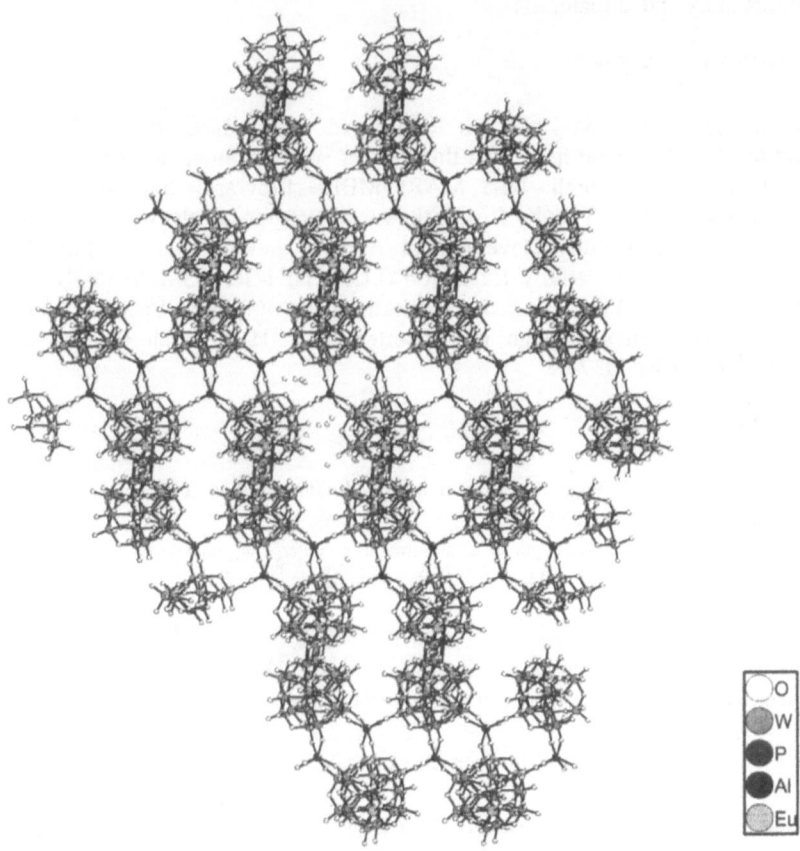

○	O
◐	W
●	P
◓	Al
◯	Eu

Figure 6. Connection of dimeric $[Eu((H_2O)_3(\alpha\text{-}2\text{-}P_2W_{17}O_{61})]_2^{14-}$ units by Al(III) to form porous structure.

CONCLUSION

Lanthanide polyoxometalates can be used in conventional syntheses to build up two dimensional materials. Linking up lacunary subunits is accomplished by lanthanide binding to terminal oxygen atoms of an adjacent polyoxometalate, forming dimers. The appropriate cations can link the dimers into a lattice, as found in the case of Al^{3+}.

FUTURE WORK

In the preparation of materials, aqueous lanthanide and polyoxometalate chemistry is fraught with problems owing to multi-equilibria. Occluded water is often hard to remove and may preclude some applications. Additionally, in aqueous syntheses, different counterions complicate the issue as shown from our work and that of others. Therefore, it is very difficult to "predict" complexes and structures that will be formed in reactions carried out in water. We intend to branch into syntheses using organic soluble lacunary polyoxoanions and organic soluble or organometallic lanthanide precursors. We hope to establish reactions that can be depended on to produce complexes that can be reliably used in synthesis of extended materials.

ACKNOWLEDGEMENT

We are grateful to Ms. Marilyn Rampersad, Ms. Frances Perez and Ms. Kymora Scotland for their technical help with this project. We acknowledge the following sources of support for this research: NIH MARC/MBRS Program; Gertrude Elion Graduate Scholarship Award (QL), Faculty Research Award Program of the City University of New York, Eugene Lang Faculty Development Award, National Science Foundation for a CAREER Award and Creativity Extension (LCF), NIH-SO6 GM60654 (SCORE) (LCF) and NSF Grant PCM8111745 for the purchase of the 400 MHz Spectrometer. Research Infrastructure at Hunter College is supported by NIH-Research Centers in Minority Institutions Grant RR03037-08.

REFERENCES

1. A. Müller, F. Peters, M. T. Pope, D. Gatteschi, Polyoxometalates-very large clusters. Nanoscale magnets, *Chemical Reviews*, 98:239 (1998).
2. A.Müller, E. Krickemeyer, H. Bogge, M. Schmidtmann, F. Peters, Organizational forms of matter-an inorganic super fullerene and keplerate based on molybdenum oxide, *Angew. Chem. Int. Ed.* 37: 3360 (1998).
3. A. Müller, S. Sarkar, S. Q. N. Shah, H. Bogge, M. Schmidtmann, S. Sarker, P. Kogerler, B. Hauptfleish, A. X. Trautwein, V. Schunemann, Archimedean synthesis and magic numbers. Sizing giant molybdenum-oxide-based molecular spheres of the keplerate type, *Angew. Chem. Int. Ed.* 38:3238 (1999).
4. A.Müller, S. Q. N. Shah, H. Bogge, M. Schmidtmann, Molecular growth from a ^{176}Mo to a ^{248}Mo cluster, *Nature*, 397:48 (1999).
5. A. Müller, S. Polarz, S. K. Das, E. Krickemeyer, H. Bogge, M. Schmidtmann, B. Hauptfleisch, Open-and-shut for guests in molybdenum oxide-based giant spheres, baskets, and rings containing the pentagon as a common structural element, *Angew. Chem. Int. Ed.* 38:3241 (1999).
6. V. N. Molchanov, I. V. Tatjanina, E. A. Torchenkova, L. P. Kazansky, A Novel type of heteropolynuclear complex anion : X-ray crystal structure of the polymeric complex anion $[Th(H_2O)_3UMo_{12}O_{42}]_n^{4n-}$, *J.C.S. Chem. Comm.* 1981:93 (1981).

7. A. Kitamura, T. Ozeki, A. Yagasaki, β-ocatamolybdate as abuilding-block. Synthesis and structural characterization of rare-earth molybdate adducts, *Inorganic Chem.*, 36:4275 (1997).

8. T. Yamase, H. Naruke, Y. Sasaki, Crystallographic characterizaiton of polyoxotungstate $[Eu_3(H_2O)_3(SbW_9O_{33})(W_5O_{18})_3]^{18-}$ and energy-transfer in its crystalline lattices, *J.Chem. Soc. Dalton Trans.* 1687 (1990).

9. K. Wassermann, M. H. Dickman, M. T. Pople, Self-assembly of polyoxometalates. The compact, water-soluble heterpolytungstate anion $[(As^{III}_{12}Ce^{III}_{16}(H_2O)_{36}W_{148}O_{524})^{76-}$, *Angew. Chem. Int. Ed. Engl.* 36:1445 (1997).

10. M. Sadakane, M. H. Dickman, M. T. Pope, Controlled assembly of polyoxometalate chains from lacunary building-blocks and lanthanide-cation linkers, *Angew. Chem. Int. Ed.* 39:2914 (2000).

11. J. Bartis, M. Dankova, M. Blumenstein, L. C. Francesconi, Preparation of lanthanide complexes of heteropolytungstates and characterization by ^{183}W and ^{31}P nmr-spectroscopy, *Journal of Alloys and Compunds* 249:56 (1997).

12. J. Bartis, M. Dankova, J. J. Lessmann, Q.-H. Luo, W. D. Horrocks Jr., L. C. Francesconi, Lanthanide complexes of the α_1 isomer of the $[P_2W_{17}O_{61}]^{10-}$ heteropolytungstate. Preparation, stoichiometry, and structural characterization by ^{183}W and ^{31}P nmr-spectroscopy and europium(III) luminescence spectroscopy, *Inorganic Chemistry* 38:1042 (1999).

13. V. N. Molchanov, L. P. Kazanskii, E. A. Torchenkova, V. I. Simonov, Crystal structure of $[K_{16}Ce(P_2W_{17}O_{61})_2]\cdot nH_2O$ (N=50), *Sov Phys. Crystallogr.* 24:96 (1979).

14. Q. Luo, R. C. Howell, M. Dankova, J. Bartis, C. W. Williams, J. Horrocks, J. DeW, J. V. G. Young, A. L. Rheingold, L. C. Francesconi, M. R. Antonio, Coordination of Rare-Earth Elements in Complexes with Monovacant Wells-Dawson Polyoxoanions, *Inorg. Chem.* 40:1894 (2001),.

15. J. -H. Son, H. Choi, Y. -U.Kwon, Porous crystal-formation from polyoxometalate building-blocks. Single-crystal structure of $[AlO_4Al_{12}(OH)_{12}(H_2O)_{24}][Al(OH)_6Mo_6O_{18}]_2(OH)\cdot 29.5H_2O$, *J. Am. Chem. Soc.* 122:7432 (2000).

16. A. Müller, C. Serain, Soluble molybdenum blues. Des pudels kern, *Acc. Chem. Res.* 33:2. (2000).

17. C. E. Van Pelt, W. J. Crooks III, G. R. Choppin, *submitted* (2000).

18. J. P. Ciabrini, R. Contant, Mixed heteropolyanion. Synthesis and formation. Constants of cerium(III) and cerium(IV) complexes with lacunary tungstophosphates, *J. Chem. Res., Synop.* 10:391 (1993).

19. M. S. Weeks, C. L. Hill, R. F. Schinazi, Synthesis, characterization, and Anti-human-immunodeficiency-virus activity of water-soluble salts of polyoxotungstate anions with cobalently attached organic groups, *Journal of Medicinal Chemistry* 35:1216 (1992).

DYNAMICS OF ORGANOMETALLIC OXIDES : FROM SYNTHESIS AND REACTIVITY TO DFT CALCULATIONS

V. Artero[1], A. Proust[*1], M.-M. Rohmer[2] and M. Bénard[2]

[1]Chimie Inorganique et Matériaux Moléculaires, Université Pierre et Marie Curie, 4 place Jussieu, Case 42, 75252 Paris Cedex 05, France
proust@ccr.jussieu.fr
[2]Chimie Quantique, Institut le Bel, Université Louis Pasteur, 4 rue Blaise Pascal, 67000 Strasbourg, France

INTRODUCTION

Organometallic oxides lie at the border line between organometallic and coordination chemistry. The field was originally explored by F. Bottomley[1] and W. A. Herrmann[2] and co-workers, and as far as organometallic derivatives of polyoxometalates[3] are concerned by the groups of W. Klemperer,[4] R. Finke[5] and K. Isobe.[6] We came in the field while examining the comprehensive coordination chemistry of the lacunar pentamolybdate[7] $[Mo_5O_{13}(OMe)_4(NO)]^{3-}$ and we thus described supported organometallics like $[\{Mo_5O_{13}(OMe)_4(NO)\}(Cp*Rh)]^-$, $[\{Mo_5O_{13}(OMe)_4(NO)\}_2(Cp*Rh)_2(\mu-X)]$ (X = Cl, Br)[8] and $[\{Mo_5O_{13}(OMe)_4(NO)\}\{Re(CO)_3\}]^{2-}$. During these studies, we also observed some degradation-aggregation processes involving the pentamolybdate and leading to the formation of molecular organometallic oxides. Some topological and electronic considerations led us then to recognize that it should be possible to substitute appropriate organometallic moieties for metal oxo units[9] within the intimate structure of polyoxometalates.[10] This drove our impetus to further explore the chemistry of these organometallic oxides.

Organometallic derivatives of polyoxometalates are commonly divided into two classes, as initially proposed by R. Finke.[5a] In $[\{P_2W_{15}Nb_3O_{62}\}(Cp*Rh)]^{7-}$ [5] and $[\{Mo_5O_{13}(OMe)_4(NO)\}\{Rh(cod)\}]^{2-}$,[11] the organometallic moieties are considered to be supported on the polyanion framework. This is in accordance with short Mo-O bonds for those oxygens further linked to the rhodium center. In this case, Mo-O bond lengths are more characteristic of terminal rather than doubly-bridging oxygens and are comparable to the two other ones delimiting the lacuna. In both species, the nucleophily of the surface oxygens has been enhanced by the substitution of pentavalent niobium atoms for hexavalent tungsten atoms and by the presence of the lacuna in the Lindqvist related pentamolybdate.

Polyoxometalate Chemistry for Nano-Composite Design
Edited by Yamase and Pope, Kluwer Academic/Plenum Publishers, 2002

The second class of derivatives is composed of the so-called integrated organometallics. $[M_5O_{18}\{CpTi\}]^{3-}$,[12] $[Mo_5O_{18}\{Cp*Mo\}]^-$ [13] or the triple cubane $[Mo_4O_{16}\{Cp*Rh\}_4]$ [14] belong to this family. In these cases, the organometallic fragments complete the polyanion.

Organometallic derivatives of polyoxometalates are perfectly characterized at the molecular level and in turn provide models for the grafting of an organometallic catalyst on an oxide surface, especially in the case of the supported derivatives. Beyond, they also display their own reactivity and catalytic activity.[5d] Some synergy between the organometallic and oxide parts has been reported in the literature,[15] as well as bifunctional activity.[16] The mobility of organometallic fragments at an oxide surface could also be reproduced on a polyanion.[17]

ORGANOMETALLIC DERIVATIVES OF POLYOXOMETALATES RELATED TO OLEFIN METATHESIS

As alkylidene and alkylidyne complexes are still under active investigation because of their catalytic activity in metathesis type reactions, we decided to benefit from our own experience in the functionalization of polyoxometalates to try to introduce an alkylidene function in an oxide core. Our aims were two fold : (i) provide some insights for a better understanding of the reaction where $[Mo_6O_{19}]^{2-}$ itself is used as a metathesis precatalyst and try to characterize potential intermediates ;[18] (ii) provide an alkylidene function in a special oxo environment to evaluate the role of ancillary oxo ligands on the catalytic activity, since theoretical calculations[19] had suggested that π-donor ligands should favour the formation of the metallacycle proposed as an intermediate in the Herisson and Chauvin mechanism[20] for olefin metathesis. It is worth noting in this context that the catalysts developped by R. R. Schrock, some of which are commercial, bear, besides the alkylidene function, an π-donor organo-imido and peculiar alkoxo ligands.

The first strategy to introduce an alkylidene function in a polyoxometalate is to benefit from the analogy of reactivity between the {MoO} and organic {CO} functions, as it has been exploited in the formation of organo-imido derivatives of polyoxometalates.[21-23] We have thus studied the reactivity of phosphonium ylides $R_3P=C(R_1)R_2$ towards polyoxomolybdates, in organic solvents, as an extension of the Wittig reaction.[24] However, whatever the ylide used, stabilized or not, with various substituents either on the carbone or phosphorus atoms, and the polyoxomolybdate considered, $[PMo_{12}O_{40}]^{3-}$, $[Mo_6O_{19}]^{2-}$ or even $[MoO_2(dedtc)_2]$ (dedtc = diethyldithiocarbamate) the reaction only results in the reduction of the polyanion and quantitative formation of the corresponding phosphonium cation, in agreement with the following mechanism.

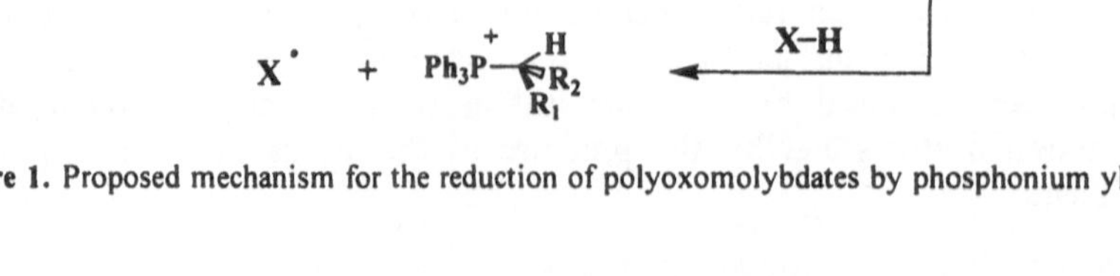

Figure 1. Proposed mechanism for the reduction of polyoxomolybdates by phosphonium ylides

The monoelectronic reduction of the polyanion, also observed by ^{31}P NMR in the case of the dodecaphosphomolybdate, suggests the intermediate formation of a radical cation $[R_3P=C(R_1)R_2]^{\bullet+}$, which further evolves to the phosphonium cation by hydrogen abstraction, for example from solvent or residual water. The radical character of the reaction is further supported by the formation of diphenyldisulphide when thiophenol is initially introduced to the reaction medium.

An alternative strategy for the functionalization of polyoxometalates relies on self-assembly processes. Up to now we failed to introduced a Schrock-type alkylidyne in the lacunar undodecaphosphotungstate. As the current trend in olefin metathesis reaction now favours ruthenium catalysts, such as the Grubbs's ones, we turn to ruthenium precursors and to more stable Lappert-type carbene fragments, stabilized in the α-position by nitrogen atoms. The reaction of $[PW_{11}O_{39}]^{7-}$ with the carbene precursor $[Ru\{=C(NMeCH_2)_2\}_4Cl_2]$ yields, after partial degradation and subsequent aggregation processes, the first carbene derivative of a polyoxometalate : $[K\{PW_9O_{34}\}_2\{WO_2\}\{Ru\{=C(NMeCH_2)_2\}_2\}]^{12-}$. Its structure is related to that of $[KP_2W_{20}O_{72}]^{13-}$ previously reported by J. Fuchs et al [25] two PW$_9$ subunits are linked through a cis-$\{Ru(carbene)_2\}$ moiety and an extra cis-$\{WO_2\}^{2+}$ center ; the whole anion encapsulates a potassium cation, interacting with 10 oxo ligands and two water molecules. The two PW$_9$ units are moved forwards from each other. According to the whole charge, the ruthenium center should be in the +III oxidation state. This was confirmed by its electrochemical behavior and its magnetic and spectroscopic properties.

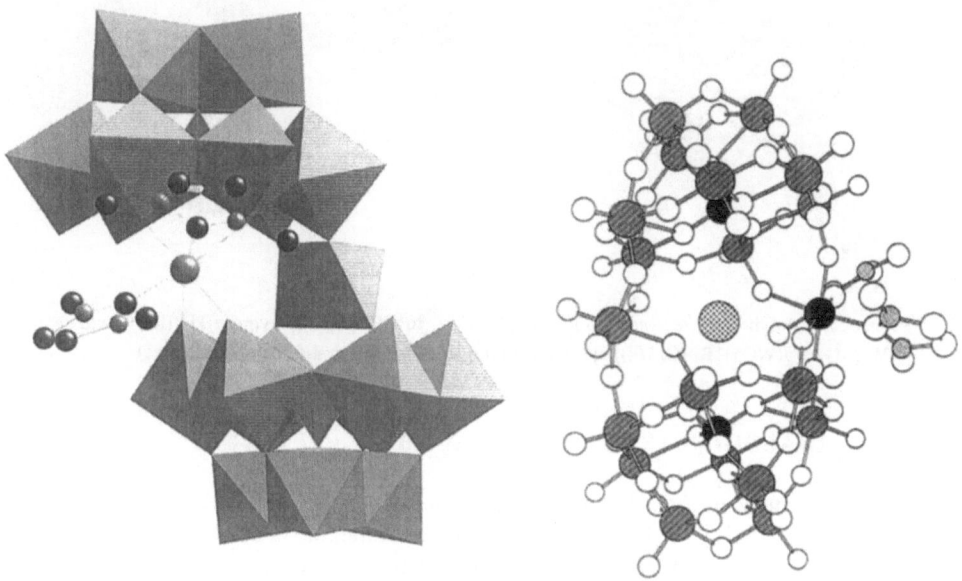

Figure 2. Polyhedral(left) and ball and stick (right) representation of $[K\{PW_9O_{34}\}_2\{WO_2\}\{Ru\{=C(NMeCH_2)_2\}_2\}]^{12-}$ (single crystal X-Ray diffraction structure)

However, the reactivity at the ruthenium center is questionable, since it seems difficult to create some unsaturation. This would be easier for a ruthenium bearing a labile ligand, as for example an arene ligand which could be displaced by photolysis. Indeed, Fürstner et al have recently reported that the compound $[Ru(p\text{-cymene})Cl_2(PCy_3)]$ can act as a metathesis catalyst after irradiation and release of the p-cymene group.[26] This prompted us to investigate self condensation-type reactions involving both $[PW_{11}O_{39}]^{7-}$ and $[Ru(\eta^6\text{-arene})]^{2+}$ precursors. Two supported ruthenium compounds $[\{PW_{11}O_{39}\}\{Ru(H_2O)(p\text{-cymene})\}]^{5-}$ and $[\{PW_{11}O_{39}\}_2(WO_2)\{Ru(C_6H_6)\}_2]^{8-}$ have been fully characterized, including single crystal X-ray diffraction. The photoactivation of $[\{PW_{11}O_{39}\}\{Ru(H_2O)(p\text{-cymene})\}]^{5-}$ has also been investigated and monitored by ^{31}P NMR. The chemical shift of the species obtained after irradiation in water is that of $[PW_{11}O_{39}\{Ru(H_2O)\}]^{5-}$, as previously reported by M. Pope *et al.*[27] Moreover, carrying out the irradiation in the presence of some DMSO yields $[PW_{11}O_{39}\{Ru(DMSO)\}]^{5-}$, also characterized by its ^{31}P chemical shift. On the other hand, if the photolysis is performed in the presence of oxygen, the Ru(III) species $[PW_{11}O_{39}\{Ru(DMSO)\}]^{4-}$ is obtained, with all spectroscopic and electrochemical features in agreement with those of the literature.[27] The $[\{PW_{11}O_{39}\}\{Ru(H_2O)(arene)\}]^{5-}$ species thus prove to be convenient intermediates for a straightforward preparation of pure $[PW_{11}O_{39}\{Ru(H_2O)\}]^{5-}$ catalyst.

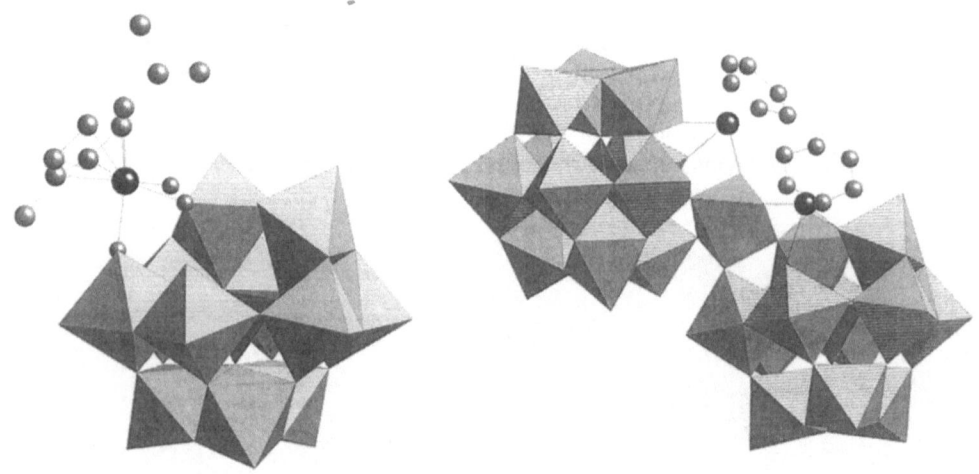

Figure 3. Polyhedral representation of $[\{PW_{11}O_{39}\}\{Ru(H_2O)(p\text{-cymene})\}]^{5-}$(left) and $[\{PW_{11}O_{39}\}_2(WO_2)\{Ru(C_6H_6)\}_2]^{8-}$ (right) (single crystal X-ray diffraction structures)

ISOLOBAL ANALOGIES AS SYNTHETIC GUIDES

We have recently proposed that the fragments $d^0\text{-}fac\text{-}\{MO_2(OR)\}^+$ (M = Mo, W) and $d^6\text{-}fac\text{-}\{M'L_3\}$ (M'L$_3$ = $Mn(CO)_3^+$, $Re(CO)_3^+$ or $Ru(arene)^{2+}$, for example) should be considered as isolobal[9] on the basis of (i) related molecular structures of some polyoxometalates and some organometallic oxides and (ii) theoretical calculations reported in the literature, which concluded to the isolobal analogy between ReO_3 and $Re(CO)_3^+$ fragments.[28] We have thus brought into light the structural relationships within pairs of compounds like $[Mo_3O_6(OMe)\{MeC(CH_2O)_3\}_2]^-$ [29] and $[Mo_2O_4\{MeC(CH_2O)_3\}_2\{Mn(CO)_3\}]^-$, $[Mo_4O_8(OEt)_2\{HOC(CH_2O)_3\}_2]$ [30] and $[Mo_2O_4\{HOC(CH_2O)_3\}_2\{Mn(CO)_3\}_2]$, $[Mo_8O_{20}(OMe)_4\{MeC(CH_2O)_3\}_2]^{2-}$ [31] and $[Mo_6O_{16}(OMe)_2\{MeC(CH_2O)_3\}_2\{Mn(CO)_3\}_2]^{2-}$, within the rhomb-like species $[Mo_4O_{10}(OMe)_4]^{2-}$ [32] and $[Mo_2O_6(OMe)_4\{Re(CO)_3\}_2]^{2-}$, or within the cubane-like species $[\{MoO_2(OMe)\}_2\{Mn(CO)_3\}_2O(OMe)_3]^-$ and $[\{Mn(CO)_3\}_4(\mu_3\text{-OMe})_4]$.[33] Two members of each of the former pairs are related by formal substitution of $\{M'(CO)_3\}^+$ fragments for $\{MoO_2(OR)\}^+$ fragments.

To further validate this concept of isolobal analogies, we have also explored the reactions of $[Ru(\eta^6\text{-arene})]^{2+}$ units, which also obey to the electronic $d^6\text{-}fac\text{-}\{M'L_3\}$ description, with oxomolybdate and oxotungstate precursors. The tetranuclear core $[Mo_2O_6(OMe)_4\{Ru(p\text{-cymene})_2\}]$, related to $[Mo_4O_{10}(OMe)_4]^{2-}$, could indeed be obtained, as it was independently shown by G. Süss-Fink and his group[34b] and by us. The nature of the products of such reactions strongly depends on the reaction conditions. In the case of the $\{Ru(C_6Me_6)\}^{2+}$ precursor and in methanol, two different products have been isolated and characterized by X-ray diffraction : $[Mo_5O_{18}\{Ru(C_6Me_6)\}_2\{Ru(C_6Me_6)(H_2O)\}]$, we will come back later on, and $[Mo_6O_{20}(OMe)_2\{Ru(C_6Me_6)\}_2]^{2-}$. The later is structurally related to the Cp^*Rh^{2+} derivative $[Mo_6O_{20}(OMe)_2\{Cp^*Rh\}_2]^{2-}$, described by K. Isobe et al,[35] which is not so surprising since Cp^*Rh^{2+} is also a $d^6\text{-}fac\text{-}\{M'L_3\}$ fragment. This compound could alternatively be described as belonging to the same familly as the γ–octamolybdate or as beeing related to the methoxo derivative $[Mo_8O_{24}(OMe)_4]^{4-}$ [36] by formal substitution of two organometallic fragments for two $\{MoO_2(OMe)\}^+$ fragments. It could then be described as two superimposed but shifted tetranuclear planar arrangements.

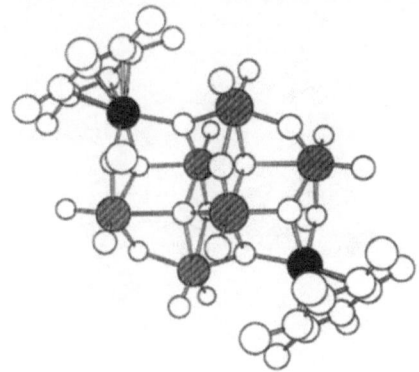

Figure 4. Molecular drawing of $[Mo_6O_{20}(OMe)_2\{Ru(C_6Me_6)\}_2]^{2-}$ (X-ray diffraction structure)

INTERPLAY OF CUBIC UNITS

These self-assembly processes have also been explored in the presence of tungstate. In this case, the products can be more easily described as cubic arrangements. In acetonitrile, the reaction between tungstate and $\{Ru(p\text{-cymene})\}^{2+}$ fragments yields $[W_4O_{16}\{Ru(p\text{-cymene})\}_4]$, a tetratungstate cubic core stabilized by four $\{Ru(p\text{-cymene})\}^{2+}$ units.[37] The molybdenum analog $[Mo_4O_{16}\{Ru(p\text{-cymene})\}_4]$, obtained in water, was initially reported by the group of G. Süss-Fink and, because of its shape, its molecular structure was compared to a windmill.[34] A central W_4O_4 cubic core stabilized by six $\{Ir(cod)\}^+$ fragments has also been described by R. Finke et al.[5c]

The $[W_4O_{16}\{Ru(p\text{-cymene})\}_4]$ complex evolves in MeCN to yield a unprecedented double-cubane species $[W_2O_{10}\{Ru(p\text{-cymene})\}_4]$,[37] and in water to yield a dodecametallic species $[W_8O_{28}(OH)_2\{Ru(p\text{-cymene})\}_2\{Ru(p\text{-cymene})(H_2O)\}_2]^{2-}$. The later can be viewed as composed of two W_3RuO_4 cubes linked by two WO_2^{2+} units and supporting each a $\{Ru(p\text{-cymene})\}^{2+}$ fragment, which coordination sphere is further completed by a water molecule. This compound could also been obtained straightforwardly in water, in good yield.[38]

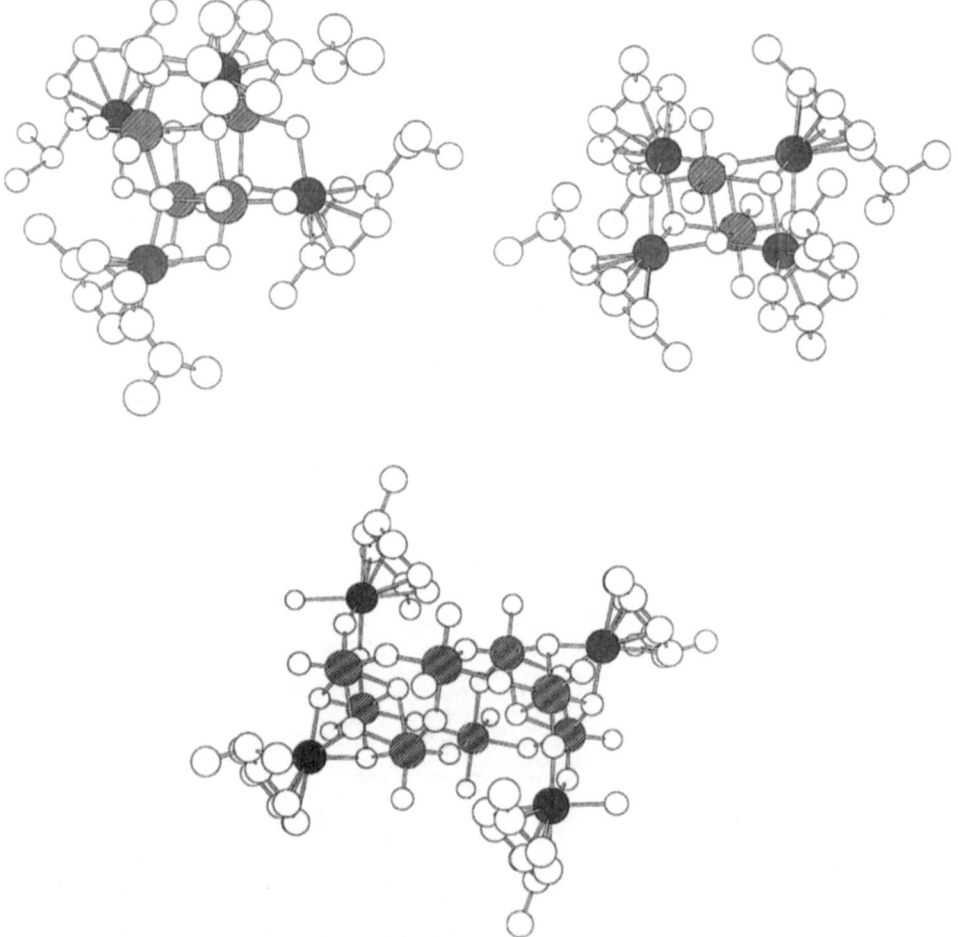

Figure 5. Molecular drawings of [W$_4$O$_{16}${Ru(p-cymene)}$_4$] (top and left), [W$_2$O$_{10}${Ru(p-cymene)}$_4$] (top and right) and [W$_8$O$_{28}$(OH)$_2${Ru(p-cymene)}$_2${Ru(p-cymene)(H$_2$O)}$_2$]$^{2-}$ (bottom) (X-ray diffraction structures)

In the case of {Ru(C$_6$Me$_6$)}$^{2+}$ fragment, the tungsten analog of the above pentamolybdate can be isolated.[38] The molecular structure of these [M$_5$O$_{18}${Ru(C$_6$Me$_6$)}$_2${Ru(C$_6$Me$_6$)(H$_2$O)}] (M = Mo, W) can alternatively be described on the basis of two edge-sharing W$_3$RuO$_4$ cubes or as a central lacunar Lindqvist-type anion stabilized by three supported organometallic cations, one bearing an extra water molecule. These two different approaches in the description of the molecular structure also apply to the description of the hexavanadate derivatives [V$_6$O$_{19}$(Cp*Rh)$_4$][39] and [V$_6$O$_{19}${Ru(*p*-cymene)}$_4$].[34b,c] In other words, this raises the question of supported *versus* integrated organometallic fragments. We failed to find a general cristallographic criterion to discriminate between these two different descriptions, which mainly relie on the preliminary identification of a parent polyanion.

NUCLEATION AND GROWTH PROCESSES

At this stage, we could also address the question of nucleation and growth processes involved in these self-assembly reactions. In some cases, the molecular structures display a planar or a rhomb-like arrangement of the coordination polyhedra, like for $[Mo_2O_4(OMe)_6\{Mn(CO)_3\}_2]$, $[Mo_2O_6(OMe)_4\{Ru(p\text{-cymene})_2\}]$ and $[Mo_6O_{20}(OMe)_2\{Ru(C_6Me_6)\}_2]^{2-}$. In some other cases, they rather display cubic arrangements like in $[\{MoO_2(OMe)\}_2\{Mn(CO)_3\}_2O(OMe)_3]^-$ or $[W_2O_{10}\{Ru(p\text{-cymene})\}_4]$. Both arrangements obey to the generic formula M_4O_{16} but they differ in the oxygen connectivities. What are the factors which could then favour one of the two types of arrangements upon the other ? To assemble four octahedra we can think of assembling first three and then adding the fourth. Starting with a close-packing arrangement of three octahedra we could then equally obtain a rhomb-like arrangement or a cubane-type. In the presence of trisalkoxo-ligands, and whatever the solvent, the triangular cavity created by the edge-sharing of the three octahedra is generally capped and this prevents the formation of cubic units. Planar arrangements are then observed. In the presence of methoxo ligands, which could only partially fill the cavity, or in other words which could behave as triply bridging ligands, both rhomb-like or cubane types have been obtained with molybdenum, as exemplified in the case of $\{Mn(CO)_3\}^+$ derivatives. A pentamolybdate core can also be recognized in $[\{Mo_5O_{16}(OMe)_2\}_2\{Mn(H_2O)_2\}\{Mn(CO)_3\}_2]^{4-}$,[39] as in $[M_5O_{18}\{Ru(C_6Me_6)\}_2\{Ru(C_6Me_6)(H_2O)\}]$. In the case of tungsten, methoxo derivatives are less common and no product other than $[W_5O_{18}\{Ru(C_6Me_6)\}_2\{Ru(C_6Me_6)(H_2O)\}]$ could be definitively characterized from reaction with $\{Ru(arene)\}^{2+}$ fragments in methanol, even if we suspect the formation of a mono-cubane. On the other hand, in solvents other than methanol, the isolated species, both for molybdenum and tungsten do not display any planar arrangement but either a cubic or, more generally, a more compact arrangement, which could thus tentatively be seen as more stable.

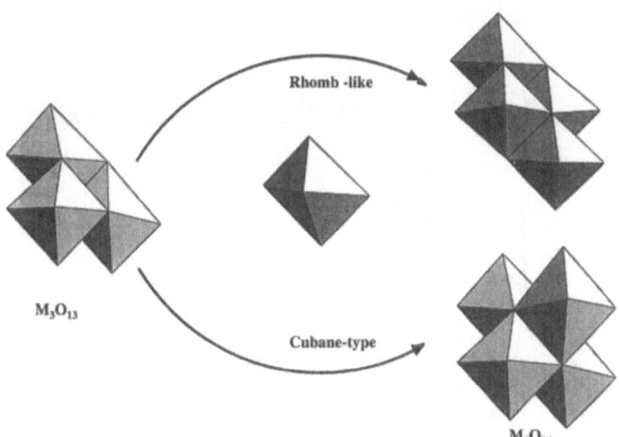

Figure 6. Relationship between rhomb-like and cubane-type M_4O_{16} arrangements

STEREOCHEMICAL NON RIGIDITY OF $[M_4O_{16}\{Ru(arene)\}_4]$

All the above described species have been characterized in solution by multinuclear NMR, in collaboration with R. Thouvenot. The ^{17}O NMR spectrum of $[W_4O_{16}\{Ru(p\text{-cymene})\}_4]$, recorded in chloroform, is thus reproduced below. It is fully consistent with the molecular structure as determined by X-ray diffraction, so demonstrating that it is retained in solution, with four sets of oxygen signals respectively assigned, from low to high field, to terminal oxygens, to the two types of doubly bridging and to quadruply bridging oxygens. However, it is at variance with the results reported by G. Süss-Fink *et al* on the molybdenum analog [Mo_4O_{16}{Ru(*p*-cymene)}$_4$], but with an NMR study carried out in dichloromethane.[34b] This led us to reprepare the molybdenum-windmill and to reinvestigate its NMR features.[37] The spectrum obtained in dichloromethane is indeed in agreement with that published by G. Süss-Fink *et al*. It displays three main signals, none of which in the region of the doubly bridging oxygens, from which it could be inferred that the structure in solution departs from the windmill-form. It can however be compatible with a triple-cubane form of the same generic formula [Mo_4O_{16}{Ru(*p*-cymene)}$_4$]. Such a form had indeed been characterized by the group of K. Isobe as a Cp*Rh^{2+} derivative in [Mo_4O_{16}(Cp*Rh)$_4$][14] and its ^{17}O NMR spectrum is similar to that reported by G. Süss-Fink.

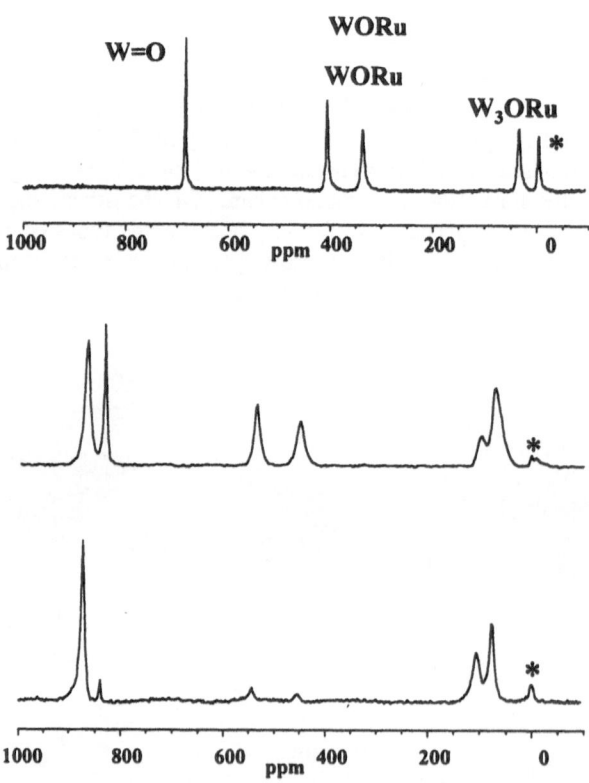

Figure 7. ^{17}O NMR spectra of enriched [W_4O_{16}{Ru(*p*-cymene)}$_4$] in CHCl$_3$ (top), [Mo_4O_{16}{Ru(*p*-cymene)}$_4$] in CHCl$_3$ (middle) and [W_4O_{16}{Ru(*p*-cymene)}$_4$] in CH$_2$Cl$_2$ (bottom)

However, minor signals, especially in the region of the doubly bridging oxygens, were also detected in the spectrum of [Mo_4O_{16}{Ru(*p*-cymene)}$_4$] and were shown to grow on going from dichloromethane to chloroform. This behaviour was checked to be reversible on going back from chloroform to dichloromethane. A similar evolution can also be detected by ^{1}H and ^{95}Mo NMR experiments. These observations led us to the conclusion that the windmill and the triple cubane forms are very probably in equilibrium in solution, eventhough only the windmill-form is isolated in the solid state. The equilibrium is driven to the right in dichloromethane and the two forms are in equivalent proportions in chloroform. This stereochemical change probably involves a concerted motion of only two {Ru(p-cymene)}$^{2+}$ units.

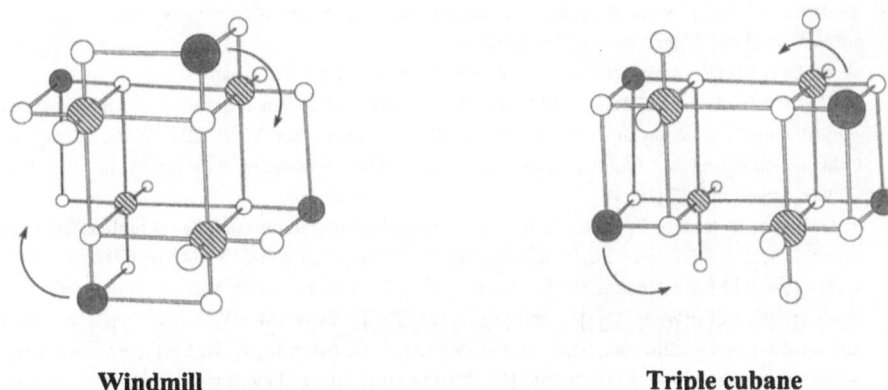

Windmill   **Triple cubane**

Figure 8. Schematic representation of the interconversion between the windmill-like and the triple cubane form of [M$_4$O$_{16}${Ru(arene)}$_4$] (postulated mechanism)

In addition to this equilibrium, which is slow on the NMR time scales, a faster dynamic process has been detected as the temperature is raised : the signals attributed to the doubly-bridging oxygens become broader and hard to observe above 313 K. This coalescence very likely reflects a concerted motion of the four organometallic units along the diagonal faces of the central cubic core.

From a careful examination of the spectra of all the species [M$_4$O$_{16}${Ru(arene)}$_4$] (M = Mo, W, arene = p-cymene, C$_6$Me$_6$) we have characterized, we can draw the conclusions that the equilibrium windmill-triple cubane depends on (i) the solvent, as we have seen previously, (ii) the metal since [W$_4$O$_{16}${Ru(p-cymene)}$_4$] was found to be rigid, (iii) the arene ligand since [Mo$_4$O$_{16}${Ru(C$_6$Me$_6$)}$_4$] was also found to be rigid.
Some calculations at the DFT level are under current investigation in the group of M.Bénard and M.-M. Rohmer in Strasbourg for a better understanding of the determining parameters. Preliminary results indicate the windmill-form as the more stable in the case of M = Mo and that the energy gap between the two forms, always in the case of molybdenum, depends on the nature of the arene group.

CONCLUSION

Several familliés of organometallic oxides have been described and especially the first carbene derivative of polyoxometallates. The concept of isolobality between some oxo and organometallic fragments has been extended to {Ru(arene)}$^{2+}$- containing species, which could turn out to be convenient precursors for catalytic purposes. Stereochemical non rigidity has been evidenced in the system [M$_4$O$_{16}${Ru(arene)}$_4$], which is under further investigation.

REFERENCES

1. a) F. Bottomley, L. Sutin, Organometallic compounds containing oxygen atoms, *Adv. Organomet. Chem.* 28:339 (1988) ; b) F. Bottomley, Cyclopentadienylmetal oxides, *Polyhedron* 11:1707(1992).

2. W. A. Herrmann, Essays on organometallic chemistry .7. Laboratory curiosities of yesterday, catalysts of tomorrow-organometallic oxides, *J. Organomet. Chem.* 500:149 (1995).

3. P. Gouzerh, A. Proust, Main-group element, organic, and organometallic derivatives of polyoxometalates, *Chem. Rev.* 98:77 (1998).

4. a) V. W. Day, W. G. Klemperer, Metal oxide chemistry in solution: The early transition metal polyoxoanions, *Science* 228:533 (1985). b) For a review of the extensive studies of Klemperer and co-workers on polyoxoanion-supported organometallic complexes, see: V. W. Day, W. G. Klemperer in *Polyoxometalates: From Platonic Solids to Anti-Retroviral Activity*, M. T. Pope, A. Müller,eds., Kluwer, Dordrecht, 87 (1994).

5. a) R. G. Finke, B. Rapko, P. J. Domaille, Trisubstituted heteropolytungstates as soluble metal oxide analog.2. 1, 2, 3-β-$SiW_9V_3O_{40}^{7-}$ supported $CpTi^{3-}$, $(Bu_4N)_4[CpTi \cdot SiW_9V_3O_{40}]$, *Organometallics* 5:175 (1986); b) For references for the extensive studies of Finke and co-workers on the synthesis and characterization of $[PW_9M_3O_{37}]^{7-}$- and $[P_2W_{15}M_3O_{62}]^{9-}$-based (M = V^{5+}, Nb^{5+}) polyoxoanion-supported organometallic complexes, see: H. Weiner, J. D. Aiken III, R. G. Finke, Polyoxometalate catalyst precursors-improved synthesis, H^+-titration procedure, and evidence for ^{31}P NMR as a highly sensitive support-site indicator for the prototype polyoxoanion-organometallic-support system [(η-$C_4H_9)_4N]_9P_2W_{15}Nb_3O_{62}$, *Inorg. Chem.* 35:7903 (1996); c) Y. Hayashi, F. Müller, Y. Lin, S. S. Miller, O. P. Anderson, R. G. Finke, $CH_3CN \subset (n-Bu_4N)_2[\{Ir(1,5-COD)\}_6W_4O_{16}] \cdot 2CH_3CN$: A hybrid inorganic-organometallic, flexiblecavity host, acetonitrile-guest complex composed of a $[W_4O_4]^{n+}$ tetratungstate cube and 6 polyoxoanion-supported $(1, 5-COD)Ir^+$ organometallic groups, *J. Am. Chem. Soc.* 119:11401 (1997) ; d) N. Mizuno, H. Weiner, R. G. Finke, Cooxidative epoxidation of cyclohexene with molecular-oxygen, isobutylaldehyde reductant, and the polyoxoanion-supported catalyst precursor $[(\eta-C_4H_9)_4N]_5Na_3[(1,5-COD)Ir \cdot P_2W_{15}Nb_3O_{62}]$-the importance of key control experiments including omitting the catalyst and adding radical-chain initiators, *J. Mol. Catal., A Chem.*, 114:15 (1996) ; J. D. Aiken III, Y. Lin, R. G. Finke, A perspective on nanocluster catalysis-polyoxoanion and $(n-C_4H_9N)^+$ stabilized Ir(0) (similar-to-300) Nanocluster soluble heterogenous catalysts, *J. Mol. Catal., A Chem.*, 114:29 (1996)

6. a) Y. Hayashi, Y. Ozawa, K. Isobe, Site-selective oxygen-exchange and substitution of organometallic groups in an amphiphilic quadruple-cubane-type cluster-synthesis and molecular-structure of $[(RhCp^*)_4V_6O_{19}]$, $[(IrCp^*)_4V_6O_{19}]$, *Inorg. Chem.* 30:1025 (1991); b) K. Isobe, A. Yagasaki, Cubane-type clusters as potential models for inorganic solid-surfaces, *Acc. Chem. Res.* 26:524 (1993); c) R. Xi, B. Wang, K. Isobe, T. Nishioka, K. Toriumi, Y. Ozawa, Isolation and X-ray crystal-structure of a new octamolybdate-$[(RhCp^*)_2(\mu_2-SCH_3)_3]_4(Mo_8O_{26}) \cdot 2CH_3CN$ ($RhCp^* = \eta^5-C_5Me_5$), *Inorg. Chem.* 33:833 (1994).

7. a) P. Gouzerh, Y. Jeannin, A. Proust, F. Robert, 2 Novel polyoxomolybdates containing the $(MoNO)^{3+}$ unit-$[Mo_5Na(NO)O_{13}(OCH_3)_4]^{2-}$ and $[Mo_6(NO)O_{18}]^{3-}$, *Angew. Chem., Int. Ed. Eng.* 28:1363 (1989) ; b) A. Proust, P. Gouzerh, F. Robert, Molybdenum oxo nitrosyl complexes.1. Defect lindqvist compounds of the type $[Mo_5O_{13}(Or)_4(NO)]^{3-}$ (R=CH_3, C_2H_5)-solid-state interactions with alkali-metal cations, *Inorg. Chem.* 32:5291 (1993) ; c) A. Proust, R. Thouvenot, F. Robert, P. Gouzerh, Molybdenum oxo nitrosyl complexes. 2. ^{95}Mo NMR-studies of defect and complete lindqvist-type derivatives-crystal and molecular-structure of $(n-Bu_4N)_2[Mo_6O_{17}(OCH_3)(NO)]$, *Inorg. Chem.* 32:5299 (1993).

8. A. Proust, P. Gouzerh, F. Robert, Organometallic oxides-lacunary lindqvist-type polyanion-supported cyclopentadienylrhodium complex fragments, *Angew. Chem., Int. Ed. Engl.* 32:115 (1993); b) R. Villanneau, A. Proust, F. Villain, F. Robert, M. Verdaguer, P. Gouzerh, manuscript in preparation.

9. R. Villanneau, R. Delmont, A. Proust, P. Gouzerh, Merging organometallic chemistry with polyoxometalate chemistry, *Chem. Eur. J.* 6:1184 (2000).

10. M. T. Pope in *Hetero and Isopolyanions*, Springer-Verlag, Berlin, New-York, (1983) ; (b) M. T. Pope, A. Müller, Polyoxometalate chemistry-an old field with polyoxometalate chemistry, *Angew. Chem., Int. Ed. Engl.* 30:34 (1991). c) *Polyoxometalates: From Platonic Solids to Anti-Retroviral Activity*, M. T. Pope, A. Müller, eds, Kluwer Academic Publishers, Dordrecht, (1994) ; d) Special issue of *Chem. Rev.* 98 (1998).

11. E. Marceau, A. Proust, P. Gouzerh, unpublished results.

12. a) V. W. Day, M. F. Fredrich, M. R. Thompson, W. G. Klemperer, R.-S. Liu, W. Shum, Reactivity of the $[(\eta^5\text{-}C_5H_5)Ti(Mo_5O_{18})]^{3-}$ anion: Synthesis and structure of MoO_2Cl^+ and $Mn(CO_3)^+$ adducts, *J. Am. Chem. Soc.* 103:3597 (1981).

13. A. Proust, R. Thouvenot, P. Herson, Revisitig the sysnthesis of $[Mo_6(\eta^5\text{-}C_5Me_5)O_{18}]$-X-ray structural-analysis, UV-visible, electrochemical and multinuclear NMR characterization, *J. Chem. Soc., Dalton Trans.* 51 (1999).

14. a) Y. Hayashi, K. Toriumi, K. Isobe, Novel triple cubane-type organometallic oxide clusters: $[MC_p{}^*MoO_4]_4 \cdot nH_2O$ (M=Rh and Ir; $C_p{}^*=C_6Me_5$; n=2 for Rh and 0 for Ir), *J. Am. Chem. Soc.* 110:3666 (1988); b) Y. Do, X.-Z. You, C. Zhang, Y. Ozawa, K. Isobe, Trishomocubane-type methoxide cluster as a novel mediator in the extension of cube size in organometallic oxide clusters-synthesis and structures of $[(RhC_p{}^*)_2Mo_3O_9(OMe)_4] \cdot MeOH$ and a linear quadruple cubane-type cluster $[(RhC_p{}^*)_4Mo_6O_{22}] \cdot CH_2Cl_2$ ($C_p{}^*=\eta^5\text{-}C_5Me_5$), *J. Am. Chem. Soc.* 113:5892 (1991).

15. C. Zhang, Y. Ozawa, Y. Hayashi, K. Isobe, Oxidation of cyclohexene with tetra-butyl hydroperoxide catalyze by transition-metal oxide clusters, *J. Organomet. Chem.* 373:C21 (1989).

16. A. R. Siedle, C. G. Markell, P. A. Lyon, K. O. Hodgson, A. L. Roe, Bifunctional rhodium-oxometalate catalysts, *Inorg. Chem.* 26:219 (1987).

17. T. Nagata, M. Pohl, H. Weiner, R. G. Finke, Polyoxoanion-supported Organometallic complexes-carbonyls of rhenium(I), iridium(I), and rhodium(I) that are soluble analogs of solid-oxide-supported $M(CO)N^+$ and that exhibit novel $M(CO)N^+$ mobility, *Inorg. Chem.* 36:1366 (1997).

18. M. McCann, D. McDonnell, Ring-opening polymerization of norbornene using a single-crystal of $(Bu^n{}_4N)_2(Mo_6O_{19})$ as a heterogeneous catalyst, *J. Chem. Soc., Chem. Commun.* 1718 (1993).

19. A. K. Rappé, W. A. Goddard III, Olefine metathesis. A mechanistic study of high-valent group 6 catalysts, *J. Am. Chem. Soc.* 104:448 (1982).

20. J.-L. Herisson, Y. Chauvin, Catalysis of olefin transferomations by tungsten complexes. II. Telomerization of cyclic olefins in the presence of acyclic olefins, *Makromol. Chem.* 141:161 (1970).

21. a) A. Proust, R. Thouvenot, M. Chaussade, P. Gouzerh, F. Robert, Phenylimido derivatives of $[Mo_6O_{19}]^{2-}$-syntheses, X-ray structures, vibrational, electrochemical, ^{95}Mo and ^{14}N NMR-studies, *Inorg. Chim. Acta* 224:81 (1994) ; b) A. Proust, S. Taunier, V. Artero, F. Robert, R. Thouvenot, P. Gouzerh, The unexpected reactivity of p-tolylisocyanate towards the keggin anion $\alpha\text{-}[PMo_{12}O_{40}]^{3-}$, *J. Chem. Soc., Chem. Commun.* 21895 (1996).

22. W. Clegg, R. J. Errington, K. A. Fraser, S. A. Holmes, A. Schäfer, Functionalization of $[Mo_6O_{19}]^{2-}$ with aromatic-amines-Synthesis and structure of a hexamolybdate building-block with linear difunctionality, *J. Chem. Soc., Chem. Commun.* 455 (1995).

23. a) J. B. Strong, R. Ostrander, A. L. Rheingold, E. A. Maatta, Ensheathing a polyoxometalate-convenient systematic introduction of organoimido ligands at terminal oxo sites in $[Mo_6O_{19}]^{2-}$, *J. Am. Chem. Soc.* 116:3601 (1994) ; b) Y. Du, A. L. Rheingold, E. A. Maatta, A polyoxometalate incorporating an organoimido ligand-preparation and structure of $[Mo_5O_{18}(MoNC_6H_4CH_3)]^{2-}$, *J. Am. Chem. Soc.* 114:345 (1992) ; c) J. L. Stark, A. L. Rheingold, E. A. Maatta, Polyoxometalate clusters as building blocks: Preparation and structure of bis(hexamolybdate) complexes covalently bridged by organodiimido ligands, *J. Chem. Soc., Chem. Commun.* 1165 (1995) ; d) J. B. Strong, B. S. Haggerty, A. L. Rheingold, E. A. Maatta, A superoctahedral complex derived from a polyoxometalate-The hexakis(arylimido)hexamolybdate anion $[Mo_6(NAr)_6O_{13}H]^-$ *J. Chem. Soc., Chem. Commun.* 1137 (1997) ; e) J. B. Strong, G. P. A. Yap, R. Ostrander, L. M. Liable-Sands, A. L. Rheingold, R. Thouvenot, P. Gouzerh, E. A. Maatta, A new class of functionalized polyoxometalates-Synthetic, structural, spectroscopic, and electrochemical studies of organoimido derivatives of $[Mo_6O_{19}]^{2-}$, *J. Am. Chem. Soc.* 122:639 (2000).

24. V. Artero, A. Proust, Reduction of the phosphododecamolybdate ion by phosphonium ylides and phosphanes, *Eur. J. Inorg. chem.* 2393 (2000).

25. J. Fuchs, R. Palm, Structure and vibrational spectrum of heteropoly compound potassium tungstophosphate hydrate ($K_{13}[KP_2W_{20}O_{72}]\cdot xH_2O$), *Z. Naturforsch.* 39B:757 (1984).

26. A. Hafner, A. Mühlebach, P. A. van der Schaff, One-componenet catalysts for thermal and photoinduced ring-opening metathesis polymerization, *Angew. Chem., Int. Ed. Engl.* 36:2121 (1997); A. Fürstner, L. Ackermann, A most user-friendly protocol for ring closing metathesis reactions, *Chem. Commun.* 95 (1999).

27. C. Rong, M. T. Pope, lacunary polyoxometalate anions are π-acceptor ligands-Characterization of some tungstoruthenate (II, III, IV, V) heteropolyanions and their atoms-transfer reactivity, *J. Am. Chem. Soc.* 114:2932 (1992).

28. T. Szyperski, P. Schwerdtfeger, On the stability of trioxo(η^5-cyclopentadienyl) compounds of manganese, technetium, and phenium-An abinitio SCF study, *Angew. Chem., Int. Ed. Engl.* 28:1228 (1989).

29. K. H. van Dijik, J. van der Haar, D. J. Stufkens, and A. Oskam, Metal to ligand charge-transfer photochemistry of metal-metal-bonded complexes.7. Photochemistry of $(CO)_4CoM(CO)_3(bpy)$ (M=Mn, Re; bpy=2, 2'-Bipyridine): Photochatalyic disproportionation of the manganese complex in the presence of PR_3, *Inorg. Chem.* 28:75 (1989).

30. A. J. Wilson, W. T. Robinson, C. J. Wilkins, 1, 3-Diethoxy-1,2;1,4; 2,3,4:2,3;3,4;2,4-bis-μ_4-{2-hydroxymethyl-2-methyl-1,3-propanediolato^{3-}]-μ-O, μ-O', μ-O"}-tetrakis[cis-diocomolybdenum(VI)], $C_{14}H_{28}Mo_4O_{16}$, *Acta Crystallogr. Sect. C.* 39:54 (1983).

31. L. Ma; S. Liu, J. Zubieta, The first isolation of an intermediate in the formation of a hexaruthenium carbido-cluster from the reaction of $[Ru_3(CO)_{12}]$: X-ray structure analyses of $[Ru_6(\eta^2-\mu_4-CO)_2(CO_{13}(\eta^6-C_6H_3Me_3)]$ and $[HRu_6(\eta^2-\mu_4-CO)(CO)_{13}(\eta^7-\mu_2-C_6H_3Me_2CH_2)]$, *J. Chem. Soc., Chem. Commun.* 440 (1989).

32. H. Kang, S. Liu, S. N. Shaikh, T. Nicholson, J. Zubieta, Synthesis and structural investigation of polyoxomolybdate coordination-compounds displaying a tetranuclear core crystal and molecular-structures of $(n\text{-}Bu_4N)_2[Mo_4O_{10}(OMe)_4X_2]$ (X=-OMe, -Cl) and their relationship to the catecholate derivative $(n\text{-}Bu_4N)_2[Mo_4O_{10}(OMe)_2(OC_6H_4O)_2]$ and to the diazenido complexes of the o-aminophenolate and the naphthalene-2, 3-diolate derivatives (n-$Bu_4N)_2[Mo_4O_{10}(OMe)_2(HNC_6H_4O)_2(NNC_6H_5)_4]$ and (n-$Bu_4N)_2[Mo_4O_{10}(OMe)_2(C_{10}H_6O_2)_2(NNC_6H_5)_4]$-Comparison to the structrue of a binuclear complex with the $[Mo_2(OMe)_2(NNC_6H_5)_4]^{2+}$ core, $[Mo_2(OMe)_2(HNC_6H_4O)_2(NNC_6H_5)_4]^{4+}$, *Inorg. Chem.* 28:920 (1989).

33. C. E. Holloway, M. Melnik, Manganese carbonyl and organometallic compounds-Analysis and classification of crystallographic and structural data, *J. Organomet. Chem.* 396:129 (1990).

34. a) G. Süss-Fink, L. Plasseraud, V. Ferrand, H. Stoeckli-Evans, $[(p\text{-}Pr^iC_6H_4Me)_4Ru_4Mo_4O_{16}]$: an amphiphilic organoruthenium oxomolybdenum cluster presenting a unique framework geometry, *J. Chem. Soc., Chem. Commun.* 1657 (1997); b) G. Süss-Fink, L. Plasseraud, V. Ferrand, S. Stanislas, H. Stoeckli-Evans, M. Henry, G. Laurenczy, R. Roulet, Amphiphilic organoruthenium oxomolybdenum and oxovanadium clusters, *Polyhedron* 17:2817 (1998); c) L. Plasseraud, H. Stoechli-Evans, G. Süss-Fink, $[(\eta^6\text{-}p\text{-}Pr^iC_6H_4Me)_2Ru_2Mo_2O_6(OMe)_4]$-A new tetranuclear mixed-metal oxo cluster presenting a cube-based chair structure, *Inorg. Chem. Commun.* 2:44 (1999).

35. S. Takara, T. Nishioka, I. Kinoshita, K. Isobe, A Novel organometallic oxide cluster with multi-valley sites - Synthesis and structure of $(nBu_4N)_2[\{(\eta^5C_5Me_5)Rh\}_2Mo_6O_{20}(OMe)_2]$ and its framework transformations, *J. Chem. Soc., Chem. Commun.* 891 (1997).

36. E. M. McCarron III, R. L. Harlow, Synthesis and structure of $Na_4[Mo_8O_{24}(OCH_3)_4]\cdot 8MeOH$: A novel isopoly molybdate that decomposes with the loss of formaldehyde, *J. Am. Chem. Soc.* 105:6179 (1983).

37. V. Artero, A. Proust, P. Herson, R. Thouvenot, P. Gouzerh, (η^6-arene) ruthenim oxomolybdenum and oxotungsten clusters - Stereochemical nonrigidity of $[\{Ru(\eta^6\text{-}p\text{-}MeC_6H_4Pr^i)\}_4Mo_4O_{16}]$ and crystal-structure of $[\{Ru(\eta^6\text{-}p\text{-}MeC_6H_4Pr^i)\}_4W_2O_{10}]$, *Chem. Commun.* 883 (2000).

38. V. Artero, A. Proust, P. herson, P. Gouzerh, Interplay of cubic building blocks in (η^6-arene) ruthenium-containing tungsten and molybdenum oxides, *Eur. J. Chem.* 7:3901 (2001).

39. a) H. K. Chae, W. G. Klemperer, V. W. Day, An organometal hydroxide route to [(C$_5$Me$_5$)Rh]$_4$[V$_4$O$_{19}$], *Inorg. Chem.* 28:1423 (1989); b) Y. Hayashi, Y. Ozawa, K. Isobe, The first vanadate hexamer capped by 4 pentamethylcyclopentadienyl-rhodium or pentamethylcyclopentadienyl-iridium groups, *Chem. Lett.* 425 (1989).

40. R. Villaneau, PhD thesis, Université Pierre et Marie Curie (1996).

AN ORGANORHODIUM TUNGSTEN OXIDE CLUSTER WITH A WINDMILL-LIKE SKELETON: SYNTHESIS OF [(Cp*Rh)$_4$W$_4$O$_{16}$] AND DIRECT OBSERVATION BY ESI-MS OF AN UNSTABLE INTERMEDIATE [Cp*RhClWO$_4$]

Koji Nishikawa, [1] Koichi Kido, [1] Jun'ichi Yoshida, [1] Takanori Nishioka, [1] Isamu Kinoshita, [1] Brian K. Breedlove, [1] Yoshihito Hayashi, [2] Akira Uehara, [2] and Kiyoshi Isobe*[1]

[1]Department of Material Science, Graduate School of Science, Osaka City University, 3-3-138 Sugimoto, Sumiyoshi-ku, Osaka 558-8585, Japan
[2]Department of Chemistry, Faculty of Science, Kanazawa University, Kakuma, Kanazawa 920-1192, Japan

INTRODUCTION

Organometallic oxide clusters containing cubic or incomplete cubic frameworks are very useful for homogeneous or heterogeneous catalysts in the oxidation and metathesis of unsaturated hydrocarbons.[1]   Theoretical studies modeling catalytic hydrocarbon transformations on integrated cubic metal oxide clusters show that coordinatively unsaturated metal sites and valley sites play an important role in the activation.[2]   We recently reported a novel organometallic oxide cluster with multiple valley sites, [n-Bu$_4$N]$_2$[(Cp*Rh)$_2$Mo$_6$O$_{20}$(OMe)$_2$],[3] having a so-called double-book shelf-like structure.   It converts into different frameworks such as the face-sharing triple-cubane-type cluster [(Cp*Rh)$_4$Mo$_4$O$_{16}$], the incomplete double-cubane-type cluster [(Cp*Rh)$_2$Mo$_3$O$_9$(OMe)$_4$], and the face-sharing quadruple-cubane-type cluster [(Cp*Rh)$_4$Mo$_6$O$_{22}$] under different conditions.   The mechanism for the formation of the double-book-shelf cluster was investigated by electrospray ionization mass spectrometry (ESI-MS),[4] which aided in the detection of unstable species generated in solution.[5]   In that study, an ESI mass spectrum showed an isotopic envelope for an intermediate in the formation of the cluster, [Cp*RhMo$_3$O$_8$(OMe)$_5$]$^-$ generated by the reaction of [{Cp*RhCl(μ–Cl)}$_2$] and [n-Bu$_4$N]$_2$[Mo$_2$O$_7$] in methanol at –78 °C.   Understanding the mechanism for the formation of these clusters may provide strategies to synthesize desired oxide clusters for use as catalysts in the transformation of hydrocarbons.

RESULTS AND DISCUSSION

Here, we report the synthesis and characterization of an organorhodium tungsten oxide cluster having a windmill-like skeleton $[(Cp^*Rh)_4W_4O_{16}]$ (**3**) and mechanism for the formation of **3** by ESI-MS in acetonitrile. A modified sprayer for ESI-MS was used to aid in the detection of any unstable species in the reactions which may form on a millisecond time scale.

Cluster **3** can be synthesized from the reaction of $[\{Cp^*RhCl(\mu\text{-}Cl)\}_2]$ (**1**) with two equivalents of $[n\text{-}Bu_4N]_2[WO_4]$ (**2**) in acetonitrile under a nitrogen atmosphere.‡ Orange crystals were obtained by recrystallization with methanol-acetone. Single crystal X-ray analysis of **3** revealed it consists of a tungsten tetranuclear framework of $[W_4O_{16}]^{8-}$, forming a distorted cubic core as a center capped by four $[Cp^*Rh]^{2+}$ moieties each binding to a triply bridging oxygen atom and to two terminal oxygen atoms (Fig. 1).§ The four tungsten atoms are coordinated to three oxygen atoms, of which two are bound to two different rhodium atoms and another is free. The W–O bonds can be classified into two different categories. The first one consists of W—O bonds with double bond character and includes the W—O bonds with terminal oxygen atoms (1.713(10) Å) and with oxygen atoms bridging two rhodium atoms (1.78(1)–1.812(9) Å). The second has single bond character and involves quadruply bridging oxygen atoms (O1, O1*, O1', and O1"). The W—O bond distance is 2.374(8) Å and is much longer than those in the first category. This suggests that the bonds might to be cleaved to produce $[(Cp^*Rh)_2W_2O_8]$, which has a ladder-type

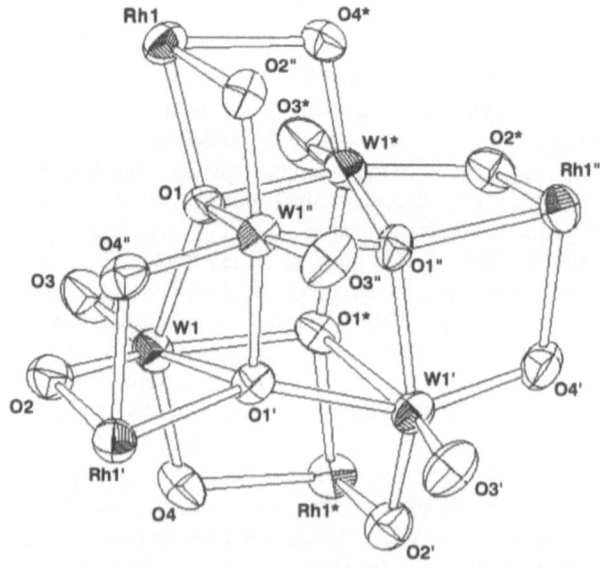

Figure 1. ORTEP drawing of the $Rh_4W_4O_{16}$ framework of cluster **3**. Selected bond distances (Å) and angles (°): Rh1–O1, 2.128(9); Rh1–O2", 2.14(1); Rh1–O4*, 2.104(10); W1–O1, 2.080(9); W1–O1', 2.374(8); W1–O1*, 2.105(9); W1–O2, 1.78(1); W1–O3, 1.713(10); W1–O4, 1.812(9); O1–Rh1–O2", 77.6(4); O1–Rh1–O4*, 74.1(4); O2"–Rh1–O4*, 81.5(5); O1–W1–O1*, 71.3(4); O1–W1–O1', 71.5(4); O1*–W1–O1', 71.1(3); W1–O1–W1*, 105.0(4); W1–O1–W1", 106.8(4); W1*–O1–W1", 105.9(4).

framework. The O—Rh—O bond angles are in the range of 74.1(4)–81.5(5) °, resulting in the rhodium groups having a "piano stool" type arrangement. As well, each Cp*Rh group is located at a vertex of a distorted incomplete cubic framework. The structure is very similar to $[\{Ru(\eta^6-p-MeC_6H_4Pr^i)\}_4Mo_4O_{16}]$.[6] They also contain valley sites that may be active sites for the activation of hydrocarbons. The chemical behavior of **3** in different solvents was investigated and it was found that **3** decomposes to **1** in $CHCl_3$ and converts to the triple cubane type cluster in water.**

Fig. 2 shows an ESI mass spectrum involving the mixing of two acetonitrile solutions of **1** and **2**, respectively.‖ In the measurement, only a negative ion peak at $m/z = 521$ appeared as a new peak when the two solutions were mixed at the tip of the mixing sprayer

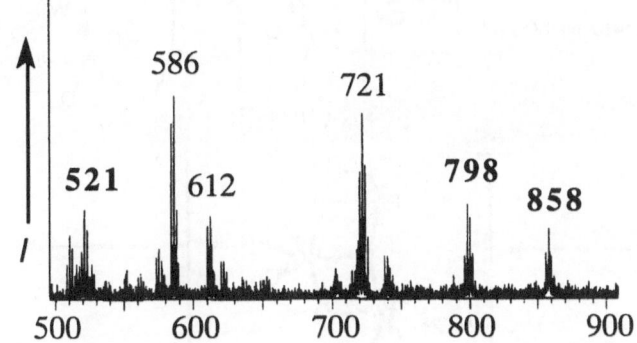

Figure 2. Negative-ion mixing ESI mass spectrum of a reaction mixture of each acetonitrile solution of **1** and **2**. Assignment of the peaks is as follows: $[Cp*RhClWO_4]^-$: m/z 521; $[(n-Bu_4N)Cp*RhCl_3-H]^-$: m/z 586; $[(n-Bu_4N)W_2O_7]^-$: m/z 722; $[(n-Bu_4N)Cp*RhCl_2WO_4]^-$: m/z 798; $[Cp*RhW_2O_7(OH)(MeCN)_3]^-$: m/z 858. See text for details. I = relative intensity.

and the effluent from the tip was introduced directly into a sector type mass spectrometer. The peak had a very low intensity and a characteristic distribution of isotopomers that matches well with the calculated isotopic distribution for $[Cp*RhClWO_4]^-$ (3_{im}). Moreover, the two signals at $m/z = 798$ and 858 can be assigned to $[(n-Bu_4N)Cp*RhCl_2WO_4]^-$ and $[Cp*RhW_2O_7(OH)(MeCN)_3]^-$, respectively, from calculated distributions of isotopomers. These two peaks also appeared when an acetonitrile solution containing both **1** and **2** in a 1 : 1 ratio was stirred for 10 min at room temperature, and injected into a sprayer. However, the peak for 3_{im} was not observed in the spectrum, suggesting that it has a short life time during the reaction of **1** and **2**.

Scheme 1 shows a proposed mechanism for the formation of **3** through the intermediates 3_{im} and 3_{ladder}. In the ESI-MS experiments, 3_{im} was found to be very reactive during the condensation of **1** and **2**. The intermediate 3_{im} may dimerize to form 3_{ladder} as indicated by the peak at $m/z = 858$ which represents $[3_{ladder} -Cp*Rh + MeCN]^-$ 3_{ladder} was not directly detected in the mass spectrometry experiments due to its neutral charge. Moreover, the molecular ion peak for **3** was not observed because of its low solubility in acetonitrile.

The windmill-like framework of **3** can be formed by connecting parallel rectangles of two 3_{ladder} complexes, with the faces of their molecular planes perpendicular to each other. On the other hand, the triple cubane framework*** seems to form by the formation of bonds between two 3_{ladder} complexes with the faces of their molecular planes parallel to each other.   Such a mechanism is also supported by the fact that differences in the reaction temperatures of the reactions lead to the clusters with different frameworks.   3 was isolated selectively at 60 °C or above, and the triple cubane type cluster was formed at temperatures below 30 °C.   Moreover, the triple cubane type cluster isomerizes to 3 in acetonitrile at 60 °C or above.   The investigation of this isomerization is in progress.

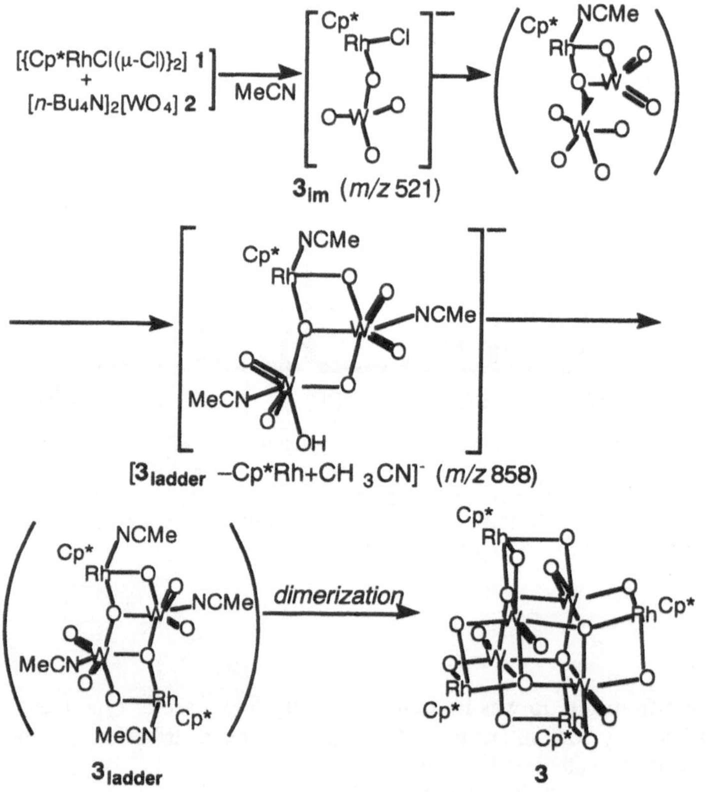

Scheme 1.   Proposed mechanism for the formation of **3** via the unstable intermediates 3_{Im} and 3_{ladder} generated from a 1 : 1 reaction of **1** and **2** in CH_3CN.

Acknowledgment

We thank the financial support by a Grant-in-Aid for Scientific Research on Priority Areas (No. 10149101 "Metal-assembled Complexes") and on (C) (No. 11682201)   from Ministry of Education, Science, Sports and Culture, Japan.

Endnotes

‡ Synthesis of **3**: To a solution of [Cp*RhCl(μ–Cl)]$_2$ (0.50 g, 0.81 mmol) in CH$_3$CN (5.0 cm^3) was added a solution of [n-Bu$_4$N]$_2$[WO$_4$] (1.20 g, 1.60 mmol) in CH$_3$CN (20 cm^3) with stirring at r.t. under Ar. The mixture was heated at 60 °C for 6 h and the resulting orange solid was collected by filtration and washed with a large amount of acetonitrile and acetone. (yield: 0.67 g, 85.0% based on Rh)   Anal. Calcd for C$_{40}$H$_{60}$O$_{16}$Rh$_4$W$_4$: C, 24.72; H, 3.11%.   Found: C, 25.03%; H, 3.22.   IR (KBr, ν (W–O)), cm^{-1}: 931(s), 812(s), 742(s), 588(m), and 493(m).

§ *Crystal data* for C$_{40}$H$_{60}$O$_{16}$Rh$_4$W$_4$, **3**: single crystals of **3** suitable for X-ray crystallography were obtained by recrystallization from methanol/acetone.   An orange crystal of dimensions 0.24 x 0.18 x 0.10 mm was sealed in a capillary.   Cluster **3** crystallizes in the tetragonal space group I $\bar{4}$ (no. 82) with a = 13.0710(6), c = 14.8142(6) Å, V = 2531.0(2) Å^3, Z = 2, and D_c = 2.551 gcm^{-3}.   A total number of 2917 reflections were collected on a Rigaku RAXIS-RAPID Imaging Plate with graphite-monochromated Mo-Kα radiation in the limit of 2θ ≤ 55.0 ° at 296 K.   The reflections with I > 3.00σ(I) were corrected for Lorentz-polarization factors but not for extinction.   An absorption correction was applied using an empirical method.   All crystallographic calculations were carried out using *teXsan* programs.   The positions of the rhodium and molybdenum atoms were obtained by SIR92 direct methods.   The remaining atoms were located on different Fourier maps.   Hydrogen atoms were fixed at calculated positions and used for structure-factor calculations.   145 variables were refined by full-matrix least-squares techniques using 2523 independent reflections and converted to R = 0.035, R_w = 0.054 and GOF = 1.40.   A final difference Fourier map yielded $\Delta\rho_{max}$ = 1.17 e Å^{-3}, $\Delta\rho_{min}$ = -2.11 e Å^{-3}.

|| Mass spectrometric experiments were performed on a JMS-700S spectrometer equipped with a modified mixing sprayer.   The sprayer consists of triaxial stainless steel pipes of dimensions (inner pipe: 150 μm I.D., 250 μm O.D.; middle pipe: 300 μm I.D., 400 μm O.D.; outer pipe: 600 μm I.D., 800 μm O.D.), and the inner and middle layers can flow different solutions introduced using a syringe pump and an automated pump system (HP, Agilent 1100 Series).   N$_2$ gas was flowed through the outer pipe to assist in liquid nebulization.   Spraying can occur within ca. 20 ms after mixing the reactants.   Reactive species generated between ca. 20–100 ms can be tracked.   The three layers have similar sectional areas at the tip of the sprayer, allowing for the control of the ratio of two solutions by controlling only the flow rate of the solutions.   In addition, the middle pipe can be moved within ±2 mm in parallel with the others, and contributes to controlling the time scale of mixing.   In these experiments, each acetonitrile solution (10^{-5} M) of **1** and **2** was introduced into the inner and middle, respectively.   The sprayer was held at a potential of –6.0 kV, and the orifice potential was maintained at 0 kV.   The flow rates of the solutions were 100 μL/min.   The negative-ion ESI mass spectra were measured in the range of m/z 0 to 3000.

** A detailed discussion will be given in another report.

*** It is an isomer of **3** with a composition of [(Cp*Rh)$_4$W$_4$O$_{16}$], and its structure is very similar to that of the molybdenum analog [(Cp*Rh)$_4$Mo$_4$O$_{16}$].   A detailed structure of the cluster and its reactivity will be reported elsewhere.

REFERENCES

1. a) C. Zhang, Y. Ozawa, Y. Hayashi, K. Isobe, Oxidation of cyclohexene with t-butyl hydroperoxide catalyzed by transition metal oxide clusters, *J. Organomet. Chem.*. 373: C21 (1989); b) K. Isobe, A. Yagasaki, Cubane-type clusters as potential models for inorganic solid surfaces, *Acc. Chem. Res.* 26: 524 (1993); c) K. Takahashi, M. Yamaguchi, T. Shido, H. Ohtani, K. Isobe, M. Ichikawa, Molecular modeling of supported metal catalysts: SiO$_2$-grafted [{(η^3-C$_4$H$_7$)$_2$Rh}$_2$V$_4$O$_{12}$] and [{Rh(C$_5$Me$_5$)}$_4$V$_6$O$_{19}$] are catalytically active in the selective oxidation of propene to acetone, *J. Chem. Soc. Chem. Commun.* 1301 (1995); d) M. Ichikawa, W. Pan, Y. Imada, M. Yamaguchi, K. Isobe, T. Shido, Surface-grafted metal oxide clusters and metal carbonyl clusters in zeolite micropores; XAFS/FTIR/TPD characterization and catalytic behavior, *J. Mol. Catal. A* 107: 23 (1996).

2. a) K. Sawabe, N. Koga, K. Morokuma, Y. Iwasawa, An *ab initio* molecular orbital study on adsorption at the MgO surface. I. H$_2$ chemisorption on the (MgO)$_4$ cluster, *J. Chem. Phys.* 97: 6871 (1992); b) R. Cain, L.J. Matienzo, F. Emmi, Potential energy hypersurface for ammonia adsorbing onto nickel oxide, *J. Phys. Chem.* 94: 4985 (1990).

3. S. Takara, T. Nishioka, I. Kinoshita, K. Isobe, A novel organometallic oxide cluster with multi-valley sites: synthesis and structure of [NBu$^n_4$]$_2$[{(η^5-C$_5$Me$_5$)Rh}$_2$Mo$_6$O$_{20}$- (OMe)$_2$] and its framework transformations, *Chem. Commun.* 891 (1997).

4. S. Takara, S. Ogo, Y. Watanabe, K. Nishikawa, I. Kinoshita, K. Isobe, Direct observation by electrospray ionization mass spectrometry of [Cp*RhMo$_3$O$_N$(OMe)$_5$]$^-$, a key intermediate in the formation of the double-bookshelf-type oxide cluster [(Cp*Rh)$_2$Mo$_6$O$_{20}$(OMe)$_2$]$^{2-}$, *Angew. Chem. Int. Ed. Engl.* 38: 3051 (1999).

5. a) J. W. Sam, X. J. Tang, J. Peisach, Electrospray mass spectrometry of iron bleomycin: demonstration that activated bleomycin is a ferric peroxide complex, *J. Am. Chem. Soc.* 116: 5250 (1994); b) J. Kim, Y. Dong, E. Larka, L. Que, Jr., Electrospray ionization mass spectral characterization of transient iron species of bioinorganic relevance, *Inorg. Chem.* 35: 2369 (1996); c) C. Hinderling, D.A. Plattner, P. Chen, Direct observation of a dissociative mechanism for C—H activation by a cationic iridium(III) complex, *Angew. Chem. Int. Ed. Engl.* 36: 243 (1997); d) D. Feichtinger, D.A. Plattner, Direct proof for O=Mn^V(salen) complexes, *Angew. Chem. Int. Ed. Engl.* 36: 1718 (1997).

6. a) G. Süss-Fink, L. Plasseraud, V. Ferrand, H. Stoeckli-Evans, [(p-Pr^{i}C$_6$H$_4$Me)$_4$Ru$_4$-Mo$_4$O$_{16}$]: an amphiphilic organoruthenium oxomolybdenum cluster presenting a unique framework geometry, *Chem. Commun.* 1657 (1997); b) V. Artero, A. Proust, P. Herson, R. Thouvenot, P. Gouzerh, (η^6-Arene)ruthenium oxomolybdenum and oxotungsten clusters. Stereochemical non-rigidity of [{(η^6-p-MeC$_6$H$_4$Pr^i)Ru}$_4$Mo$_4$O$_{16}$] and crystal structure of [{(η^6-p-MeC$_6$H$_4$Pr^i)Ru}$_4$W$_2$O$_{10}$], *Chem. Commun.* 883 (2000).

ROLE OF ALKALI-METAL CATION SIZE IN ELECTRON TRANSFER TO SOLVENT-SEPARATED 1:1 [(M$^+$)(POM)] (M$^+$ = Li,$^+$ Na$^+$, K$^+$) ION PAIRS

Ira A. Weinstock,[1,2] Vladimir A. Grigoriev,[2] Danny Cheng[2] and Craig L. Hill[2]

[1]USDA Forest Service, Forest Products Laboratory
One Gifford Pinchot Drive
Madison, WI   53705

[2]Emory University
Chemistry Department
1515 Pierce Drive
Atlanta, GA   30322

INTRODUCTION

Numerous published reports show that additions of electrolytes or salts to solutions of charged electron-donor or acceptor complexes dramatically alter rates of charge- or electron-transfer processes.[1-10]   In electron-transfer oxidations of organic or inorganic substrates by negatively charged acceptors, additions of alkali-metal cations generally result in increases in electron-transfer rates.   While examples include reductions of classical coordination complexes, such as $Fe(CN)_6^{3-}$,[11-14] $Ru(CN)_6^{3-}$[15] and $IrCl_6^{3-}$,[15] polyoxometalates (POMs)[16-19] stand out as a large and increasingly useful class of anionic electron-acceptors.[20,21]   Such POM acceptor anions are typified by α-SiVW$_{11}$O$_{40}^{5-}$ (1), a POM of the Keggin structural class (Fig. 1).   In both solid-state POM-salt structures and in solution, counter cations are always present.   However, while uses of POM-based materials and solutions in diverse applications from materials chemistry to catalysis grow, the effects of counter cations are often ignored.   At the same time, in studies that have addressed the roles of counter cations, substantial effects on synthesis, structure[22] and

Polyoxometalate Chemistry for Nano-Composite Design
Edited by Yamase and Pope, Kluwer Academic/Plenum Publishers, 2002

reactivity[20,21,23,24] have been reported.[†]   Of these, alkali-metal-cation catalysis of homogeneous electron-transfer reactions is central to the use of soluble POMs in oxidation and catalysis.[20,21]

Alkali-metal cation catalysis of electron transfer to Acceptor-anions is generally attributed (1) to the formation of stable association complexes (ion pairs) between cation and Acceptor-anion or (2) to formation of stable ternary association complexes that additionally include the Donor as well.[20] (These are distinct from the ternary "activated complexes" formed later along the reaction coordinate.)[10,11]   In case (1), increase in k_{obs} for electron transfer is often attributed to a positive shift in the reduction potential of the acceptor complex upon cation association.   Such positive shifts in reduction potential are

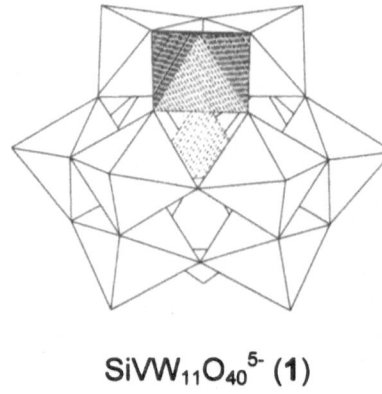

$$SiVW_{11}O_{40}{}^{5-} \ (1)$$

Figure 1.   Drawing (in polyhedral notation) of α-$SiVW_{11}O_{40}{}^{5-}$ (1).   The central tetrahedron represents the $Si^{IV}O_4$ unit and the shaded octahedron represents V^VO_6.   The unshaded octahedra represent $W^{VI}O_6$ units.

[†] Additional reports concerning electron Donor-Acceptor interactions in POM systems, many involving POMs and their counter cations: (a)   C. M. Prosser-McCartha, M. Kadkhodayan, M. M. Williamson, D. A. Bouchard, C. L. Hill, Photochemistry, spectroscopy, and X-ray structure of an intermoolecular charge-transfer complex between an organic substrate and a polyoxometallate, α-$H_3PMo_{12}O_{40}$•6(tetrametylurea), *J. Chem. Soc., Chem. Commun.* **1986**, 1747; (b) M. M. Williamson, D. A. Bouchard, C. L. Hill, Characterization of a weak intermolecular photosensitive complex between an organic substrate and a polyoxometalate. Crystal and molecular structure of α-$H_3PMo_{12}O_{40}$•6DMA•CH_3CN•0.5H_2O (DMA=N, N-dimethlacetamide), *Inorg. Chem.*, **1987**, *26*, 1436; (c) J. A. Schmidt, E. F. Hilinski, D. A. Bouchard, C. L. Hill, Electron donor-accepter complexes of polyoxometalates with organic molecules. Picosecond spectroscopy of $[(n\text{-methylpyrrolidinone})_2H^+PW_{12}O_{40}{}^{3-}$, *Chem. Phys. Lett.* **1987**, *138*, 346; (d) C. L. Hill, D. A. Bouchard, M. Kadkhodayan, M. M. Williamson, J. A. Schmidt, E. F. Hilinski, Catalytic photochemical oxidation of organic substrates by polyoxometalates. Picosecond spectroscopy, photochemistry, and structural properties of charge-transfer complexes between heteropolytungstic acids and dipolar organic compounds, *J. Am. Chem. Soc.* **1988**, *110*, 5471 (e) P. Le Maguerès, L. Ouahab, S. Golhen, D. Grandjean, O. Pena, J. C. Jégaden, C. J. Gómez-García, P. Delhaès, Diamagnetic and paramagnetic keggin polyoxometalate salts containing 1-D and 2-D decamethlferrocenium networks-preparation, crystal-structures, and magenetic-properties of $[Fe(C_5Me_5)_2]_4(POM)(solv)_n(POM=[SiMo_{12}O_{40}]^{4-}$, $[SiW_{12}O_{40}]^{4-}$, $[PMo_{12}O_{40}]^{4-}$, $[HFeMo_{12}O_{40}]^{4-}$, solv=H_2O, C_3H_7OH, CH_3CN), *Inorg. Chem.* **1994**, *33*, 5180 (f) L. Ouahab, Organic/inorganic supramolecular assemblies and synergy between physical-properties, *Chem. Mater.* **1997**, *9*, 1909 (g) D. Attanasio, M. Bonamico, V. Fares, P. Imperatori, L. Suber, Weak charge-transfer plyoxoanion salts-The reaction of Quinolin-8-ol(Hquin) with phosphotungstic acid and the crystal and molecular-structrue of $(H_2quin)_3[PW_{12}O_{40}]$•4EtOH•2$H_2O$, *J. Chem. Soc., Dalton Trans.* **1990**, 3221; (h) X. -M. Zhang, B. -Z. Shan, Z. -P. Bai, X. -Z. You, C. -Y. Duna, Electrochromism and X-ray crystal structure of a mixed-valence charge-transfer complex $[(CH_3)_2NC_6H_4NH(CH_3)_2]_2[(C_4H_9)_4N]SiMo_{12}O_{40}$, *Chem. Mater.* **1997**, *9*, 2687 (i) P. Le Maguerès, S. M. Hubig, S. V. Lindeman, P. Veya, J. K. Kochi, Novel charge-transfer materials via cocrystallization of planear aromatic donors and spherical polyoxometalate acceptors, *J. Am. Chem. Soc.* **2000**, *122*, 10073-10082.

frequently observed upon addition of cations to solutions of Acceptor-anions.[14,25-28]

More precise definition of the physicochemical role of the cation, however, requires that three generic issues be addressed:[7] (1) the stoichiometry of association complex formation; (2) the association constants, K_{IP}, for formation of these complexes; and (3) the physicochemical (structural and electronic) properties of different M^+–Acceptor complexes (i.e., stoichiometrically identical complexes differing only in the nature of M^+).[‡] Once stoichiometry (issue 1), has been established, it is then possible to differentiate between K_{IP} values (2), and the physical properties of specific M^+-containing complexes (3). In polar, coordinating solvents, for example, alkali-metal cations typically accelerate rates of electron transfer to anionic Acceptors in the order $Li^+ < Na^+ < K^+$.[10] In labile systems such as these, however, difficulties encountered in assigning precise stoichiometries have hampered efforts[29] to separate trends in K_{IP} values from often parallel trends in physicochemical properties of different M^+-containing complexes.

We herein summarize data that address issues (1) to (3) and provide new information regarding the physicochemical role of alkali-metal cation (M^+) size in electron-transfer to 1:1 $[(M^+)(POM)]$ ($M^+ = Li,^+ Na^+, K^+$) ion pairs. To achieve this, $\alpha\text{-}SiV^VW_{11}O_{40}{}^{5-}$ (1, Fig. 1), was used as a stoichiometric 1e⁻ acceptor in the 2e⁻ oxidation of a phenolic electron donor,   3,3',5,5'-tetra-*tert*-butylbiphenyl-4,4'-diol   (BPH$_2$)   to 3,3',5,5'-tetra-*tert*-butyldiphenoquinone (DPQ) (eq 1; each equiv. of 1 is reduced by 1e⁻ to 1$_{red}$).[†] Before addressing the role of alkali-metal cations, the conditions required for

---

[‡] In summarizing the effects of added electrolyte and ion pairing, Wherland (ref. 7) concludes that, "study of the effects of ion pairing would be greatly facilitated if the structures, concentrations and lifetimes of ... ion paired species could be more directly evaluated ... ."

[†] Electron and proton transfer in formally H-atom transfer processes has been examined in some detail. The following are representative informative articles: (a) R. A. Binstead, T. J. Meyer, H-atom transfer between metal complex ions in solution, *J. Am. Chem. Soc.* **1987**, *109*, 3287-3297; (b) C. M. Che, K. Lau, T. C. Lau, C. K. Poon, Proton-coupled elecron-transfer reactions-mechanisms if 2-electron reduction of trans-dioxoruthenium(VI) to trans-aquooxoruthenium(VI) and disproportionation of trans-dioxorythenium(V), *J. Am. Chem. Soc.* **1990**, *112*, 5176-5181; (c) H. H. Thorp, Photophysical studies of copper phenanthrolines bound to DNA, *Chemtracts: Inorg. Chem.* **1991**, *3*, 171-184 (d) R. Manchanda, H. H. Thorp, G. W. Brudvig, R. H. Crabtree, Proton-coupled electron-transfer in High-Valent oxomanganese dimers-role of the ancillary ligands, *Inorg. Chem.* **1991**, *30*, 494-497 (e) R. A. Binstead, M. E. McGuire, A. Dovletoglou, W. K. Seok, L. E. Roecker, T. J. Meyer, Oxidation of hydroquinones by $[(bpy)_2(py)Ru(IV)(O)]^{2+}$ and $[(bpy)_2(py)Ru(III)(O)]^{2+}$ - Proton-coupled electron transfer, *J. Am. Chem. Soc.* **1992**, *114*, 173-186 (f) R. I. Cukier, Mechanism for proton-coupled electron-transfer reactions, *J. Phys. Chem.* **1994**, *98*, 2377-81 (g) G. K. Cook, J. M. Mayer, C-H bond activation by metal oxo species-oxidation of cyclohexane by chromyl chloride, *J. Am. Chem. Soc.* **1994**, *116*, 1855-1868 (h) G. K. Cook, Mayer, J. M. *J. Am. Chem. Soc.* **1995**, *117*, 7139-7156; (i) K. A. Gardner, J. M. Mayer, Understanding C-H bond oxidations-H•H-transfer in the oxidation of toluene by permanganate, *Science (Washington, D. C.)*, **1995**, *269*, 1849-1851 (j) M. J. Baldwin, V. L. Pecoraro, Energetics of proton-coupled electron-transfer in high-valent Mn$_2$(μ-O)$_2$ systems-models for water oxidation by the oxygen-evolving complex of photosystem. II., *J. Am. Chem. Soc.* **1996**, *118*, 11325-11326. (k) M. S. Graige, M. L. Paddock, J. M. Bruce, G. Feher, M. Y. Okamura, Mechanism of proton-coupled electron-transfer for Quinone(Q$_B$) reduction in reaction centers of rhodobacter-sphaeroides, *J. Am. Chem. Soc.* **1996**, *118*, 9005-9016; (l) K. Wang, J. M. Mayer, Oxidation of hydrocarbons by $[(phen)_2Mn(\mu$-O)$_2Mn(phen)_2]^{3+}$ via hydrogen-atom abstraction, *J. Am. Chem. Soc.* **1997**, *119*, 1470-1471 (m) J. M. Mayer, Hydrogen-atom abstraction by metal-oxo complexes-understanding the analogy with organic radical reactions, *Acc. Chem. Res.* **1998**, *31*, 441-450 (n) S. A. Trammell, J. C. Wimbish, F. Odobel, L. A. Gallagher, P. M. Narula, T. Meyer, Mechanisms of surface electron transfer. Proton-coupled electron transfer, J. *J. Am. Chem. Soc.* **1998**, *120*, 13248-13249 (o) R. I. Cukier, D. G. Nocera, Proton-coupled electron-transfer, *Annu. Rev. Phys. Chem.* **1998**, *49*, 337-369 (p) S. J. Spencer, J. K. Blaho, J. Lehnes, K. A. Goldsby, Ph-dependent metal-based redox couples as models for proton-coupled electron-transfer reactions, *Coord. Chem. Rev.* **1998**, *174*, 391-416; (q) S. M. Hubig, R. Rathore, J. K. Kochi, Steric control of electron-transfer -changeover from outer-sphere to inner-sphere mechanisms in arene/quinon redox pairs, *J. Am. Chem. Soc.* **1999**, *121*, 617-626 (r) J. P. Roth, S. Lovell, J. M. Mayer, Intrinsic barriers for electron and

kinetically well behaved reaction between **1** and BPH$_2$ were determined, and the mechanism of electron transfer was established at constant ionic strength, μ, and constant Li$^+$ concentration, [Li$^+$].

The formation of 1:1 ion pairs $[(M^+)(SiVW_{11}O_{40}^{5-})]^{4-}$ ($M^+\mathbf{1}$, M^+ = Li$^+$, Na$^+$, K$^+$; eq 2) is then demonstrated.

$$M^+ + SiVW_{11}O_{40}^{5-}\ (\mathbf{1}) \rightleftharpoons [(M^+)(SiVW_{11}O_{40}^{5-})]^{4-}\ (M^+\mathbf{1}) \qquad (2)$$

Finally, the physicochemical role of alkali-metal cation size in electron transfer from BPH$_2$ to $M^+\mathbf{1}$ ion pairs (eq 3) is established by correlation of kinetic data (changes in k_{obs} values as a function of [M$^+$]) with data obtained by paramagnetic ^{7}Li NMR spectroscopy, cyclic voltammetry and single-potential step chronoamperometry.[21]

$$BPH_2 + 2\ (M^+\mathbf{1}) \longrightarrow DPQ + 2\ (M^+\mathbf{1}_{red}) + 2\ H^+ \qquad (3)$$

RESULTS AND DISCUSSION

Mechanism of Oxidation of BPH$_2$ to DPQ at Constant [Li$^+$].

The stoichiometry shown in eq 1, mechanism and conditions necessary for kinetically well behaved electron transfer from BPH$_2$ to **1** at constant [Li$^+$] were rigorously established using GC-MS, ^{1}H, ^{7}Li and ^{51}V NMR and UV-vis spectroscopy. All reactions were carried out using α-Li$_5$SiVW$_{11}$O$_{40}^{5-}$ (Li$_5\mathbf{1}$) in lithium acetate buffered 2:3 (v/v) H$_2$O:*t*-BuOH at 60 °C. The concentration of Li$^+$ was held constant by addition of LiCl.

hydrogen-atom transfer-reactions of viomimetic iron complexes, *J. Am. Chem. Soc.* **2000**, *122*, 5486-5498 (s) M-H. V. Huynh, P. S. White, T. J. Meyer, Proton-coupled electron-transfer from sulfur: A S-H/S-D kinetic isotope effect of greater-than or equal to 31.1, *Angew. Chem., Int. Ed.* **2000**, *39*, 4101-4104; (t) Y. Georgievskii, A. A. Stuchebrukhov, Concerted electron and proton-transfer-transition from nonadiabatic to adiabtic proton tunneling, *J. Chem. Phys.* **2000**, *113*, 10438-10450; (u) H. Decornez, S. Hammes-Schiffer, Model proton-coupled electron-transfer reactions in solution-Predictions of rates, mechanisms, and kinetic isotope effects, *J. Phys. Chem. A* **2000**, *104*, 9370-9384.

At constant $[Li^+]$ and $[H^+]$ (lithium acetate buffer) the reaction rate is first order in $[BPH_2]$ (0.3 to 3.6 mM) and in [1] (0.34 to 5.0 mM Li_51), zeroth order in $[1_{red}]$ (0.1 to 3.2 mM) and in [acetate] (LiOAc; 10 to 200 mM at constant $[H^+]$), and is independent of ionic strength, μ. In order to vary the ionic strength without changing $[Li^+]$, either LiCl or $Li_6Al(AlOH_2)W_{11}O_{39}$ $(Li_62)^{30}$ was used as electrolyte. Doubling of the ionic strength from 16.9 mM (10 mM LiCl) to 40.4 mM (1.60 mM Li_62), had little effect on the reaction rate.[†] Using UV-vis to measure the rate of formation of 1_{red} (initial-rate method), taking into account the 2:1 1:BPH_2 stoichiometry shown in eq 1, and defining the reaction rate as $d[DPQ]/dt = 1/2(d[1_{red}]/dt)$, the empirical rate law for reaction of 1 with BPH_2 at constant $[Li^+]$ and $[H^+]$ is given by eq 4.

$$\frac{1}{2}\frac{d[1_{red}]}{dt} = k_{obs}[1]^1[BPH_2]^1[OAc^-]^0[1_{red}]^0 \tag{4}$$

The $[H^+]$ dependence of the reaction rate (6.5×10^{-6} to 1.3×10^{-7} M H^+, corresponding to the pH range 5.19 to 6.89) was investigated at constant $[Li^+]$ (0.1 M LiOAc) by varying the concentration of HOAc. Acid dissociation of BPH_2 to BPH^- becomes kinetically significant at pH values larger than 5.5 (i.e., at $[H^+]$ values smaller than 3.15×10^{-6} M). At pH values greater than 5.5, linear plot of k_{obs} values ($k_{obs} \equiv (d[1_{red}]/dt)/([Li_51][BPH_2])$) versus $1/[H^+]$ reveals an inverse dependence on $[H^+]$. The reaction rate becomes independent of $[H^+]$ at high $[H^+]$ ($[H^+]$ values of ca. 3×10^{-6} M and larger, which correspond to pH values below 5.5).

Temperature-dependence data gave activation parameters of $\Delta H^{\ddagger} = 8.5 \pm 1.4$ kcal mol^{-1} and $\Delta S^{\ddagger} = -39 \pm 5$ cal mol^{-1} K^{-1}. No evidence of pre-association[31] between BPH_2 and 1 was observed in combined ^1H and ^{51}V NMR studies. In addition, pH (pD)-dependent deuterium kinetic-isotope data indicated that the O–H bond in BPH_2 remains intact during rate-limiting electron transfer from BPH_2 to 1. At pH 5.45 and pD 5.44 (D_2O and d-tert-butanol), a k_H/k_D value of 2.0 ± 0.3 was obtained. At pH 3.90 and pD 3.90, however, a k_H/k_D value of 1.2 ± 0.2 was observed. The decrease in k_H/k_D values as the pH (pD)

$$BPH_2 + AcO^- (H_2O) \underset{\text{rapid pre-equilibrium}}{\overset{Ka_1}{\rightleftharpoons}} BPH^- + AcOH (H_3O^+) \tag{5}$$

$$BPH_2 + 1 \xrightarrow[\text{slow}]{k_{BPH2}} BPH_2^{+\bullet} + 1_{red} \tag{6}$$

$$BPH^- + 1 \xrightarrow[\text{slow}]{k_{BPH^-}} BP\overset{\bullet}{H} + 1_{red} \tag{7}$$

$$BPH_2^{+\bullet} + AcO^- (H_2O) \xrightarrow{\text{fast}} BP\overset{\bullet}{H} + AcOH (H_3O^+) \tag{8}$$

$$BP\overset{\bullet}{H} + 1 + AcO^- (H_2O) \xrightarrow{\text{fast}} DPQ + 1_{red} + AcOH (H_3O^+) \tag{9}$$

[†] Formal molal-scale μ values are approximated here by molar-scale values calculated by using molarities in place of molalities, m, in the Debye-Hückel relationship, $0.5\Sigma z^2 m$.

decreases is not due to changes in the extent of O–H (O-D) bond breaking in the bimolecular reaction of BPH_2 (BPD_2) with 1, but rather, is attributed to a slightly larger acid-dissociation constant for BPH_2 than for BPD_2.[32]   A reaction mechanism consistent with all the kinetic and mechanistic data is provided in eqs 5-9.

Steady-state approximation applied to intermediates in eqs 5-9 ($d[BPH_2^{+\bullet}]/dt = d[BPH^{\bullet}]/dt = 0$), gives the following expression:

$$\frac{1}{2}\frac{d[1_{red}]}{dt} = \left(\frac{k_{BPH_2} \bullet [H^+] + k_{BPH^-} \bullet K_{a_1}}{K_{a_1} + [H^+]}\right)[1][BPH_2] \tag{10}$$

in which k_{BPH_2} and k_{BPH^-} are rate constants for reaction of 1, respectively, with BPH_2 and with mono-anionic BPH^-, and K_{a_1} is the first acid-dissociation constant of BPH_2. Given that $K_{a_1} \ll [H^+]$ (K_{a_1} values are estimated to be ca. 10^{-14} to 10^{-15} M;[33-36] experimental $[H^+]$ values ranged from ca. 10^{-6} to 10^{-7} M), eq 10 simplifies to eq 11, a more general form of the empirical rate law (eq 4).

$$\frac{1}{2}\frac{d[1_{red}]}{dt} = \left(k_{BPH_2} + \frac{k_{BPH^-} \bullet K_{a_1}}{[H^+]}\right)[1][BPH_2] \tag{11}$$

Electron-Transfer to Solvent-Separated 1:1 Ion-Pairs, $[(M^+)(SiVW_{11}O_{40}^{5-})]^{4-}$ ($M^+ = Li^+$, Na^+, K^+)

The data presented above describe the reaction conditions necessary for obtaining selective oxidation of BPH_2 to DPQ by kinetically and mechanistically well defined outer-sphere electron transfer to 1 at constant $[Li^+]$.   In the work presented below, concentrations of anionic BPH^- are kept at kinetically insignificant levels by using alkali-metal acetate MOAc/HOAc buffers to keep solution pH values at 4.76.   Under these conditions, the reaction rate is described by eq 12, in which k_{obs} from eq 4 is now equal to k_{BPH_2} (i.e., the $[H^+]$-dependent term in eq 11 is not kinetically significant).

$$\frac{1}{2}\frac{d[1_{red}]}{dt} = k_{BPH_2}[1][BPH_2] \tag{12}$$

We now establish the 1:1 stoichiometry of the ion pairs, $[(M^+)(SiVW_{11}O_{40}^{5-})]^{4-}$ (M^+1, $M^+ = Li^+$, Na^+, K^+), determine the formation constants of each, and investigate in detail the role of alkali-metal-cation size on ion-pair structure and on the energy and rate of electron transfer from BPH_2 to the M^+1 pairs.

Electron Transfer in the Absence of Ion Pairing. A value for k_{BPH_2} in the absence of ion pairing was obtained by using an effectively non-associating quaternary

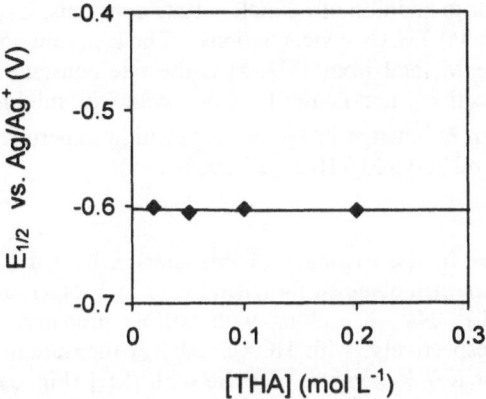

Figure 2. Reduction potential, $E_{1/2}$, of the $1/1_{red}$ couple (1.0 mM THA$_5$1 in 2:3 (v/v) H$_2$O:t-BuOH at 60 °C) in the presence of 25 to 200 mM tetra-n-hexylammonium nitrate, THAN, electrolyte. $E_{1/2}$ values are relative to a Ag/Ag$^+$ reference electrode (see Experimental).

alkylammonium cation (R_4N^+), as counter cation to **1**. Of the alkylammonium ions investigated (R = ethyl, n-propyl, n-butyl, n-pentyl, n-hexyl), only tetra-n-hexylammonium gave salts of **1** (n-hexyl$_4N^+$)$_5$**1** (THA$_5$**1**) (obtained from Li$_5$**1** by cation-exchange) that were sufficiently soluble in 2:3 (v/v) H$_2$O:t-BuOH at 60 °C. The n-hexyl$_4N^+$ cation is sufficiently large and hydrophobic that it, effectively, does not form ion pairs with **1** in the low dielectric and substantially hydrophobic solvent-system, 2:3 (v/v) H$_2$O:t-BuOH, at 60 °C (the dielectric constant is estimated from literature data to be 23.9).[37] Evidence in support of this conclusion is provided by cyclic voltammetry. If ion pairing between n-hexyl$_4N^+$ and **1** or **1**$_{red}$ occurred, increase in the concentration of n-hexyl$_4N^+$ would result in a positive shift in the potential of the $1/1_{red}$ couple.[38] However, no change in $E_{1/2}$ values of solutions of **1** (1 mM) were observed as concentrations of n-hexyl$_4N^+$NO$_3^-$ (tetra-n-hexylammonium nitrate, THAN) were varied from 25 to 200 mM (Fig. 2).[†]

[†] The Ag/Ag$^+$ reference electrode used consisted of 0.01 M AgNO$_3$ and 0.1 M THAN in 2:3 (v/v) H$_2$O:t-BuOH. Acceptable results (at least quasi-reversible electrode kinetics and acceptably small liquid-liquid junction potentials) required electrolyte concentrations of at least 25 mM.

Initial-rate methods were then used to measure k_{BPH_2} for oxidation of BPH_2 by THA_51. Reactions were carried out in 2:3 (v/v) H_2O:t-BuOH at 60 °C, using 0.5 mM THA_51 and 2.0 mM BPH_2. To obtain a suitable buffer without introducing additional cations, tetra-n-hexylammonium hydroxide and acetic acid (THAOH and HOAc; 1 mM OAc⁻) were combined in water prior to addition of t-BuOH. Rate constants, k_{obs}, were also determined at larger (50 and 100 mM) THAN concentrations. The k_{obs} value observed in the absence of added THAN (2.0 mM total from THA_51) is the rate constant for oxidation of BPH_2 (k_{BPH_2} in eq 12) by effectively non-paired **1**. Consistent with minimal association between THA and **1** in solution, no change in k_{obs} values (within experimental error) is observed after additions of 100 and 200 mM THAN (Fig. 3, below).

Electron Transfer in the Presence of Association by $[M^+]$ ($M^+ = Li^+$, Na^+, K^+). Reaction rates, k_{obs}, were then obtained for reactions of Li_51, Na_51, and K_51 as a function of added $[MCl]$ ($M^+ = Li^+$, Na^+, K^+, along with buffers prepared by combining LiOAc, NaOAc and KOAc, respectively, with HOAc; 20 k_{obs} measurements in all). For each cation, M^+ (Li^+, Na^+ or K^+), k_{obs} values increase with $[M^+]$ (Fig. 3). Because k_{obs} values are insensitive to changes in ionic strength (cf. plot of initial rate versus [THAN] in Fig. 3), changes in k_{obs} values as a function of $[M^+]$ are associated with association between M^+ and the 5- anion, **1**. The challenge presented by this data, however, is to differentiate between contributions to the ordering in k_{obs} values ($Li^+ < Na^+ < K^+$) attributable to ion-association stoichiometries, to K_{IP} values, and to the physicochemical properties of stoichiometrically identical association complexes.

At infinite dilution, the dissolved POM salts are fully dissociated into counter cations, M^+, and POM anions. At finite $[M^+]$ values, however, association between M^+ and the POM occurs. Based on electrostatic arguments[39-42] and experimental data,[38,43-45] ion association occurs sequentially: K_{IP} for formation of the 1:1 pair in eq 13 (K_{M1}) is larger than K_{IP} values for the formation of higher order, i.e., 2:1, 3:1, etc., association complexes.

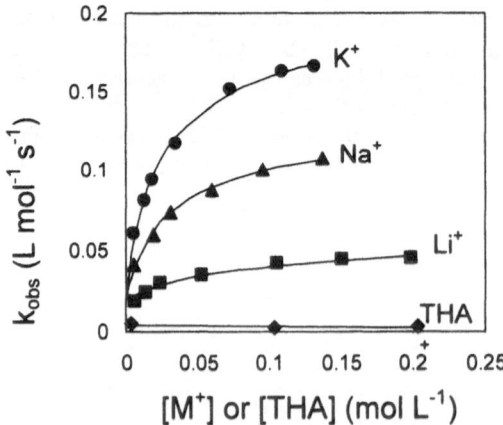

Figure 3. Initial rates (d[1_{red}]/dt) for the oxidation of BPH_2 by **1** (2.0 mM BPH_2 in 2:3 (v/v) H_2O:t-BuOH at 60 °C) in the presence of 0, 100 and 200 mM THAN (0.5 mM THA_51 in THAOAc/HOAc buffer, 2mM in OAc⁻), and initial rates versus $[M^+]_{total}$ ($M^+ = Li^+$, Na^+ and K^+, where $[M^+]_{total} = [MCl]$ (0.0-0.2 M) + 5[M_51] (1.0 mM) + [MOAc], and [MOAc] = [HOAc] = 1 mM).

$$M^+ + 1 \; \underset{}{\overset{K_{M1}}{\rightleftharpoons}} \; M^+1 \tag{13}$$

In eq 13, the equilibrium concentration of M^+1 is given by mass-balance as $K_{M1} = [M^+1]/([M^+][1])$. Expansion of eq. 13 to include rate constants, k_1 and k_{M1}, associated with unpaired 1 and with 1:1 M^+1 pairs, respectively, gives:

$$\frac{1}{2}\frac{d[1_{red}]}{dt} = k_1[1][BPH_2] + k_{M1}[M^+1][BPH_2] \tag{14}$$

Substitution of $K_{M1}[M^+][1]$ for $[M^+1]$ and rearrangement gives:

$$\frac{1}{2}\frac{d[1_{red}]}{dt} = \left(\frac{k_1 + k_{M1}K_{M1}[M^+]\gamma_M}{1 + K_{M1}[M^+]\gamma_M} \right)[1][BPH_2] \tag{15}$$

The complex rate constant in eq 15 is a rectangular hyperbolic function. It is exactly equal to k_1 at infinite dilution (the low-$[M^+]$ limit) and asymptotically approaches k_{M1} as $[M^+]$ increases to large values.[46,47]

The solid curves in Fig. 3 were calculated by simultaneous non-linear least-squares regression of reaction-rate data at all alkali-metal cation concentrations (20 k_{obs} values using three different cations) to the complex rate constant in eq 15. The calculation was performed subject to the stipulation that all three curves converge to a single k_1 value. (Convergence to a single k_{obs} value, k_1, at infinite dilution is implicit to derivation of eq 15.) Simultaneous analysis of the three curves (one for each of the cations, Li^+, Na^+ and K^+) provides unique values for rate and equilibrium constants, k_{M1} and K_{M1}, i.e., for k_{Li1}, k_{Na1}, k_{K1}, K_{Li1}, K_{Na1} and K_{K1}.

The rate constant in eq 15 is formally a function of activities of all species present. However, 1 and BPH_2 possess small initial concentrations, and these concentrations are identical at the outset of each experiment. As a result, variation in the activities of these components are much less significant than changes in the activities of the alkali-metal cations themselves (the experimental variable). The cation concentrations vary from 5 to over 200 mM (based on grams of MCl salt added per liter of solution). Therefore, to obtain meaningful k_1, k_{M1} and K_{M1} values by non-linear regression of the kinetic data, the number of adjustable parameters was restricted to the most significant activity coefficient, γ_M (for M^+).[38,48†]

[†] The $[M^+]$ values shown in Fig. 3 and used to fit the data to eq. 13 are the sum of $[M^+]$ from: added MCl, POM counter–cations and buffer. However, association between M^+ and Cl^- is also possible. The association constant for the formation of M^+Cl^- pairs, K_{MCl}, would be explicitly included in eq. 13 by substituting $[MCl]/K_{MCl}[Cl^-]$ for $[M^+]$. However, self association of the added 1:1 metal halide salts is but one of several phenomena that contribute to values of γ_M and is one reason why γ_M is retained in eq 13. Values used for the activity coefficient, γ_M, were calculated using an extended Debye-Hückel law with a linear empirical-correction term: $\log\gamma_z = -Az^2[\mu^{1/2}/(1 + Ba\mu^{1/2})] + b\mu$, in which γ_z is the activity coefficient of a single ion of charge z, A ($1.825 \; 10^6(\varepsilon T)^{-3/2} \; mol^{-1/2} \; L^{1/2} \; K^{3/2}$) and B ($50.29(\varepsilon T)^{-1/2} \; Å^{-1} \; mol^{-1/2} \; L^{1/2} \; K^{1/2}$) are constants that change with the temperature and dielectric constant of the solvent, and a and b are adjustable parameters. Values of A and B used to generate the solid curves in Fig. 3 were, A = 2.55 $mol^{-1/2} \; L^{1/2} \; K^{3/2}$ and B = 0.563 $Å^{-1} \; mol^{-1/2} \; L^{1/2} \; K^{1/2}$. Values for "a" were approximated by use of standard values commonly

The stoichiometry in eq 13 was established by recourse to three lines of evidence (see Fig. 3): (1) agreement between k_{obs} values and the complex rate constant in eq 13 over a statistically required range of $[M^+]$ values,[49,50] (2) convergence of the three curves (for Li^+, Na^+ and K^+) at the low-$[M^+]$-limit to a single k_1 value, and (3) near identity between the single low-$[M^+]$-limit k_1 value and the k_{obs} value determined using THA$_5$1 (absence of kinetically significant ion association). In the effective absence of ion association, a k_{obs} value of $3.4 \pm 0.7 \times 10^{-3}$ L mol^{-1} s^{-1} (Fig. 3, determined using THA$_5$1 and THAN) was obtained. (The two systems—THA and alkali-metal cations—differ in subtle yet potentially significant ways from one another. First, the THA system uses a THA acetate buffer whose degree of self association in 2:3 H_2O/t-BuOH differs from that of the alkali-metal acetate buffers. Secondly, the extent and possible effect of interaction between THA and BPH$_2$ has not been quantified. Finally, in the effective absence of association between THA and 1 or 1_{red}, the reduction potential of the $1/1_{red}$ couple is sufficiently negative that rapid back reaction between 1_{red} and BPH$_2 \bullet^-$ prior to irreversible fragmentation of the successor complex, [BPH$_2 \bullet^-$–1_{red}] can no longer be excluded.)

Having established the stoichiometry in eq 13, values for K_{M1} and k_{M1} were reliably assessed (eq. 15). Calculated values for K_{M1} increase in the order $K_{Li1} = 21 \pm 10$, $K_{Na1} = 54 \pm 10$ and $K_{K1} = 65 \pm 6$ L mol^{-1} (uncertainties are 95% confidence intervals). The K_{M1} values increase as the crystallographic radii of the cations become larger. Although electrostatic arguments based on contact ion pairs between hard spheres dictate that K_{M1} values should *decrease* as the radii of the associated ions increase,[51] the ordering of K_{M1} values reported here is consistent with the formation of solvent-separated ion pairs (see ^{7}Li NMR investigation of ion-pair structure, below). As the crystallographic radii of the cations increase, their charge densities decrease as e/r. Accordingly, the radii of the solvated-cations decrease from Li^+ to Na^+ to K^+.[21,22,52,53] Having established ion-pair stoichiometry, the rectangular hyperbolic functional dependence of k_{obs} values on $[M^+]$ also allows us to assign unique rate constants, k_{M1}, to the three M^+1 pairs. The rate constants increase in the order $k_{Li1} = 0.065$, $k_{Na1} = 0.137$ and $k_{K1} = 0.225$ L mol^{-1} s^{-1}. Using this information as a starting point, it is now possible to assess the energetic and structural role of alkali-metal-cation size in electron transfer. This is achieved by combined use of cyclic voltammetry, ^{7}Li NMR and chronoamperometry.

Cyclic voltammetry was used to assess the dependence of the reduction potentials ($E_{1/2}$ values) of solutions of 1 (1 mM) on the concentrations of added alkali-metal salts, MCl, M^+ = Li^+, Na^+ and K^+. Reaction conditions were identical to those used to obtain the k_{obs} data in Fig. 3. For each cation, $E_{1/2}$ values for 1-e$^-$ reduction of 1 to 1_{red} (reduction of V(V) in 1 to V(IV)) increased with $[M^+]$ (Fig. 4; 22 $E_{1/2}$ values).

assigned to Li^+, Na^+ and K^+ in water (Butler, J. N.; Cogley, D. R. *Ionic Equilibrium: Solubility and pH Calculations*; John Wiley & Sons: New York, 1998, p. 47) and "b" was set equal to zero.

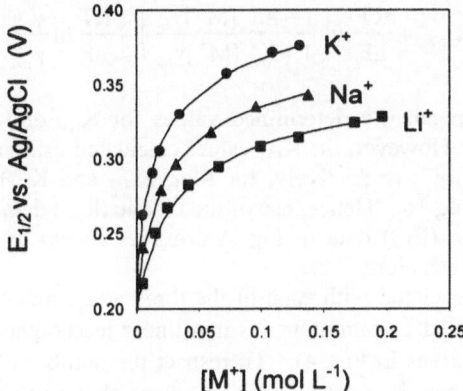

Figure 4. Reduction potentials, $E_{1/2}$, versus $[M^+]_{total}$ (M^+ = Li^+, Na^+ and K^+; $[M^+]_{total}$ = $[MCl]$ + $5[M_51]$ + $[MOAc]$, $[MCl]$ = 0.0-0.2 M, $[M_51]$ = 1.0 mM, $[MOAc]$ = $[HOAc]$ = 1 mM), at 60 °C, referenced to a Ag/AgCl electrode. (Solid curves were calculated by non-linear least-squares fitting of the data to eq 16.)

The kinetic data in Fig. 3 are fundamentally distinct from the thermodynamic data in Fig. 4. As established above (see eq 4), the rate of oxidation of BPH_2 is independent of $[1_{red}]$. As a result, the formation constant, K_{IP}, for pairing between M^+ and 1_{red} is neither directly pertinent to, nor accessible from, k_{obs} data. The situation is quite different, however, when measuring the reduction potentials of solutions of **1** in the presence of M^+. In this case, the standard Gibbs free energies of formation (ΔG_f°) of all reactants and products of the heterogeneous electrode reaction (electron transfer from the electrode to **1** in solution) contribute to the value of the $1/1_{red}$ couple.[28,54] Therefore, the concentrations of **1** and M^+**1**, 1_{red} and M^+**1**$_{red}$[55] must all be considered (Scheme 1). This is evident in the Nernstian expression that describes the dependence of $E_{1/2}$ on $[M^+]$, which includes both K_{ox} and K_{red} (eq 16;[38†] by convention, K_{IP} for formation of M^+**1**, K_{M1}, is labeled K_{ox}, while K_{IP} for formation of M^+**1**$_{red}$, K_{M1red}, is labeled K_{red}).

Scheme 1

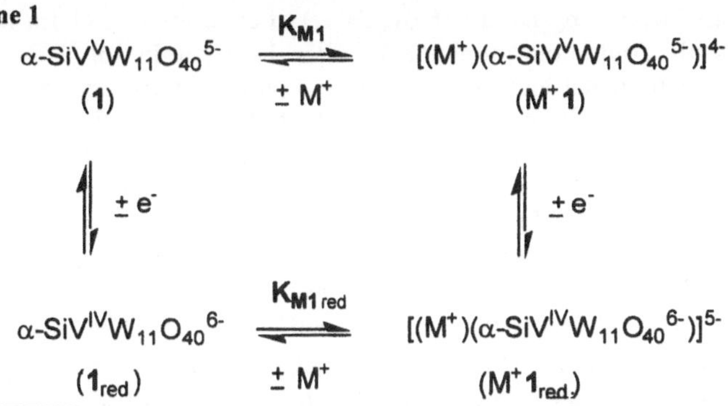

$$\alpha\text{-SiV}^V W_{11}O_{40}{}^{5-} \quad \xrightleftharpoons[\pm\, M^+]{K_{M1}} \quad [(M^+)(\alpha\text{-SiV}^V W_{11}O_{40}{}^{5-})]^{4-}$$

$$\textbf{(1)} \qquad\qquad\qquad\qquad\qquad\qquad\qquad (M^+\textbf{1})$$

$$\updownarrow \pm\, e^- \qquad\qquad\qquad\qquad\qquad \updownarrow \pm\, e^-$$

$$\alpha\text{-SiV}^{IV} W_{11}O_{40}{}^{6-} \quad \xrightarrow{K_{M1\,red}} \quad [(M^+)(\alpha\text{-SiV}^{IV} W_{11}O_{40}{}^{6-})]^{5-}$$

$$\textbf{(1}_{red}\textbf{)} \qquad\qquad \pm\, M^+ \qquad\qquad\qquad (M^+\textbf{1}_{red})$$

† In previous work (V. A. Grigoriev, C. L. Hill, I. A. Weinstock, Role of cation size in the energy of electron-transfer to 1/1 polyoxometalate ion-pairs $[M^+][X^{n+}VW_{11}O_{40}]^{(8-n)-}$ (M=Li, Na, K), *J. Am. Chem. Soc.* **2000**, *122*, 3544-45), a rectangular-hyperbolic function was used to model the change in $E_{1/2}$ as a function of $K_{M1}[M^+]$. While not rigorously correct, use of the rectangular-hyperbolic function provided a practically useful and internally consistent method for establishing association stoichiometry. However, eq 16 is not only rigorously correct, but provides values for both K_{M1} and K_{M1red}.

$$E_{1/2} = E_{1/2}^{\circ} + \frac{RT}{nF}\ln\left(\frac{1 + K_{red}[M^+]\gamma_M}{1 + K_{ox}[M^+]\gamma_M}\right) + \frac{RT}{nF}\ln\frac{\gamma_{ox}}{\gamma_{red}} \qquad (16)$$

In the absence of independently determined values for K_{red} or K_{ox}, eq 16 can, at best, provide K_{red}/K_{ox} ratios. However, the K_{M1} values calculated using the data in Fig. 3 and eq 15 (21, 54 and 65 L mol^{-1}, respectively, for K_{Li1}, K_{Na1} and K_{K1}) correspond to the K_{ox} values in Scheme 1 and eq 16. Hence, use of the kinetic (k_{obs}) data in Fig. 3 in conjunction with the thermodynamic ($E_{1/2}$) data in Fig. 4 provides access to K_{red} values not readily obtained by either approach alone.

The K_{red} values associated with each of the three alkali-metal cations, K_{Li1red}, K_{Na1red} and K_{K1red}, were calculated by simultaneous non-linear least-squares fitting of the data in Fig. 4 to eq 16 (solid curves in Fig. 4). To restrict the number of adjustable parameters, only the most essential activity coefficients, i.e., those shown[‡] in eq 16 (γ_M, γ_{ox} and γ_{red}), were used. Furthermore, the same Debye-Hückel parameters used to calculate γ_M in eq 15 were used here to calculate γ_M, γ_{ox} and γ_{red}. In addition, the non-linear regression was performed with the stipulation that the three curves in Fig. 4 converge to a single value, $E_{1/2}$ = $E_{1/2}^{\circ}$, as $[M^+]$ approaches zero. Excellent fit (solid curves in Fig. 4) lends independent support for the K_{M1} values calculated from the kinetic data in Fig. 3. Finally, calculated K_{M1red} values are 130 ± 30, 570 ± 120 and 2000 ± 300 L mol^{-1}, respectively, for M^+ = Li$^+$, Na$^+$ and K$^+$ (uncertainties are 95% confidence intervals).

Formation constants K_{M1} increase with the crystallographic radii of the M^+ ions. According to the Eigen-Fuoss model[51] for contact ion pairs between hard spheres in an unstructured dielectric continuum, K_{IP} values possess an inverse exponential dependence on the distance, d, (more exactly $d^3 e^{-1/d}$; $K = (4\pi N d^3/3000)\exp(-U_{(d)}/RT)$, where $U_{(d)}$ = $z_1 z_2 e^2/\{D_s d(1 + \chi d)$; $\chi = (8\pi N e^2 \mu/1000 D_s RT)^{1/2})$[51] between the centers of the two associated ions. The observed increase in K_{M1} values from Li$^+$ to K$^+$ thus suggests that d values decrease as the crystallographic radii of the cations become larger. To the extent that the electrostatic arguments used to derive the Eigen-Fuoss model apply, this increase in K_{M1} values implies the formation of solvent-separated ion pairs within which the *solvated radii* of the cations decrease from Li$^+$ to K$^+$. Upon reduction of 1 to 1_{red}, calculated formation constants increase by factors (K_{M1red}/K_{M1}) of ca. 6 for Li$^+$, 11 for Na$^+$ and 31 for K$^+$. Consistent with Coulombic attraction between ions within solvent-separated pairs, increase in K with incremental increase in the charge product, $z_1 z_2$, from 5- to 6-, is greater for the more tightly associated ion pairs of the less highly solvated cations.

Ion pairing between Li$^+$ and 1 and 1_{red} was also investigated by analysis of the functional dependence of ^{7}Li NMR chemical-shift values on the total concentrations of Li$_5$1 and of 1e$^-$ reduced Li$_6$1$_{red}$ (Fig. 5). Evidence for the formation of solvent-separated pairs was provided by the small value of the molar paramagnetic contact shift (δ_c) in ^{7}Li NMR spectra of paramagnetic solutions of Li$_6$1$_{red}$ (Fig. 5b).[56,57]

Eight solutions of Li$_5$1 in 2:3 (v/v) H$_2$O/*t*-BuOH (0.3 to 15 mM in 1; 1.5 to 75 mM Li$^+$) were heated to 60.0 °C in the NMR probe. Exchange between free and paired Li$^+$ ions is rapid on the NMR time scale and a single ^{7}Li NMR signal is observed (rapid-exchange limit). The chemical shift of the ^{7}Li NMR signal moves downfield as the fraction of Li$^+$ paired to 1 increases with the concentration of Li$_5$1. The data in Fig. 5a show effectively

---

[‡] The coefficient γ_M, is associated with $[M^+]$, which is varied over a wide range; both γ_{ox} and γ_{red} are experimentally significant and, from a mathematical perspective as well, $E_{1/2}$ values possess a significant functional dependence on the term $(RT/F)\ln(\gamma_{ox}/\gamma_{red})$.

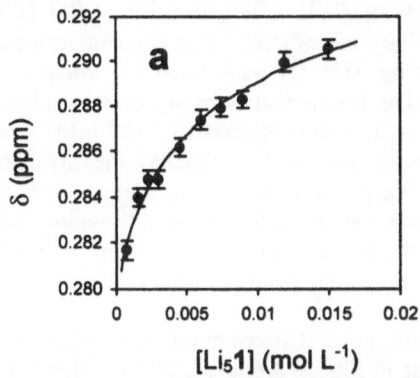
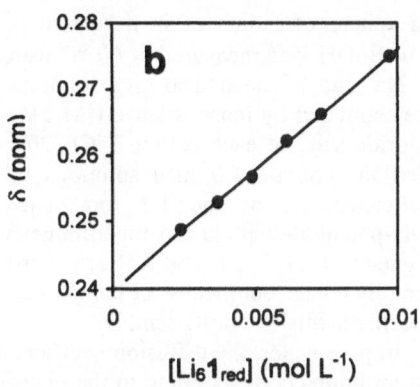

Figure 5. a. Chemical shifts of ^{7}Li NMR signals (referenced to 1.0 M LiCl in D$_2$O) as [Li$_5$1] is increased from 0.3 to 15 mM (1.5 to 75 mM Li$^+$) in 2:3 (v/v) H$_2$O:t-BuOH at 60 °C. Due to the relatively small changes in chemical-shift values observed, a fine digital resolution of 0.061 Hz (0.0004 ppm) was used. Solid curve: non-linear least-squares fit (see eq 34 in Experimental); δ_o = 0.287 ppm, δ_1 = 0.336 ppm, K_{Li1} = 65 ± 35 M^{-1} (95% confidence interval). b. Chemical shifts of ^{7}Li NMR signals as [Li$_6$1$_{red}$] is increased from 0.5 to 10 mM in 2:3 (v/v) H$_2$O:t-BuOH at 60 °C. The slope (δ_{obs}) is 3.62 ppm L mol^{-1} (R^2 = 0.998).

no pairing at the low-concentration limit (1.5 mM Li$^+$ and 0.3 mM **1**) and the formation of 1:1 ion pairs, Li$^+$**1**, as the concentration of Li$_5$**1** increases. At each [Li$_5$**1**], the chemical shift of the single, coalesced ^{7}Li NMR signal is the weighted mean of contributions from free and paired Li$^+$ ions. Non-linear least-squares regression of the data to a model describing the change in chemical shift upon 1:1 association as a function of [Li$_5$**1**] gives a formation constant, K_{Li1}, of 65 ± 35 L mol^{-1} (uncertainty is a 95% confidence interval). Despite a very small range of δ values, this K_{Li1} value is within statistical uncertainty of the value obtained independently from k_{obs} data (i.e., 21 ± 10 L mol^{-1} from eq 15).

Once convinced that the ^{7}Li NMR technique gave a reasonable value for K_{Li1}, spectra of paramagnetic solutions of Li$_6$**1**$_{red}$ were obtained (Fig. 5b). Here, the chemical shift, δ_{obs}, increases linearly with the concentration of Li$_6$**1**$_{red}$. The δ_{obs} value[56] is a linear combination of contributions from the bulk susceptibility (δ_{bulk}, a concentration-dependent result of macroscopic interactions) and the contact shift, δ_c, that arises from spin-density transfer from the paramagnetic center, V(IV), in **1**$_{red}$:

$$\delta_{obs} = \delta_{bulk} + 2\delta_c \tag{17}$$

At the rapid exchange limit (our case), δ_c is also concentration dependent. Therefore, both δ_{bulk} and δ_c must be evaluated as molar values (ppm L mol^{-1}).[57] The data in Fig. 5b give δ_{obs} = 3.62 ppm L mol^{-1}, from which values of δ_{bulk} = 4.72 and δ_c = -0.55 ppm L mol^{-1} are obtained (333 K and s = 1/2, i.e., a single unpaired electron).[56] The very small contribution of δ_c to δ_{obs} suggests that the associated ions (whose formation is indicated in Fig. 5a) are substantially electronically insulated from one another by solvent, i.e., they exist in solution as solvent-separated ion pairs.[57]

Molecular sizes of the 1:1 pairs were compared as effective hydrodynamic radii, estimated using the Stokes-Einstein equation: D = kT/6$\pi\eta$r, (D is the diffusion coefficient of a sphere of radius r in a solvent of viscosity η).[58,59†] The viscosity of 2:3 (v/v) H_2O:*t*-BuOH was measured at 60 °C using a capillary viscometer. The concentrations of Li^+, Na^+ and K^+ needed to give solutions containing 93% 1:1 paired and 7% unpaired **1** were estimated by mass balance ($[M^+\mathbf{1}]/[\mathbf{1}] = K_{M1}[M^+]$; K_{M1} values from Table 1, below). (Chloride salts of each cation, LiCl (202 mM), NaCl (110 mM) and KCl (85 mM) were added to separate 1.0 mM solutions of $Li_5\mathbf{1}$, $Na_5\mathbf{1}$ and $K_5\mathbf{1}$.) Finally, the diffusion coefficients, D, of the 1:1 ion pairs, $Li^+\mathbf{1}$, $Na^+\mathbf{1}$, and $K^+\mathbf{1}$ were determined by single-potential-step chronoamperommetry.[21] While the absolute values of the ionic radii assigned to $Li^+\mathbf{1}$, $Na^+\mathbf{1}$, and $K^+\mathbf{1}$ are approximate, the Stokes-Einstein equation provides an internally consistent means for comparing relative molecular volumes, each associated with a specific diffusion coefficient, D.

In pure water, the diffusion coefficients of solutions of **1** (even at relatively large MCl concentrations corresponding to the plateau regions in Fig. 3) are identical for added LiCl, NaCl or KCl, and give the same Stokes-Einstein radius (5.6 ± 0.2 Å) as that reported for unpaired Keggin anions in water (i.e., 5.6 Å for $PW_{12}O_{40}^{3-}$ and for $SiW_{12}O_{40}^{4-}$).[60] In the present case (2:3 (v/v) H_2O:*t*-BuOH at 60 °C), however, 1:1 ion association occurs. Now, the diffusion coefficients (and corresponding Stokes-Einstein radii) *decrease* as the crystallographic radii of the associated cations become larger (Table 1, below). While the crystallographic (Shannon and Prewitt) radii of hexacoordinate Li^+, Na^+ and K^+ ions increase, respectively, from 0.90 to 1.16 to 1.52 Å,[61] their solvated radii decrease with charge density, *e*/*r*, in the same order[52] (radii of hydrated Li^+, Na^+ and K^+ ions in water decrease from ca. 3.40 to 2.76 to 2.32 Å as the waters of hydration decrease from ca. 25.3 to 16.6 to 10.5).[62] The effective hydrodynamic radii of the solvent-separated 1:1 ion pairs thus *decrease* accordingly as the radii of the "naked" alkali-metal cations become larger.

Outer-sphere electron-transfer from BPH_2 to **1** involves ion pairing, local ordering of solvent molecules (coordination of water to M^+), steric (*t*-butyl groups in BPH_2) and orientational (specific orientations of both Donor and Acceptor) constraints. These phenomena are not consistent with the simplifying assumptions used to calculate essential parameters in theoretical models, such as that of Marcus, that relate standard free-energies of reaction, ΔG°, to electron-transfer rates, k_{et}.[1,7,63] In particular, changes in nuclear coordinates, λ (both λ_{in} and λ_{out}) associated with the alkali-metal cation and with the water molecules within the cation's solvation sphere must be evaluated. However, there is no readily applicable quantitative model for doing this. Moreover, even in stoichiometrically well characterized systems such as ours, the precise location of the associated cation within the Donor–Acceptor complex as a function of progress along the reaction coordinate is unknown. Due to these uncertainties, no comprehensive model for the role of associated cations in electron-transfer to ion pairs such as $M^+\mathbf{1}$ is available.

The absence of such a model[63] is due, in part, to the problems encountered in determining the structure and physicochemical properties of specific ion pairs and in correlating these properties with reaction-rates. This problem is addressed by the data in Table 1, which establish relationships between the physicochemical properties of the solvent-separated $M^+\mathbf{1}$ pairs and electron-transfer rates. In Scheme 2, these relationships and properties are associated with the key species (stable $M^+\mathbf{1}$ pairs as well as transient precursor and successor complexes) pertinent to assessing the rate and energy ($\Delta G°_{et}$) of electron transfer from BPH_2 to **1**. As the crystallographic radii of the cations increase from

---

† Effective hydrodynamic (Stokes-Einstein) radii of Keggin anions obtained in water from diffusion-coefficient data match those determined in water by velocity ultracentrifugation and from viscosity and density data.

Table 1. Solution Properties of 1:1 Solvent-Separated Ion Pairs[a] $[(M^+)(SiV^VW_{11}O_{40})]^{4-}$ (M^+1) and $[(M^+)(SiV^{IV}W_{11}O_{40})]^{5-}$ (M^+1_{red})

M^+	r_{eff}, Å[b]	K_{M1} (L mol^{-1})	k_{M1} (L mol^{-1} s^{-1})	$E_{1/2}$ (mV)[bc]	K_{M1red} (L mol^{-1})
Li$^+$	8.3 ± 0.4	21 ± 10	0.065 ± 0.013	327	130 ± 30
Na$^+$	7.7 ± 0.2	54 ± 10	0.137 ± 0.011	338	570 ± 120
K$^+$	6.8 ± 0.3	65 ± 6	0.225 ± 0.010	362	2000 ± 300

[a]$H_2O:t$-BuOH (2:3, v:v) at 60 °C. [b]93% 1:1 paired and 7% unpaired **1** (202 mM LiCl, 110 mM NaCl and 85 mM KCl, added, respectively to 1.0 mM solutions of Li$_5$**1**, Na$_5$**1** and K$_5$**1**). [c]Calculated from the data in Fig. 4 (202 mM LiCl, 110 mM NaCl and 85 mM KCl); vs. Ag/AgCl (3 M NaCl).

0.90 Å (Li$^+$) to 1.16 Å (Na$^+$) to 1.52 Å (K$^+$), the Stokes-Einstein radii of the solvent-separated M^+1 pairs *decrease* from 8.3 Å (Li$^+$) to 7.7 Å (Na$^+$) to 6.8 Å (K$^+$) (see discussion above). Additionally, as expected based on Coulombic arguments, the decrease in ion-pair size is associated with an *increase* in formation constants, K_{M1}. This trend in K_{M1} values alone could explain the commonly encountered observation that effectiveness in specific-cation catalysis increases with cation size: At equal [M^+] values, a larger fraction of the Acceptor anions is paired when $M^+ = K^+$ than when $M^+ = Li^+$.

However, we are now in a position to compare the relative rates of electron transfer to fully formed ion pairs that differ from one another only in the nature of M^+. Each ion pair is associated with a unique rate constant, k_{M1}. These rate constants are theoretical limiting values that represent extrapolation of the kinetic data (Fig. 3) to the hypothetical case in which all the anions, **1**, exist in solution as 1:1 pairs with M^+ cations. These calculated rate constants, k_{M1}, increase in the order k_{Li1} (0.065 L mol^{-1} s^{-1}) $< k_{Na1}$ (0.137 L mol^{-1} s^{-1}) $< k_{K1}$ (0.225 L mol^{-1} s^{-1}). At the same time, the reduction potentials of solutions containing 93% paired and 7% unpaired **1** (Table 1) increase from Li$^+$ to K$^+$. As depicted in Scheme 2, these thermodynamic values include contributions from energies associated with the formation of M^+1 pairs, with reduction of V(V) in **1** to V(IV) (electron transfer), and with the simultaneous formation of M^+1_{red} pairs. Also, as M^+ increases in size (crystallographic

Scheme 2

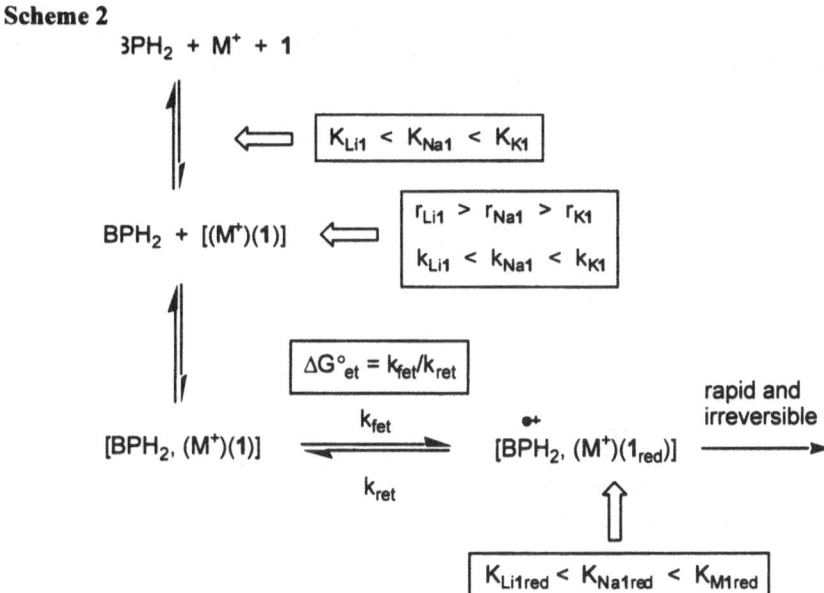

radius), the hydrodynamic radii of the 1:1 M^+1 pairs (Table 1) decrease. Based on Coulombic arguments, reduction of the smaller, more "intimate" ion pairs should be more energetically favorable. Furthermore, electrochemical data reveal that, as M^+ increases in size (crystallographic), $K_{M1_{red}}$ values increase substantially. These increases in $K_{M1_{red}}$ values are attributable to favorable increases in the energy of formation ($\Delta G°_f$ = $-RTlnK_{M1_{red}}$) of the M^+1_{red} pairs. Thus, as the crystallographic radii of the M^+ ions increase, the energies associated with ion-pair formation make larger (and more favorable) contributions to the standard free energy of reaction, $\Delta G°_{rxn}$. These increases in ion-pairing energy are likely reflected in the relative energies of reductions of M^+1 to M^+1_{red} (i.e., in the electronic transitions $[BPH_2, M^+1] \rightarrow [BPH_2^{•+}, M^+1_{red}]$; Scheme 2). Thus, as M^+1 is reduced to M^+1_{red}, contributions to $\Delta G°_{et}$ (Scheme 2) associated with ion-pair formation itself (proportional to $-RTln(K_{M1_{red}}/K_{M1})$ increase with the crystallographic radii of M^+. Using the $K_{M1_{red}}$ and K_{M1} values in Table 1, contributions from ion-pair formation energy to $\Delta G°_{et}$ increase (as $-RTln(K_{M1_{red}}/K_{M1})$ from -1.2 kcal mol^{-1} (Li$^+$), to -1.5 kcal mol^{-1} (Na$^+$), to -2.3 kcal mol^{-1} (K$^+$).

Clearly, the changes in ion-pair formation energy observed upon reductions of M^+1 to M^+1_{red} include changes in solvation energy that are not readily assessed. In addition, the precise effect of change in $\Delta G°_{et}$ on the rate of electron-transfer requires detailed information about changes in nuclear coordinates (i.e., the reorganization energies, λ, associated with local changes in both solvent structure and ion position). Ready calculation of λ values associated with changes in these complex systems is beyond the current level of theory. However, for modestly favorable (negative) $\Delta G°_{et}$ values, increase in $|\Delta G°_{et}|$ is generally associated with a decrease in the effect of λ on the magnitude of $\Delta G^{\ddagger}$ (eq 18)[20,64,65] and a resultant increase in the rate of electron transfer.

$$\Delta G^{\ddagger} = W(r) + \frac{\lambda}{4} + \frac{\Delta G°_{et}}{2} + \frac{(\Delta G°_{et})^2}{4\lambda} \tag{18}$$

EXPERIMENTAL

Materials and Methods. 3,3',5,5'-tetra-*tert*-butylbiphenyl-4,4'-diol (BPH$_2$; Polysciences, Inc., 99%) was analyzed by GC-MS and ^{1}H NMR and used as received. Other chemicals, *tert*-butyl alcohol (*t*-BuOH; 99.5%), deuterium oxide (99.9% D), 2-methyl-2-propan(ol-*d*) (>98% D), lithium acetate dihydrate (99.999%), sodium acetate (99%), potassium acetate (99%), tetra-*n*-hexylammonium nitrate (THAN; 99%) and tetra-*n*-hexylammonium hydroxide (40% in water), were of the highest purity available from commercial sources and were used as received. All reactions were carried out under prepurified grade Argon. The POMs α-K$_5$SiVW$_{11}$O$_{40}$•14.5H$_2$O (K$_5$1),[66] α-K$_6$SiVW$_{11}$O$_{40}$•7H$_2$O (K$_6$1$_{red}$)[66] and α-K$_6$Al(AlOH$_2$)W$_{11}$O$_{39}$•13H$_2$O (K$_6$2),[30] were prepared and purified according to published procedures. Purities of K$_5$1, K$_6$1$_{red}$ and K$_6$2 were verified by ^{29}Si and ^{51}V (K$_5$1 and K$_6$1$_{red}$—after oxidation by Br$_2$) or ^{27}Al (K$_6$2) NMR. Li$^+$ and Na$^+$ POM salts were prepared by cation exchange from corresponding K$^+$ salts (Amberlite IR-120 (plus) ion-exchange resin converted to Li$^+$ or Na$^+$ forms). Purities of the Li$^+$ or Na$^+$ forms of the POMs were confirmed by elemental analysis and spectroscopically (^{29}Si, ^{51}V and ^{27}Al NMR and UV-vis). Anal. Calc. (found) for α-Na$_5$SiVW$_{11}$O$_{40}$•10H$_2$O (Na$_5$1): Na 3.79 (3.64), Si 0.92 (1.10), V 1.68 (1.81), W 66.6

(67.1), H 0.66 (0.66), K 0.00 (<0.01). ^{29}Si NMR, δ, ppm: -81.87; ^{51}V NMR, δ, ppm: -551.3. Anal. Calc. (found) for α-Li$_5$SiVW$_{11}$O$_{40}$•12H$_2$O (Li$_5$1): Li 1.16 (1.07), Si 0.94 (0.79), V 1.70 (1.70), W 67.6 (68.0), H 0.81 (0.80), K 0.00 (<0.01). ^{51}V NMR, δ, ppm: -551.0; ^{29}Si NMR, δ, ppm: -81.94. Anal. Calc. (found) for α-Li$_6$SiVW$_{11}$O$_{40}$•20H$_2$O (Li$_6$1$_{red}$): Li 1.28 (1.25), Si 0.89 (0.87), V 1.62 (1.59), W 64.3 (64.2), H 1.28 (1.26), K 0.00 (<0.01). ^{29}Si NMR, δ, ppm: -81.92; ^{51}V NMR, δ, ppm: -551.0 (after oxidation in D$_2$O by Br$_2$ to Li$_5$1). UV-vis, λ_{max} = 496 nm (ε = 627 L mol^{-1} cm^{-1}). Anal. Calc. (found) for Li$_6$Al(AlOH$_2$)W$_{11}$O$_{39}$•8H$_2$O (Li$_6$2): Li 1.43 (1.37), Al 1.86 (1.76), W 69.6 (68.1), H 0.62 (0.58), K 0.00 (<0.01). ^{27}Al NMR, δ, ppm: 72.06 and 8.78 (1:1 intensity ratio). α-THA$_5$SiVW$_{11}$O$_{40}$ (THA$_5$1) was prepared by extracting aqueous solutions of 1 into an organic solvent containing tetra-n-hexylammonium nitrate (THAN). Typically, K$_5$1 (3.6 g, 1.13 mmol) was dissolved in 20 mL deionized water and extracted using 40 mL of a CHCl$_3$ solution containing 5 equiv. of THAN (2.06 g, 5.65 mmol). The organic layer (yellow) was separated and the aqueous phase extracted a second time using 20 mL of CHCl$_3$. The two organic portions were combined and washed 3 times with 40 mL water. The organic solution was then concentrated to dryness (rotary evaporation) and the solid obtained was dried overnight in vacuum at room temperature. Anal. Calc. (found) for α-((C$_6$H$_{13}$)$_4$N)$_5$SiVW$_{11}$O$_{40}$•0.4[(C$_6$H$_{13}$)$_4$N][NO$_3$] (THA$_5$1•0.4THAN): C 33.25 (33.36), H 6.04 (6.07), N 1.74 (1.69). ^{51}V NMR (in CDCl$_3$), δ, ppm: -540.7.

NMR Spectroscopy. A 400 MHz (Inova) instrument was used to acquire ^{1}H, ^{7}Li, ^{27}Al, ^{29}Si and ^{51}V NMR spectra. POM solutions used for NMR study (except ^{7}Li NMR) were prepared in pH 4.76 lithium-, sodium- or potassium-acetate buffers. POM stability was monitored by NMR spectroscopy after CHCl$_3$ extraction (removal) of organic reaction products. Paramagnetic α-SiV^{IV}W$_{11}$O$_{40}^{6-}$ (1$_{red}$) formed during reactions with BPH$_2$ was oxidized by Br$_2$ to diamagnetic 1. External NMR references for ^{7}Li, ^{27}Al, ^{29}Si and ^{51}V were, respectively, 1.0 M LiCl in D$_2$O (δ = 0 ppm), 0.10 M AlCl$_3$ ([Al(H$_2$O)$_6$]$^{3+}$ (δ = 0 ppm), 50 vol% Me$_4$Si in CDCl$_3$ (δ = 0 ppm) and 10 mM H$_4$PVMo$_{11}$O$_{40}$ in 0.60 M NaCl (δ = -533.6 ppm relative to neat VOCl$_3$; chemical shifts are reported relative to VOCl$_3$ at δ = 0 ppm).

Cyclic Voltammetry. Cyclic voltammetric (CV) measurements of POM solutions (Li$^+$, Na$^+$ and K$^+$ salts in acetate-buffered 2:3 (v/v) water/t-BuOH; the water component was first adjusted to pH 4.76 using HOAc and MOAc, M$^+$ = Li$^+$, Na$^+$, or K$^+$) were carried out at 25 or 60 ^{o}C under argon using a BAS CV-50W. A three-electrode cell (glassy-carbon working electrode, platinum auxiliary electrode and Ag/AgCl reference electrode) was used. Unless stated, sweep rates of 100 mV/s were used. Cation, M$^+$, concentrations were varied by additions of LiCl, NaCl or KCl. CV measurements of solutions containing tetra-n-hexylammonium (THA) cations required use of a Ag/Ag$^+$ reference electrode (a Ag wire was immersed in 2:3 (v/v) H$_2$O:t-BuOH containing AgNO$_3$ (10 mM) and tetra-n-hexylammonium nitrate (THAN, 100 mM)).

UV-vis Absorbance Measurements. Kinetic data were obtained by electronic absorption (UV-vis) spectroscopy (Hewlett-Packard 8451A diode array spectrometer). A quartz cuvette equipped with a gas-tight side arm was filled with either POM or BPH$_2$ solution and the side arm and mouth were sealed with rubber septa. POM solutions and solid BPH$_2$ were degassed (3 freeze/pump/thaw cycles), and the cuvette was placed in the sample compartment, equipped with thermostat ($\pm$ 0.01°) and stirrer, of the UV-vis spectrometer. Once at temperature, the reaction was initiated by injecting BPH_2 (or POM) stock solutions into the POM (or BPH_2) solution with constant stirring (800 RPM). For prolonged reactions (>1 h), a quartz cuvette with a stopcock side arm and a Teflon stopper were used for better protection against gas leaks. The extinction coefficient, ε, of α-$Li_6SiV^{IV}W_{11}O_{40}$ (Li_61_{red}) at $\lambda = 520$ nm, used to obtain kinetic data by UV-vis spectroscopy, was determined by linear regression of absorbance versus POM concentration data. (Although $\lambda_{max} = 496$ nm, 520 nm was used in order to minimize overlapping with tailing from the strongly absorbing $O^{2-} \rightarrow W^{6+}$ charge-transfer band located in the UV region.) Values for ε were obtained at 25 and 60 °C.

Product Yields and Reaction Stoichiometry. For product analysis studies, Li_51 (0.02 mmol) was dissolved in 4.0 mL of 0.1 M aqueous LiOAc buffer solution (pH 4.76) in a 25-mL Schlenk flask. Next, *t*-BuOH (6.0 mL) was added and the solution degassed. An aliquot of degassed BPH_2 stock solution (0.005 to 0.02 mmol in *t*-BuOH) was injected by gas-tight syringe through a septum into the flask charged with the POM solution, which was kept under Ar at 60 °C with constant stirring (800 RPM). After reaction, the organic products were extracted using three 5-mL portions of $CHCl_3$ (the POM salts remained in the aqueous phase). Roto-evaporation of the combined $CHCl_3$ extracts left a solid residue. This residue was dissolved in $CDCl_3$ for ^1H NMR measurements (3',4'-dichloroacetophenone as an internal quantitative standard) and GC-MS analysis (Hewlett Packard 5890). The aqueous solution (after extraction by $CHCl_3$) was diluted with water to precisely 25 mL to give a POM solution of known concentration. The concentration of 1_{red} present in the aqueous solution was quantified by both UV-visible spectrometry ($\lambda = 520$ nm, $\varepsilon = 619$ L mol^{-1} cm^{-1}) and oxidative titration using standardized $(NH_4)_2Ce^{IV}(NO_3)_6$ as oxidant and ferroin as indicator.[67] In 2:3 (v/v) H_2O:*t*-BuOH at 60 °C, BPH_2 is cleanly oxidized by 2 equiv. of **1** to 3,3',5,5'-tetra-*tert*-butyldiphenoquinone (DPQ, 100% by ^1H NMR and GC-MS; see eq 1). The stoichiometry in eq 1 was established by using a range of BPH_2-to-**1** ratios. BPH_2 and DPQ are the only organic compounds observed by ^1H NMR, and no intermediates of partial oxidation of BPH_2 are observed.

Reaction-Rate Data. Kinetic data (initial reaction rates) were determined by recording the absorbance at 520 nm once every 2 seconds for 0.5 h (POM conversion < 5%). The timer was started upon initiation of each reaction (injection of BPH_2 or POM) and was controlled by the Time-Based Measurement program in the HP UV-Visible Chemstation General Scanning Software. For each experiment, at least 3 reproducible measurements of reaction rate (or rate constant) were obtained. pH values in water and in water/*t*-BuOH mixtures were measured using a Corning model 240 pH meter, equipped with a Semi-Micro Combination electrode. Unless otherwise noted, pH values listed are those of the aqueous buffer solutions (pH = 3.90 to 5.76) prior to mixing with *t*-BuOH. For experiments designed to determine the pH dependence of the reaction rate, pH values were measured directly in the water/*t*-BuOH mixtures just before initiation of reaction. Due to gradual drift in pH-meter readings in the mixed-solvent system, a reproducible measurement protocol was used.[34] The glass electrode of the pH meter was calibrated using aqueous standards (pH 4.00 and 7.00), immersed in the mixed-solvent solution, and left to equilibrate for 2 min. Three readings were taken (one immediately, and the next two after 2-min intervals), and the average of the 3 readings was taken as the apparent pH.

Rate Law. Orders of reaction with respect to the concentrations of **1**, BPH_2, 1_{red}, H^+ and OAc^- (investigation of general-base catalysis) and ionic strength (μ) were established by initial-rate methods using Li^+ salts ($Li_5$1 and $Li_6$1$_{red}$) in LiOAc buffered H_2O:t-BuOH (2:3, v/v) under Ar at 60 °C. In each case, stock POM solutions were prepared in LiOAc buffered water and stock BPH$_2$ solutions were prepared in t-BuOH. Unless noted, ionic strength was kept constant by addition of LiCl. All solutions were degassed thoroughly, combined quantitatively under Ar, and heated to 60 °C. Reactions were initiated by injections of stock POM or BPH$_2$ solutions. To determine rate dependence on [OAc⁻], relative LiOAc/HOAc concentrations were varied while keeping the ionic strength constant. To determine the dependence of reaction rate on ionic strength (μ) at constant [Li^+], μ was varied by adding LiCl or α-$Li_6Al(AlOH_2)W_{11}O_{39}$ ($Li_6$2). Kinetic isotope data were obtained using BPH$_2$ in 2:3 (v/v) H_2O:t-BuOH and BPD$_2$ in 2:3 (v/v) D_2O/t-BuOD.

Cation-Anion Association. To derive the functional dependencies of rate constant and POM redox potential on cation concentration, it is assumed that **1** and **1**$_{red}$ are in rapid equilibrium with the corresponding ion pairs (eqs 19 and 20).

$$M^+ + \mathbf{1} \quad \xrightleftharpoons{K_{MI}} \quad (M^+\mathbf{1}) \tag{19}$$

$$M^+ + \mathbf{1}_{red} \quad \xrightleftharpoons{K_{MIred}} \quad (M^+\mathbf{1}_{red}) \tag{20}$$

Respective equilibrium constants are shown in eqs 21 and 22.

$$K_{MI} = \frac{[M^+\mathbf{1}]\gamma'_{ox}}{[M^+]\gamma_M[\mathbf{1}]\gamma_{ox}} \tag{21}$$

$$K_{MI\,red} = \frac{[M^+\mathbf{1}_{red}]\gamma'_{red}}{[M^+]\gamma_M[\mathbf{1}_{red}]\gamma_{red}} \tag{22}$$

The activity coefficients, γ, correspond to those of unpaired and paired **1** and **1**$_{red}$ and of M^+ ions, and must be included in order to correct for non-ideal behavior resulting from significant changes in ionic strength (μ) upon variation in cation, M^+, concentration.

An expression describing the dependence of the reaction rate ($d[\alpha$-$SiV^{IV}W_{11}O_{40}{}^{6-}]/dt$ or $d[\mathbf{1}_{red}]/dt$), and of the observed rate constant on [M^+], consistent with the proposed mechanism of oxidation of BPH$_2$ by **1** (eqs 5-9), was derived using eqs 23-25:

$$\mathbf{1} + BPH_2 \ \square \quad \xrightarrow{k_1} \quad Products \tag{23}$$

$$M^+\mathbf{1} + BPH_2 \ \square \quad \xrightarrow{k_{MI}} \quad Products \tag{24}$$

$$rate = k_1[\mathbf{1}]\gamma_{ox}[BPH_2]\gamma_0 + k_{MI}[M^+\mathbf{1}]\gamma'_{ox}[BPH_2]\gamma_0 \tag{25}$$

(γ_0 is the activity coefficient for BPH$_2$). Use of eqs 21 and 25 and rearrangement give eq 26.

$$\text{rate} = (k_1 + k_{M1}K_{M1}[M^+]\gamma_M)\frac{[BPH_2]\gamma_{ox}\gamma_o}{1 + K_{M1}[M^+]\gamma_M\dfrac{\gamma_{ox}}{\gamma'_{ox}}}([1]+[M^+1])$$

(26)

Substitution of initial rate (rate$_o$) and initial concentrations ([1]$_o$ and [BPH$_2$]$_o$) gives eq 27:

$$\text{rate}_o = \frac{k_1 + k_{M1}K_{M1}[M^+]\gamma_M}{1 + K_{M1}[M^+]\gamma_M\dfrac{\gamma_{ox}}{\gamma'_{ox}}}[1]_o[BPH_2]_o\gamma_{ox}\gamma_o$$

(27)

It is apparent from inspection of eq 27 that the observed rate constant, k$_{obs}$, possesses a rectangular hyperbolic functional dependence on [M$^+$]:

$$k_{obs} = \frac{k_1 + k_{M1}K_{M1}[M^+]\gamma_M}{1 + K_{M1}[M^+]\gamma_M\dfrac{\gamma_{ox}}{\gamma'_{ox}}}\gamma_{ox}\gamma_o$$

(28)

The dependence of the **1/1**$_{red}$ half-wave potentials on the [M$^+$] was analyzed using a Nernstian equation modified to include two association constants (by convention, K$_{ox}$ = K$_{M1}$ and K$_{red}$ = K$_{M1_{red}}$):[38]

$$E_{1/2} = E^\circ + \frac{RT}{nF}\ln\left(\frac{1 + K_{red}\dfrac{\gamma_M\gamma_{red}}{\gamma'_{red}}[M^+]}{1 + K_{ox}\dfrac{\gamma_M\gamma_{ox}}{\gamma'_{ox}}[M^+]}\right) + \frac{RT}{nF}\ln\frac{\gamma_{ox}}{\gamma_{red}}$$

(29)

E$^\circ$ is the standard potential for unpaired **1** at zero ionic strength, $\mu = 0$; the other symbols are conventional or described above.

The activity coefficients in eqs 28 and 29 were calculated using an extended Debye-Hückel law with an empirical linear correction term:[48]

$$\log\gamma_z = -Az^2\frac{\sqrt{\mu}}{1 + Ba\sqrt{\mu}} + b\mu$$

(30)

The activity coefficient, γ_z, is that of a single ion of charge z. The terms A (1.825 10^6(εT)$^{-3/2}$ mol$^{-1/2}$L$^{1/2}$K$^{3/2}$) and B (50.29(εT)$^{-1/2}$ Å^{-1}mol$^{-1/2}$L$^{1/2}$K$^{1/2}$) are temperature-dependent constants that change with the dielectric constant of the solvent, while a and b are adjustable parameters.

Kinetic and electrochemical data, respectively, were fitted to eqs 28 and 29. Non-linear least-square fits of the observed rate constant and the formal redox potential versus [M$^+$] were carried out using the Solver Function in Microsoft Excel-98. Sums of deviation-squared values were minimized by varying k$_1$, k$_{M1}$, K$_{ox}$, K$_{red}$, E$_o$, a and b in eqs 28-30. The ratios γ_{ox}/γ'_{ox} and $\gamma_{red}/\gamma'_{red}$ were assigned a value of 1. Additional limitations and constraints imposed on the adjustable parameters to improve fitting are discussed above.

^{7}Li NMR Studies. Solutions of $Li_5\mathbf{1}$ (0.3 to 20 mM) or of $Li_6\mathbf{1_{red}}$ (0.5 to 10 mM) were prepared in 2:3 H_2O:t-BuOH inserted into an NMR tube. A coaxial tube containing 1.0 M LiCl in D_2O, used as an external reference, was inserted and the samples heated to 60 °C in the NMR probe. NMR spectra were recorded using fine digital resolution (0.061 Hz; 0.0004 ppm).

Experimentally-verified solution behavior was used to established boundary conditions needed to derive an expression for the functional dependence of the chemical shift on the total concentration of $Li_5\mathbf{1}$ ($[Li_5\mathbf{1}]_{total}$). These conditions are as follows: (1) In very dilute solutions, $Li_5\mathbf{1}$ is completely dissociated into Li^+ and **1**, (2) at larger $[Li_5\mathbf{1}]$ (within experimental limits), 1:1 ion pairing occurs (eq 31), and (3) exchange between freely solvated and associated Li^+ cations is rapid on the ^{7}Li NMR time scale. As a result, the observed chemical shift, δ, is the weighted mean of the chemical shifts of the signals due to unpaired, δ_o, and paired, δ_1, Li^+ ions (eq 32).[68-70]

$$Li^+ \quad + \mathbf{1} \qquad\qquad \underset{\longleftarrow}{\overset{K_{Li1}}{\longrightarrow}} \qquad\qquad Li\mathbf{1} \qquad\qquad (31)$$

$$\delta = \delta_0 \frac{[Li^+]}{5[Li_5\mathbf{1}]_{total}} + \delta_1 \frac{[Li\mathbf{1}]}{5[Li_5\mathbf{1}]_{total}}$$

$$(32)$$

An expression for the concentration of paired species, $[Li\mathbf{1}]$, as a function of the *total concentration* of $Li_5\mathbf{1}$ in terms of the association constant, K_{Li1}, was derived using eq 31 and appropriate mass-balance expressions:

$$[Li\mathbf{1}] = 3[Li_5\mathbf{1}]_{total} +$$

$$\frac{1}{2K_{Li1}\gamma_{Li}}\frac{\gamma_{Li1}}{\gamma_1} - \frac{1}{2}\sqrt{\left(6[Li_5\mathbf{1}]_{total} + \frac{1}{K_{Li1}\gamma_{Li}}\frac{\gamma_{Li1}}{\gamma_1}\right)^2 - 20[Li_5\mathbf{1}]_{total}^2} \qquad (33)$$

Combining eqs 32 and 33, an expression for the observed chemical shift (δ) as a function of the total POM concentration ($[Li_5\mathbf{1}]_{total}$) is obtained:

$$\delta = \delta_0 + (\delta_1 - \delta_0) \times$$

$$\frac{3[Li_5\mathbf{1}]_{total} + \frac{1}{2K_{Li1}\gamma_{Li}}\frac{\gamma_{Li1}}{\gamma_1} - \frac{1}{2}\sqrt{\left(6[Li_5\mathbf{1}]_{total} + \frac{1}{K_{Li1}\gamma_{Li}}\frac{\gamma_{Li1}}{\gamma_1}\right)^2 - 20[Li_5\mathbf{1}]_{total}^2}}{5[Li_5\mathbf{1}]_{total}} \qquad (34)$$

Experimental chemical-shift values, δ, as a function of $[Li_5\mathbf{1}]_{total}$ were fitted to eq 34 by non-linear least-square regression by varying the parameters δ_o, δ_1 and K_{Li1}. The activity coefficient of Li^+ (γ_{Li}) was calculated using the parameters a and b obtained from non-linear least-squares fit of the kinetic data. The ratio γ_{Li1}/γ_1 was assigned a value of 1.

Acknowledgments. We thank Dr. Carl J. Houtman for assistance with error-analysis (non-linear-regression calculations), and the DOE (DE-FC36-95GO10090) (I. A. W. and C. L. H.) and NSF (CHE-9412465) (C. L. H.) for support.

REFERENCES

1.   P. Chen and T. J. Meyer, Medium effects on charge-transfer in metal-complexes, *Chem. Rev.*, **1998**, *98*, 1439-1477.

2.   N. D. Stalnaker, J. C. Solenberger, A. C. Wahl, Electron-transfer between iron, ruthenium, and osmium complexes containing 2, 2'-bipyridyl, 1, 10-phenanthroline, or their deriatives. Effects of electrolytes on rates, *J. Phys. Chem.*, **1977**, *81*, 601-604.

3.   J. A. Goodwin, D. M. Stanbury, L. J. Wilson, C. W. Eigenbrot, W. R. Scheidt, Molecular structures and electron-transfer kinetics for some pentacoordinate Cu^I/Cu^{II} redox-acitve pairs, *J. Am. Chem. Soc.*, **1987**, *109*, 2979-2991.

4.   H. Doine, T. W. Swaddle, Pressure effects on the rates of electron transfer between tris(hexafluoroacetylacetonato)ruthenium(II) and -(III) in Different solvents, *Inorg. Chem.*, **1988**, *27*, 665-670.

5.   T. G. Braga, A. C. Wahl, Rates of Electron transfer from osmium(II) to iron(III) complex ions containing 2, 2'-bipyridine or its derivatives as ligands. Effects of electrolytes at low concentrations and reactant-separation distance, *J. Phys. Chem.*, **1989**, *89*, 5822-5828.

6.   M. A. Murguia, S. Wherland, Volumes of activation for electron-transfer between a series of cobalt clathrochelates and ferrocenes as a function of solvent and added electrolyte, *Inorg. Chem.*, **1991**, *30*, 139-144.

7.   S. Wherland, Nonaueous, outer-sphere electron-transfer kinetics of transition-metal complees, *Coord. Chem. Rev.*, **1993**, *123*, 169-199.

8.   C. P. Andrieux, M. Robert, and J. -M. Savéant, Role of environmental-factors in the dynamics of intramolecular dissociative electron-transfer-effect of solvation and ion-pairing on cleavage rates of anion-radicals, *J. Am. Chem. Soc.*, **1995**, *117*, 9340-9346.

9.   Y. S. Fu, T. W. Swaddle, Electrochemical kinetics of cyanometalate complexes in aqueous-solution at high-pressures, *Inorg. Chem.*, **1999**, *38*, 876-880.

10.   P. D. Metelski, T. W. Swaddle, Cation catalysis of anion-anion electron-transfer in aqueous-solution-self-exchange reaction-kinetics of some hexacyanometalate and octacyanometalate couples at variable-pressure, *Inorg. Chem.*, **1999**, *38*, 301-307.

11.   M. Shporer, G. Ron, A. Loewenstein, G. Navon, Study of some cyano-metal complexes by nuclear magnetic resonance. II. Kinetics of electron transfer between ferri- and ferrocyamide ions, *Inorg. Chem.*, **1965**, *4*, 361-364.

12.   R. J. Campion, N. Purdie, N. Sutin, The kinetics of some related electron-transfer reactions, *Inorg. Chem.*, **1964**, *3*, 1091-1094.

13.   R. J. Campion, C. F. Deck, P. J. King, A. C. Wahl, Kinetics of electron exchange between hexacyanoferrate(II) and -(III) ions, *Inorg. Chem.*, **1967**, *6*, 672-681.

14.   G. Gritzner, K. Danksagmüller, V. Gutmann, Outer-sphere coordination effects on the redox behavior of the $Fe(CN)_6^{3-}/Fe(CN)_6^{4-}$ couple in non-aqueous solvents, *J. Electroanal. Chem.*, **1976**, *72*, 177-185.

15.   J. M. A. Hoddenbagh, D. H. Macartney, Kinetics of electron-transfer reactions involving the $Ru(CN)_6^{4-/3-}$ couple in aqueous-media, *Inorg. Chem.*, **1990**, *29*, 245-251.

16.   M. T. Pope, Heteropoly and Isopoly Oxometalates, Springer-Verlag: Berlin, **1983**.

17.   M. T. Pope, A. Müller, Polyoxometalate chemistry-An old field with new dimensions in deveral disciplines, *Angew. Chem., Int. Ed.*, **1991**, *30*, 34-48.

18.   (a) C. L. Hill, C. M. Prosser-McCartha, Homogeneous catalysis by transition-metal oxygen anion clusters, *Coord. Chem. Rev.*, **1995**, *143*, 407-455; (b) R. Neumann, Polyoxometalate complexes in organic oxidation chemistry, *Prog. Inorg. Chem.*, **1998**, *47*, 317-370.

19.   C. L. Hill, G. E., Polyoxometalates, *Chem. Rev.*, **1998**, *98*, 1-389.

20.   (a) S. K. Saha, M. Ali, P. Banerjee, Electron exchange and transfer-reactions of heteropoly oxometalates, *Coord. Chem. Rev.*, **1993**, *122*, 41-62; (b) I. A. Weinstock, Homogeneous-phase electron-transfer reactions of polyoxometarates, *Chem. Rev.*, **1998**, *98*, 113-170.

21.   V. A. Grigoriev, C. L. Hill, I. A. Weinstock, Role of cation size in the energy of Electron-transfer to 1/1 polyoxometalate ion-pairs $((M^+)(X^{n+}VW_{11}O_{40}))((8-n)-)$ (M=Li, Na, K), *J. Am. Chem. Soc.*, **2000**, *122*, 3544-3545.

22.   J. F. Kirby, L. C. W. Baker, Effects of counterions in heteroly electrolyte chemistry. 1. Evaluations of relative interactions by NMR on kozik salts, *Inorg. Chem.*, **1998**, *37*, 5537-5543.

23.   M. Misono, Heterogeneous catalysis by heteropoly compounds of molybdenum and tungsten, *Catal. Rev.-Sci. Eng.*, **1987**, *29*, 269-321.

24.   N. Mizuno, M. Misono, Heterogenous catalysis, *Chem. Rev.*, **1998**, *98*, 199-218.

25.   G. I. H. Hanania, D. H. Irvine, W. A. Eaton, P. George, Thermodynamic aspects of the potassium hexacyanoferate(III)-(II) system. II. Reduction potential, *J. Phys. Chem.*, **1967**, *71*, 2022-2030.

26.   P. Carloni, L. Eberson, Strong medium and couterion effects upon the redox potential of the 12-tungstocobaltate(III/II) couple, *Acta Chem. Scand.*, **1991**, *45*, 373-376.

27.   P. L. Boulas, M. Gomez-Kaifer, L. Echegoyen, Electronchemistry of supramolecular systems, *Angew. Chem., Int. Ed.*, **1998**, *37*, 216-247.

28.   P. D. Beer, P. A. Gale, G. Z. Chen, Electrochemical molecular recognition. Pathways between complexation and signaling, *J. Chem. Soc., Dalton Trans.*, **1999**, 1897-1910.

29.   L. Eberson, Electron-transfer reactions in organic chemistry. 4. A mechanistic study of the oxidation of p-Methoxytoluene by 12-Tungstocobalt(III)ate Ion, *J. Am. Chem. Soc.*, **1983**, *105*, 3192-3199.

30.   I. A. Weinstock, J. J. Cowan, E. M. G. Barbuzzi, H. Zeng, C. L. Hill, Equilibria between alpha-isomers of keggin heteropolytungstates, *J. Am. Chem. Soc.*, **1999**, *121*, 4608-4617.

31.   R. Neumann, M. Levin, Aerobic oxidative dehydrogenations catalyzed by the mixed-addenda heteropolyanion $PV_2Mo_{10}O_{40}^{5-}$ - a kinetic and mechanistic study, *J. Am. Chem. Soc.*, **1992**, *114*, 7278-7286.

32.   M. Fujio, R. T. McIver, Jr., R. W. Taft, Effects on the Acidities of phenols from specific substitutent-solvent interactions. Inherent substituent parameters from gas-phase acidities, *J. Am. Chem. Soc.*, **1981**, *103*, 4017-4029 and earlier research.

33.   B. D. England, D. A. House, The dissociation constant of phenol in alcoholic solvents, *J. Chem. Soc.*, **1962**, *1962*, 4421-4423.

34.   L. A. Cohen, W. M. Jones, A study of free energy relationships in hindered phenols. Linear dependence for solvation effects in ionization, *J. Am. Chem. Soc.*, **1963**, *85*, 3397-3402.

35.   J. Buckingham. *Dictionary of Organic Compounds*; 6 ed.; Chapman & Hall: London, **1996**; Vol. 1.

36.   J. Mohanty, H. Pal, A. V.Sapre, Photophysical properties of 2,2'-biphenyldiols and 4,4'-biphenyldiols, *Bull. Chem. Soc. Jpn.*, **1999**, *72*, 2193-2202.

37.   G. Akerlof, Dielectric contents of some organic solvent-water mixtures at various temperatures, *J. Am. Chem. Soc.*, **1932**, *54*, 4125-4139.

38.   J. N. Butler, D. R. Cogley. *Ionic Equilibrium: Solubility and pH Calculations*; John Wiley & Sons: New York, **1998**.

39. R. M. Fuoss, Conductance-concentration function for the paired ion model, *J. Phys. Chem.*, **1978**, *82*, 2427-2440.

40. R. M. Fuoss, Ionic association. I. Derivation of constants from conductance data, *J. Am. Chem. Soc.*, **1957**, *79*, 3301-3303.

41. R. M. Fuoss, C. A. Kraus, Ionic association. II. Several salts in dioxane-water mixture, *J. Am. Chem. Soc.*, **1957**, *79*, 3304-3310.

42. R. M. Fuoss, Ionic association. III. The equilibrium between ion pairs and free ions, *J. Am. Chem. Soc.*, **1958**, *80*, 5059-5061.

43. A. F. Lindmark, Kinetic, Equilibrium, and ion-pairing studies on 7-coodinate molybdenum(II) and tungsten(II) isocyanide complexes in acetonitrile, *Inorg. Chem.*, **1992**, *31*, 3507-3513.

44. R. Andreu, J. J. Calvente, W. R. Fawcett, M. Molero, Role of ion-pairing in double-layer effects at self-assembled monolayers containing a simple redox couple, *J. Phys. Chem.*, **1997**, *101*, 2884-2894.

45. S. G. Capewell, G. T. Hefter, P. M. May, Association constants for the $NaSO_4^-$ ion-pair in concentrated cesium-chloride solutions, *Talanta*, **1999**, *49*, 25-30.

46. K. A. Connors, *Binding Constants: The Measurement of Molecular Complex Stability*; John Wiley & Sons: New York, **1987**.

47. R. L. Blackbourn, J. T. Hupp, Optical electron transfer processes. The dependence of intervalence line shape and transition energy on chromophore concentration, *Chem. Phys. Lett.*, **1988**, *150*, 399-405.

48. R. A. Robinson,; Stokes, R. H. *Electrolyte Solutions*; Butterworths: London, **1959**.

49. D. A. Deranleau, Theory of the measurement of weak molecular complexes. I. General consideration, *J. Am. Chem. Soc.*, **1969**, *91*, 4044-4049.

50. D. A. Deranleau, Theory of the measurement of weak molecular complexes. II. Consequences of multiple equilibria, *J. Am. Chem. Soc.*, **1969**, *91*, 4050-4054.

51. R. Billing, D. Rehorek, H. Henning, In *Photoinduced Electron Transfer II*; Mattay, J., Ed.; Springer-Verlag: Berlin, **1990**; Vol. 158.

52. K. H. Stern, E. S. Amis, Ionic size, *Chem. Rev.*, **1959**, *59*, 1-64.

53. T. S. Sorensen, P. Sloth, M. Shroder, Ionic radii from experimental activities and simple statistical-mechanical theories for strong electrolytes with small Bjerrum parameters, *Acta Chem. Scand. Ser. A*, **1984**, *A38*, 735-756.

54. D. Lexa, P. Rentien, J.-M. Savéant, F. Xu, Methods for investigating the mechanistic and kinetic role of ligand exchange reactions in coordination electrochemistry. Cyclic Voltammetry of chloroion(III) tetraphenylporphyrinin dimethylformamide, *J. Electroanal. Chem.*, **1985**, *191*, 253-279.

55. (a) D. M. Way, J. B. Cooper, M. Sadek, T. Vu, P. J. Mahon, A. M. Bond, R. T. C. Brownlee, A. G. Wedd, Systematic electrochmial synthesis of reduced forms of the α-$[S_2Mo_{18}O_{62}]^{4-}$ anion, *Inorg. Chem* **1997**, *36*, 4227-4233; (b) D. M. Way, A. M. Bond, A. G. Wedd, Multielectron reduction of α-$[S_2Mo_{18}O_{62}]^{4-}$ in aprotic and protic media-voltammetric studies, *Inorg. Chem.*, **1997**, *36*, 2826-2833; (c) P. D. Prenzler, C. Boskovic, A. M. Bond, A. G. Wedd, Coupled electron-transfer and proton-transfer processes in the reduction of α-$[P_2W_{18}O_{62}]^{6-}$ and α-$[H_2W_{12}O_{40}]^{6-}$ as revealed by simulation of cyclic voltammograms, *Anal. Chem.* **1999**, *71*, 3133-3139; (d) A. M. Bond, T. Vu, A. G. Wedd, Voltammetric studies of the interaction of the lithium cation with reduced forms of the Dawson $[S_2Mo_{18}O_{62}]^{4-}$ polyoxometalate anion, *J. Electroanal. Chem.*, **2000**, *494*, 96-104

56. C. G. Screttas, G. A. Heropoulos, B. R. Steele, D. Bethell, 31P contact shifs as a measure of weak ligand affinities-interaction between alkali-metal fluorenone radical-anions and certain phosphorus(III or V) ligands, *Mag. Res. Chem.*, **1998**, *36*, 656-662.

57. M. Micha-Screttas, G. A. Heropoulos, B. R. Steele, Evidence for a concentration-dependent [6(7)]Li NMR contact shift in tetrahydrofuran solutions of lithium naphthalene radical-anion and the effect of added LiCl-[6(7)]Li, *J. Chem. Soc., Perkin Trans. 2*, **1999**, 1443-1446.

58. M. C. Baker, P. A. Lyons, S. J. Singer, Velocity ultracentrifugation and diffusion of silicotungstic acid, *J. Am. Chem. Soc*, **1955**, *77*, 2011-2012.

59. T. Kurucsev, A. M. Sargeson, B. O. West, Size and hydration of inorganic ions from viscosity and density measurements, *J. Phys. Chem.*, **1957**, *61*, 1567-1569.

60. M. T. Pope, G. M. Varga. Jr., Heteropoly blues. I. Reduction stoichiometries and reduction potentials of some 12-tungstates, *Inorg. Chem.*, **1966**, *5*, 1249-1254.

61. R. D. Shannon, C. T. Prewitt, Effective inoic radii in oxides and fluorides, *Acta. Crystallogr.* **1969**, *B25*, 925-946.

62. F. A. Cotton, G. Wilkinson, *Advanced Inorganic Chemistry*; Wiley: New York, 1980; Vol. 4th.

63. P. Piotrowiak, Specific ion-pairing effects in weakly exoergic intramolecular electron-transfer, *Inorg. Chim. Acta.*, **1994**, *225*, 269-274.

64. R. A. Marcus, Chemical and electrochemical electrron-transfer theory, *Annu. Rev. Phys. Chem.*, **1964**, *15*, 155-196.

65. L. Eberson, *Electron Transfer Reactions in Organic Chemistry*; Springer-Verlag: Berlin, **1987**.

66. P. J. Domaille, 1- and 2-dimensional [183]W and [51]V NMR characterization of isopolymetalates and heteropolymetalates, *J. Am. Chem. Soc.*, **1984**, *106*, 7677-7687.

67. I. M. Kolthoff, E. B. Sandell, E. J. Meehan, S. Bruckenstein, *Quantitative Chemical Analysis*; 4th ed.; Macmillan: London, **1969**.

68. A. V. McCormick, A. T. Bell, C. J. Radke, Evidence from alkali-metal NMR-spectroscopy for ion-pairing in alkaline silicate solutions, *J. Phys. Chem.*, **1989**, *93*, 1733-1737.

69. T. Akai, N. Nakamura, H. Chihara, Ion-pairing and microscopic structure of R_4NPF_6 in the binary-system Cd_3CN-CCl_4 studied by nuclear-magnetic resonance spectroscopy, *J. Chem. Soc., Faraday Trans.*, **1993**, *89*, 1339-1343.

70. J. Barthel, U. Ströder, L. Iberl, H. Hammer, The Temperature Dependence of the Properties of Electrolyte Solutions. IV. Determination of Cationic Transference Numbers in Methanol, Ethanol, Propanol, and Acetonitrile at Various Temperatures., *Ber. Bunsenges. Phys. Chem.*, **1982**, *86*, 636.

NEW CLASSES OF FUNCTIONALIZED POLYOXOMETALATES: ORGANO-NITROGEN DERIVATIVES OF LINDQVIST SYSTEMS

Aaron R. Moore, Haidoo Kwen, Christopher G. Hamaker,
Thomas R. Mohs, Alicia M. Beatty, Bradley Harmon,
Kale Needham, and Eric A. Maatta

Department of Chemistry
Kansas State University
Manhattan, KS 66506-3701 U.S.A.

INTRODUCTION

The compositional and structural diversity of polyoxometalates,[1] coupled with their remarkable catalytic, magnetic, medicinal, redox, and photophysical properties,[2] justifies their description as class of inorganic compounds "unmatched in terms of molecular and electronic structural versatility, reactivity and relevance".[1b] The development of rational methods for the systematic modification and functionalization of polyoxometalates is a challenging endeavor, but is one which offers the prospect of exploiting more fully the desirable attributes of these systems.

In the specific case of the Lindqvist-type hexametalates $[M_6O_{19}]^{n-}$ (Figure 1), one effective strategy for producing modified derivatives involves the formal replacement of one (or more) of their terminal oxo ligands with other isolobal ligating atoms or groups. If one considers the oxo ligands as being closed-shell $[O]^{2-}$ entities, then attractive targets for this approach include the $\{1\sigma/2\pi\}$-donor $[N]^{3-}$ nitrido, $[N\text{-}R]^{2-}$ organoimido, and $[NNCR_2]^{2-}$ diazoalkane (hydrazonato) ligands. Similar, if not always strictly identical, considerations can be applied to $[NO]^+$ nitrosyl[3] and $[NNAr]^+$ diazenido[4] π-acceptor ligands, as well as to $[NNR_2]^{2-}$ hydrazido[5] and η^5-pentamethylcyclopentadienyl[6] donors, all of which have been demonstrated to serve as oxo surrogates in Lindqvist systems. In this contribution, we will focus on organoimido and diazoalkane derivatives of the Lindqvist hexametalates $[Mo_6O_{19}]^{2-}$ and $[W_6O_{19}]^{2-}$.

Polyoxometalate Chemistry for Nano-Composite Design
Edited by Yamase and Pope, Kluwer Academic/Plenum Publishers, 2002

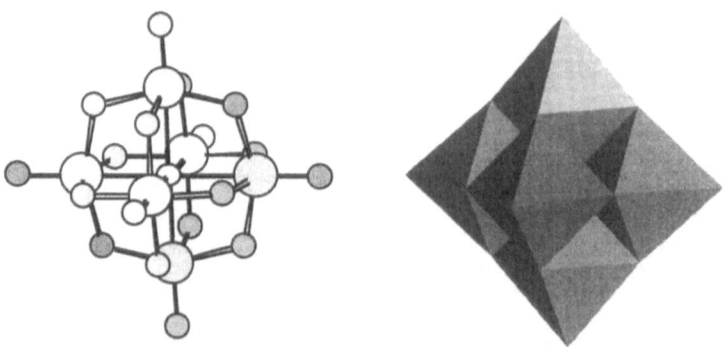

Figure 1. Lindqvist structure in conventional (left) and polyhedral (right) representations.

ORGANOIMIDO HEXAMETALATES

Mono-functional systems $[Mo_6O_{18}(NR)]^{2-}$

The p-tolylimido species $[Mo_6O_{18}(NTol)]^{2-}$ was the first member of this class to be accessed. Following its original brief mention as having arisen from an assembly reaction involving $[Mo_2O_7]^{2-}$ and $[Mo(NTol)Cl_4(THF)]$ in moist solvent,[5] a direct metathetical synthesis employing $Ph_3P{=}NTol$ and $[Mo_6O_{19}]^{2-}$ afforded the product in high yield and allowed its complete characterization, including an X-ray structural determination.[7] An analogous direct reaction employing $Ph_3P{=}NPh$ furnished the phenylimido derivative $[Mo_6O_{18}(NPh)]^{2-}$,[8] while direct metathetical reactions using isocyanates RNCO were used to prepare the corresponding $[Mo_6O_{18}(NR)]^{2-}$ systems (R = n-Bu,[9] Cy,[9] 2,6-(i-Pr)$_2$C$_6$H$_3$[9] and t-Bu[10]).

The n-butylimido species (**1-Bu**) is representative of this series and its structure is shown in Figure 2. The metrical parameters associated with the Mo-imido linkages in the mono-substituted imido-hexamolybdates feature short Mo-N bond lengths and near-linear Mo-N-C angles, consistent with the presence of considerable Mo≡N triple bond character.

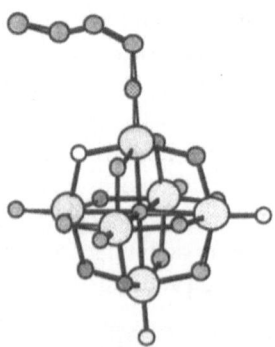

Figure 2. The molecular structure of the $[Mo_6O_{18}(NBu)]^{2-}$ anion (**1-Bu**). Adapted with permission from *J. Am. Chem. Soc.* **2000**, *122*, 639-649. © 2000 American Chemical Society.

As a class, the $[Mo_6O_{18}(NR)]^{2-}$ systems are more difficult to reduce than is the hexamolybdate parent, displaying $E_{1/2}$ values which are shifted cathodically by *ca.* −200 mV with respect to that of $[Mo_6O_{19}]^{2-}$. This fact clearly demonstrates that organoimido ligands are superior electron donors in comparison to the oxo ligand. The electronic spectra of the alkylimido $[Mo_6O_{18}(NR)]^{2-}$ species ($\lambda_{max} \approx 325$ nm; $\varepsilon \approx 6500$ $M^{-1}cm^{-1}$) are not significantly perturbed from that of $[Mo_6O_{19}]^{2-}$, but those of the mono-arylimido systems are both bathochromically shifted ($\lambda_{max} \approx 345$ nm) and considerably more intense ($\varepsilon \approx 20000$ $M^{-1}cm^{-1}$).

Multiply-functionalized systems $[Mo_6O_{(19-x)}(NR)_x]^{2-}$ ($x = 2 - 6$)

The demonstrated ability to accommodate polyfunctionalization is a noteworthy aspect of the organoimido-hexamolybdate series. To date, all of the reported examples bear arylimido ligands exclusively, but there is no reason to preclude the existence of stable (poly)alkylimido hexamolybdate species. Known systems include *cis*-$[Mo_6O_{17}(NPh)_2]^{2-}$,[8] the very interesting *trans*-$[Mo_6O_{17}(NC_6H_4NH_2)_2]^{2-}$ species,[11] and the homologous series $[Mo_6O_{(19-x)}(NAr)_x]^{2-}$ (Ar = 2,6-$(i$-Pr$)_2C_6H_3$); $x = 2, 3, 4, 5$);[9] for the latter group, a comprehensive account of their syntheses, structures, multinuclear (^{14}N, ^{17}O, ^{95}Mo) NMR and electronic spectra, and electrochemical behavior has recently appeared.[12] The protonated *hexakis*-derivative $[Mo_6O_{13}(NAr)_6H]^-$ has also been prepared and structurally characterized.[13] The *bis*-, *tris*-, and *tetrakis*-(NAr) hexamolybdate derivatives **2-Ar**, **3-Ar**, and **4-Ar** adopt *cis*-, *fac*-, and *cis*- structures, respectively (Figure 3) rather than the presumably less crowded *trans*-, *mer*-, and *trans*- alternatives, suggesting that the presence

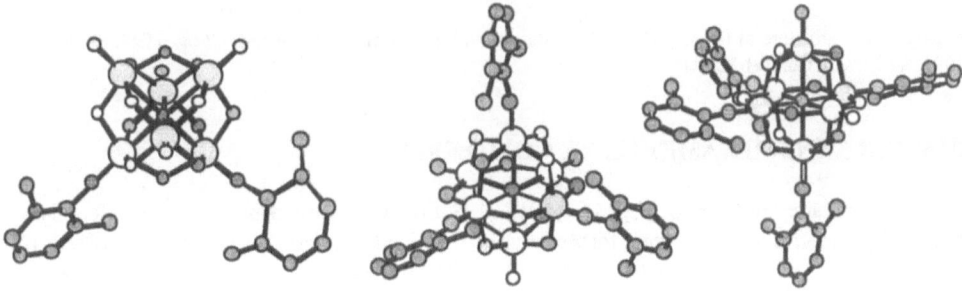

Figure 3. Structures of the *cis*-**2-Ar**, *fac*-**3-Ar** and *cis*-**4-Ar** anions. *i*-Pr methyl groups omitted for clarity. Adapted with permission from *J. Am. Chem. Soc.* **2000**, *122*, 639-649. © 2000 American Chemical Society.

of an extant [NAr] ligand preferentially activates its *cis*-sites for subsequent substitution.

Progressive substitution within the $[Mo_6O_{(19-x)}(NAr)_x]^{2-}$ series induces a slight but steady redshift in their electronic spectra λ_{max} values, accompanied by a nearly additive increase in the molar absorptivity values for these bands. The successive incorporation of [NAr] ligands into the $[Mo_6O_{19}]^{2-}$ framework likewise induces a steady decrease in the species' $E_{1/2}$ values by *ca.* -220 mV per [NAr] group.

A Hexatungstate Analogue: $[W_6O_{18}(NAr)]^{2-}$

The oxo ligands within the hexatungstate $[W_6O_{19}]^{2-}$ have proven to be inert to metathetical exchange with various $R_3P=NR'$ or RNCO imido delivery reagents. To date, only one organoimido-hexatungstate species has been reported, namely $[W_6O_{18}(NAr)]^{2-}$ (Ar = 2,6-(i-Pr)$_2$C$_6$H$_3$).[14] This yellow complex can be isolated in low yield (*ca.* 10%) from the reaction of $[Bu_4]_2[WO_4]$ and ArNCO in either 1,2-dichloroethane or pyridine solvent. The structure (Figure 4) closely resembles that of its Mo analogue, and the perturbations in the properties of the hexatungstate core induced by oxo/[NAr] substitution mirror those of the corresponding Mo systems. Thus, the $E_{1/2}$ value of $[W_6O_{18}(NAr)]^{2-}$ is more negative than that of $[W_6O_{19}]^{2-}$ by -0.21 V (*c.f.* the corresponding disparity of -0.25 V between the analogous Mo systems), and its electronic spectrum (λ_{max} = 292 nm; ε = 2.2 × 10^4) is both red-shifted and more intense than that of the hexatungstate (λ_{max} = 280 nm; ε = 1.1 × 10^4). Unlike certain $[Mo_6O_{18}(NR)]^{2-}$ systems, $[W_6O_{18}(NAr)]^{2-}$ is quite stable with respect to hydrolysis: deliberate addition of H_2O (1-10 equiv.) to its CD$_3$CN solutions at 56°C produced no detectable decomposition as monitored by ^{1}H NMR techniques over a one month period.

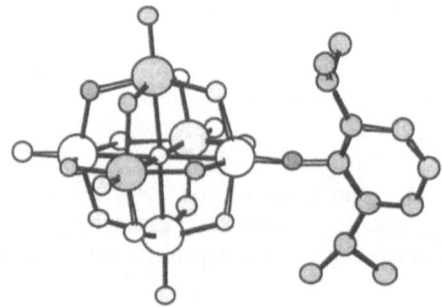

Figure 4. The structure of $[W_6O_{18}(NAr)]^{2-}$. Adapted with permission from *Inorg. Chem.* **1995**, *34*, 9. © 1995 American Chemical Society.

DIAZOALKANE-HEXAMETALATE COMPLEXES

Diazoalkane complexes, $[L_nM(N_2CRR')]$,[15] find widespread use in various catalytic and stoichiometric organic transformations.[16] Typically, such species are prepared by direct reaction of RR'CN$_2$ with low-valent metal complexes (*i. e.*, those bearing two or more *d*-electrons). As ligands in high-valent systems, however, diazoalkanes have not found the broad utility of their organoimido relatives,[17] in part because of synthetic restrictions. One reason for our interest in preparing diazoalkane-polyoxometalate systems stems from the recognition that the functionality within the [N$_2$CRR'] unit derives from a different pool (aldehydes, ketones, quinones, *etc.*) than does that of the [NR] ligand (amines or carboxylic derivatives), and thus their incorporation could broaden and complement the existing range of functional groups in polyoxometalates bearing organo-nitrogen ligands.

If the diazoalkane ligand is considered to bind as a "hydrazone-derived" [N$_2$CRR']$^{2-}$ entity, then its similarity to an organoimido [NR]$^{2-}$ species is evident; this suggested the possibility of phosphine-mediated metathetical delivery using $[R_3P=N_2CRR']$ phosphazine reagents, by analogy to the corresponding reactivity of $[R_3P=NR]$ iminophosphoranes as shown in Figure 5. This strategy has furnished examples of diazoalkane-hexametalates.

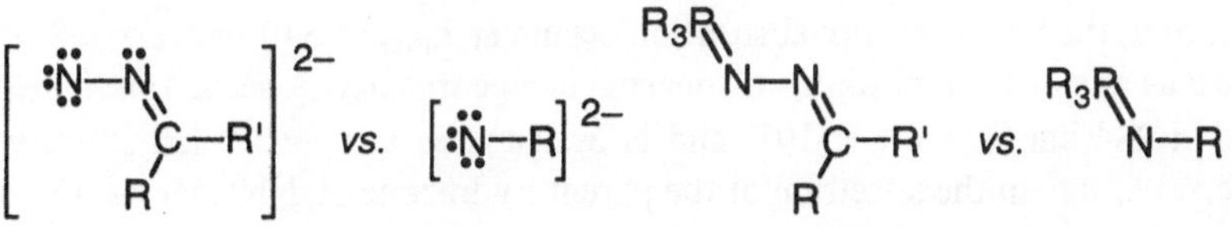

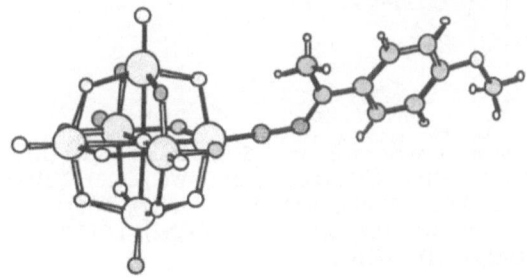

Figure 5. Relationships between diazoalkane and organoimido ligands, and their delivery reagents.

A Diazoalkane-hexamolybdate: $[Mo_6O_{18}(N_2C(Me)C_6H_4OMe)]^{2-}$

The phosphazine $Ph_3P=NN=C(Me)C_6H_4OMe$ reacts cleanly with $[Mo_6O_{19}]^{2-}$ in pyridine to afford orange $[Mo_6O_{18}(N_2C(Me)C_6H_4OMe)]^{2-}$ **6** in high yield as the $[Bu_4N]^+$ salt.[18] Its structure (Figure 6) indicates that the bound diazoalkane ligand is adequately described by the dianionic $[N_2C(Me)Ar]^{2-}$ formalism: both the short Mo-N bond length

Figure 6. Structure of the $[Mo_6O_{18}(N_2C(Me)C_6H_4OMe)]^{2-}$ anion **6**. Adapted with permission from *Angew. Chem. Int. Ed.* **1999**, *38*, 1145. © WILEY-VCH Verlag GmbH, D-69451 Weinheim, 1999.

(1.738(11)Å) and the near-linear Mo-N-N bond angle (172.0(10)°) of **6** are similar to corresponding values observed in organoimido-hexamolybdate structures, and are consistent with the presence of considerable Mo≡N triple bond character. Cyclic voltammetry studies of **6** revealed a one-electron reduction wave shifted cathodically (by –0.19 V) compared to that of $[Mo_6O_{19}]^{2-}$; the $E_{1/2}$ value of **6** (–0.894 V *vs.* Ag/Ag⁺) is comparable to that observed for $[Mo_6O_{18}(NR)]^{2-}$ systems and indicates that, as is the case for $[NR]^{2-}$ ligands, the $[N_2C(Me)Ar]^{2-}$ ligand is superior to an oxo ligand as an electron donor.

A Diazoalkane-hexatungstate: $[W_6O_{18}(N_2C(Me)C_6H_4OMe)]^{2-}$

As was true in the case of the imido-hexatungstate $[W_6O_{18}(NAr)]^{2-}$, we have not yet been able to accomplish a direct metathetical synthesis of a diazoalkane-hexatungstate through reaction of $[W_6O_{19}]^{2-}$ with a phosphazine reagent, but an assembly route was successful. Equimolar amounts of $[Bu_4]_2[WO_4]$ and $Ph_3P=NN=C(Me)C_6H_4OMe$ react in refluxing 1,2-dichloroethane over a period of three days to produce a red-orange solution; following evaporation, dissolution in CH_2Cl_2, and precipitation by addition of Et_2O, an orange residue was obtained which was recrystallized with difficulty from CH_3NO_2 by slow admission of Et_2O vapor. Several crops of $[Bu_4]_2[W_6O_{19}]$ were harvested prior to the obtention of orange crystals of $[Bu_4N]_2[W_6O_{18}(N_2C(Me)C_6H_4OMe)]$, **7**, in *ca.* 10% yield. The crystal structure† of the anion within **7** is shown in Figure 7. In the electronic spectrum of 7, the lowest energy absorption occurs at λ_{max} = 349 nm (ε = 6.2 × 10^4) and is assigned as arising from nitrogen-to-tungsten charge transfer. A second prominent band is observed at 284 nm (ε = 6.1 × 10^4) and is assigned as a $\pi \rightarrow \pi^*$ transition within the [N$_2$CMeAr] ligand: in the spectrum of the parent hydrazone H$_2$NNCMeAr, this feature is observed at λ_{max} = 271 nm (ε = 3.2 × 10^4). For comparison, the electronic spectrum of the analogous Mo system 6 displays bands at λ_{max} = 397 nm (ε = 7.7 × 10^4) and 311 nm (ε = 7.3 × 10^4).

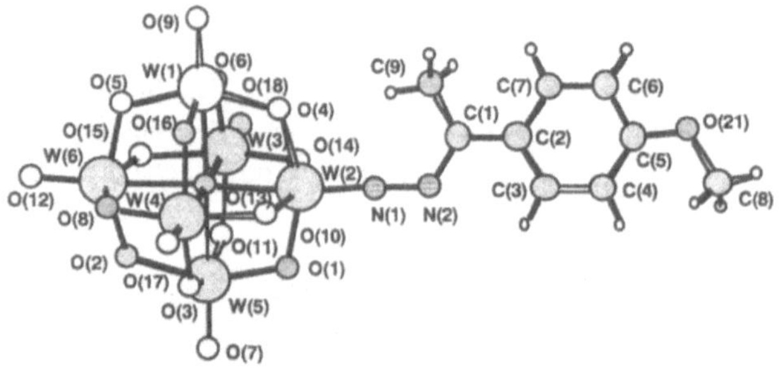

Figure 7. Anion structure in the diazo-hexatungstate 7. Selected bond lengths (Å) and bond angles (°): W(2)-N(1), 1.747(10); N(1)-N(2), 1.336(14); N(2)-C(1), 1.328(17); W(2)-O(13), 2.218(7); W(2)-O(14), 1.919(9); W(2)-O(10), 1.952(9); W(2)-O(4), 1.963(8); W(2)-O(1), 1.971(7); W(6)-O(12), 1.716(9); W(6)-O(13), 2.395(7); W(6)-O(15), 1.884(10); W(6)-O(8), 1.907(10); W(6)-O(2), 1.917(9); W(6)-O(5), 1.921(8); W(2)-N(1)-N(2), 173.9(11); N(1)-N(2)-C(1), 118.1(12).

The ligated diazoalkane of 7 exhibits metrical parameters (W(2)-N(1) = 1.747 Å; W(2)-N(1)-N(2) = 173.9°) which resemble those found for its Mo analogue 6. Distortions in the [W$_6$O$_{18}$] core resulting from diazoalkane binding are evident in the displacement of the central O(13) atom toward the nitrogen-bearing W(2) atom (W(6)-O(13) = 2.395 Å; W(2)-O(13) = 2.218 Å) and in the lengthening of the W-O$_b$ bond lengths at the diazoalkane binding site (W(2)-O$_b$ = 1.951(9) Å av) as compared with those at the *trans*-oxo site (W(6)-O$_b$ = 1.907(9) Å av).

FUNCTIONALIZED ORGANOIMIDO LIGANDS IN HEXAMETALATES

A second generation of organoimido-hexamolybdate species is beginning to emerge, in which the substituent (R) within the bound [NR] ligand bears a secondary functional group. Appropriately functionalized systems offer promising opportunities in crystal engineering (designed supramolecular chemistry), molecular materials (covalent attachment of redox-active, photo-responsive, and/or spin-carrying groups), hybrid covalent polymers, surface-adsorbed systems, *etc.* Examples of functionalized organoimido-hexamolybdates include [Mo$_6$O$_{17}$(NC$_6$H$_4$NH$_2$)$_2$]$^{2-}$,[11] the covalent donor-acceptor ferrocenylimido system [Mo$_6$O$_{18}$(NFc)]$^{2-}$,[19] and the coupled *bis*-systems [Mo$_5$O$_{18}$(Mo≡N-Z-N≡Mo)Mo$_5$O$_{18}$]$^{4-}$ (Z = p-C$_6$H$_4$, p-C$_6$H$_3$CH$_3$, cyclo-1,4-C$_6$H$_{10}$), in which a second imido-hexamolybdate unit serves as the "functionality".[20]

The covalent incorporation of polymerizable functionality into the hexamolybdate framework has been accomplished by the synthesis of the *p*-styrylimido-hexamolybdate [Bu$_4$N]$_2$[Mo$_6$O$_{18}$(NC$_6$H$_4$-CH=CH$_2$)]$^{2-}$, **8**.[21] 8 is obtained in 90% yield from the direct reaction of $[Mo_6O_{19}]^{2-}$ with $Ph_3P=NC_6H_4CH=CH_2$ in pyridine; its structure is shown in Figure 8. Solutions of **8** in acetonitrile are hydrolytically stable and exhibit an intense orange color (λ_{max} = 366 nm; ε = 50400 $M^{-1}cm^{-1}$). As expected, **8** is electrochemically active: cyclic voltammetry reveals a one-electron reduction at $E_{1/2}$ = –0.886 V (*vs.* Ag/Ag$^+$).

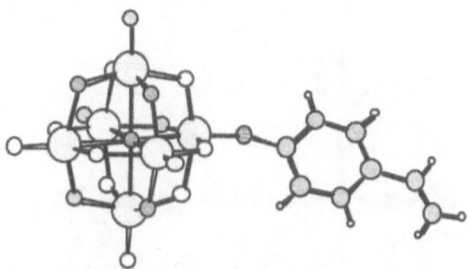

Figure 8. The *p*-styrylimido-hexamolybdate anion of **8**. Adapted from *Chem. Commun.* **2000**, 1793. Reproduced by permission of The Royal Society of Chemistry.

Initial studies employing **8** as a co-monomer in conventional free radical-induced copolymerizations with styrenes have yielded promising results. Reaction of **8** with four equivalents of 4-methylstyrene (4 -MS), in the presence of AIBN as initiator (0.18 equiv.), was conducted in 1,2-dichloroethane at 60°C for 48 h, and led to the isolation of a greenish solid (**9**) in 65% yield (by mass). In the IR spectrum of **9**, there was a complete absence of bands in the vinylic ν(C=C) region, but the 'doublet' pattern characteristic of mono-substituted imido-hexamolybdate species was observed in the ν(Mo$\equiv$O$_t$) region (975(sh), 952(s) cm^{-1}). In the ^{1}H NMR spectrum of **9**, no traces of the vinylic resonances of either precursor could be discerned, and all oberved spectral features (broadened resonances for both pendant aryl groups and for the aliphatic backbone; sharp resonances for the [Bu$_4$N]$^+$ countercations) were consistent with those of the expected copolymer. A comparison of relative integrated intensities indicated that **9** incoporates *ca.* 2.7 (4-MS) units for each unit of **8**, as stylized in Figure 9; elemental analysis results agree with this formulation.

Analogous copolymerizations using varying [4-MS]:[**8**] ratios have allowed some degree of compositional control over the resulting materials. Thus, using an eight-fold excess of 4-MS produced a composition with *ca.* 3.5 (4-MS) units per unit of **8**, while a twelve-fold excess of 4-MS afforded a composition approximating 4.5 (4-MS) units per unit of **8**. With increasing 4-MS incorporation, the materials become increasingly soluble in non-polar solvents such as benzene and toluene. Further studies of these systems, including molecular weight determinations, are underway, as are investigations of imido-hexamolybdates bearing methacryl and other polymerizable entities.

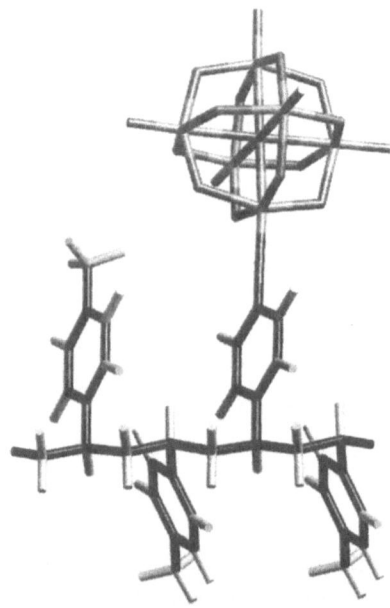

Figure 9. Idealized representation of a segment within **9** ([Bu$_4$N]$^+$ cations omitted for clarity). From *Chem. Commun.* **2000**, 1793. Reproduced by permission of The Royal Society of Chemistry.

OUTLOOK AND PROSPECTS

The organoimido and diazoalkane Lindqvist derivatives constitute a versatile class of intrinsically modified polyoxometalates, whose peripheral substituents can be used to advantage in further elaboration. In contemplating the prospects for polyoxometalate-based nanocomposite materials, it seems clear that the strategy of integrating functionality into such systems by the covalent incorporation of suitable nitrogenous ligands has much to offer. Despite frequent (and annoying) encounters with synthetic roadblocks, it should be possible to access hexametalate species bearing an array of useful remote functionality; as an example, imido-hexamolybdates currently under development in our laboratories include systems bearing carboxylic acid and various pyridyl groups.[22] Generalizing this chemistry to encompass other polyoxometalate systems, especially Keggin derivatives, remains a significant challenge, but one which is ripe with opportunity. We continue to pursue these attractive targets.

ACKNOWLEDGMENTS

We are grateful to the United States Department of Energy, Office of Basic Energy Sciences, for supporting our polyoxometalate work at Kansas State University. We also thank our colleagues Prof. Pierre Gouzerh, Prof. Anna Proust and Dr. René Thouvenot (Université Pierre et Marie Curie), Prof. Arnold Rheingold (University of Delaware), and Dr. Victor G. Young, Jr. (University of Minnesota) for their excellent and multifaceted collaborations.

NOTES

†Crystal Data for 7

$C_{41}H_{82}N_4O_{19}W_6$, irregular orange block, $M_W = 2038.21$, orthorhombic, space group *Pbca*, $a = 15.9010(8)$, $b = 16.8950(8)$, $c = 43.241(2)$ Å, $V = 11616.7(10)$ Å^3, $Z = 8$, $D_c = 2.331$ g cm^{-3}, $T = 173(2)$ K, $\mu = 11.901$ mm^{-1}, $\lambda = 0.71073$ Å. Of the79082 reflections collected ($2\theta_{max} = 56.66°$), 13829 were unique and 7566 were observed ($I > 2\sigma(I)$). An absorption correction was applied ($T_{min}/T_{max} = 0.4116$). Full-matrix least-squares on F^2 converged with $R_1 = 0.0520$ and $wR_2 = 0.1053$. Crystallographic files in .cif format are available from the author (e-mail: eam@ksu.edu).

REFERENCES

1. a) M. T. Pope, *Heteropoly and Isopoly Oxometalates*, Springer-Verlag: New York, (1983). b) M. T. Pope, A. Müller, Polyoxometalate chemistry-An old field with new dimensions in several disciplines, *Angew. Chem. Int. Ed.* 30:34 (1991). c) *Polyoxometalates: From Platonic Solids to Anti-Retroviral Activity*, M. T. Pope, A. Müller, Eds.; Kluwer Academic Publishers: Dordrecht, The Netherlands, (1994).
2. Comprehensive coverage of a broad range of topics within polyoxometalate chemistry can be found in a special thematic issue of *Chemical Reviews*: *Chem. Rev.* 98(1) (1998) C. L. Hill, guest editor.
3. a) P. Gouzerh, Y. Jeannin, A. Proust, F. Robert, Two novel polyoxomolybdates containing the $[MoNO]^{3+}$ unit-$[Mo_5Na(NO)O_{13}(OCH_3)_4]^{2-}$ and $[Mo_6(NO)O_{18}]^{3-}$, *Angew. Chem. Int. Ed. Engl.* 28:1363 (1989). b) A. Proust, R. Thouvenot, S.-G. Roh, J.-K. Yoo, P. Gouzerh, Lindqvist-type oxo-nitrosyl complexes-Synthesis, vibrational, multinuclear magnetic-resonance (^{14}N, ^{17}O, ^{95}Mo and ^{183}W) and electrochemical studies of $\{M_5O_{18}[M'(NO)]\}^{3-}$ anions (M, M'=Mo, W), *Inorg. Chem.* 34:4106 (1995).
4. a) T.-C. Hsieh, J. A. Zubieta, Synthesis and characterization of oxomolybdate clusters containing coordinatively bound organo-diazenido units: The crystal and molecular structure of the hexanuclear diazenido-oxomolybdate, $(NBu^n_4)_3[Mo_6O_{18}(N_2C_6H_5)]$, *Polyhedron* 5:1655 (1986). b) S. Bank, S. Liu, S. N. Shaikh, X. Sun, J. Zubieta, P. D. Ellis, ^{95}Mo NMR Studies of (Aryldiazenido)-and (Organohydrazido) molybdates. Crystal and molecular structure of $[n-Bu_4N]_3[Mo_6O_{18}(NNC_6F_5)]$, *Inorg. Chem.* 27:3535 (1988).
5. H. Kang, J. Zubieta, Coordination complexes of polyoxomolybdates with a hexanuclear core: synthesis and structural characterization of $(NBu_4)_2[Mo_6O_{18}(NNMePh)]$, *J. Chem. Soc., Chem. Commun.* **1988**, 1192.
6. a) J. R. Harper, A. L. Rheingold, Arsaoxanes as reversible, ligating oxygen-transfer agents in the synthesis of neutral metal oxo clusters-The X-ray structures of $Cp^*_2W_6O_{17}$ and $Cp^*_2Mo_8O_{16}$, *J. Am. Chem. Soc.* 112:4037 (1990). b) F. Bottomley, J. Chen, Organometallic oxides-oxidation of $[(\eta-C_5Me_5)Mo(Co)_2]_2$ with O_2 to form syn-$[(\eta-C_5Me_5)MoCl]_2(\mu-Cl)_2(\mu-O)$, syn-$[(\eta-C_5Me_5)MoCl]_2(\mu-Cl)(\mu-Co_3H)(\mu-O)$, and $[C_5Me_5O][(\eta-C_5Me_5)Mo_6O_{18}]$, *Organometallics.* 11:3404 (1992), c) A. Proust, R. Thouvenot, P. Herson, Revisiting the synthesis of $[Mo_6(\eta^5-C_5Me_5)O_{18}]^-$. X-ray structures of $Cp^*_2W_6O_{17}$ and $Cp^*_6Mo_8O_{16}$, *J. Chem. Soc., Dalton Trans.* 51 (1999).
7. Y. Du, A. L. Rheingold, E. A. Maatta, A polyoxometalate incorporating an organoimido ligand. Preparation and structure of $[Mo_5O_{18}(MoNC_6H_4CH_3)]^{2-}$, *J. Am. Chem. Soc.* 114:345 (1992).
8. A. Proust, R. Thouvenot, M. Chaussade, F. Robert, P. Gouzerh, Phenylimido derivatives of $[Mo_6O_{19}]^{2-}$. Syntheses, X-ray structures, vibrational, electrochemical ^{95}Mo and ^{14}N NMR-studies, *Inorg. Chim. Acta* 224:81 (1994).
9. J. B. Strong, R. Ostrander, A. L. Rheingold, E. A. Maatta, Ensheathing a polyoxometalate. Convenient systematic introduction of organoimido ligands at terminal oxo sites in $[Mo_6O_{19}]^{2-}$, *J. Am. Chem. Soc.* 116:3601 (1994).
10. R. J. Errington, C. Lax, D. G. Richards, W. Clegg, K. A. Fraser, New aspects of non-aqueous polyoxometalate chemistry, in: *Polyoxometalates: From Platonic Solids to Anti-Retroviral Activity*, M. T. Pope, A. Müller, Eds.; Kluwer Academic Publishers: Dordrecht, The Netherlands, (1994).

11. W. Clegg, R. J. Errington, K. Fraser, S. A. Holmes, A. Schäfer, Functionalization of $[Mo_6O_{19}]^{2-}$ with Aromatic-amines. Synthesis and structure of a hexamolybdate building-block with linear difunctionality, *J. Chem. Soc., Chem. Commun.* 455 (1995).

12. J. B. Strong, G. P. A. Yap, R. Ostrander, L. M. Liable-Sands, A. L. Rheingold, R. Thouvenot, P. Gouzerh, E. A. Maatta, A new class of functionalized polyoxometalates. Synthetic, structural, spectroscopic, and electrochemical studies of organimido derivatives of $[Mo_6O_{19}]^{2-}$, *J. Am. Chem. Soc.* 122:639 (2000).

13. J. B. Strong, B. S. Haggerty, A. L. Rheingold, E. A. Maatta, A superoctahedral complex derived from a polyoxometalate. The hexakis(arylimido)hexamolybdate anion $[Mo_6(NAr)_6O_{13}H]^-$, *J. Chem. Soc., Chem. Commun.* 1137 (1997).

14. T. R. Mohs, G. P. A. Yap, A. L. Rheingold, E. A. Maatta, An organoimido derivative of the hexatungstate cluster. Preparation and structure of $[W_6O_{18}(NAr)]^{2-}$ $(Ar=2,6-(^iPr)_2C_6H_3)$, *Inorg. Chem.* 34:9 (1995).

15. a) M. Dartiguenave, M. J. Menu, E. Deydier, Y. Dartiguenave, H. Siebald, Crystal and molecular-structures of transition-metal complexes with N-bonded and C-bonded diazoalkane ligands, *Coord. Chem. Rev.* 178-180:623 (1998). b) Y. Mizobe, Y. Ishii, M. Hidai, Synthesis and reactivities of diazoalkane complexes, *Coord. Chem. Rev.* 139:281 (1995). c) D. Sutton, Organometallic diazo-compounds, *Chem. Rev.* 93:995 (1993). d) W. A. Herrmann, Organometallic syntheses with diazoalkanes, *Angew. Chem. Int. Ed.* 17:800 (1978).

16. See, for example, the following sources and references therein: a) M. P. Doyle, M. A. McKervey, T. Ye, *Modern Catalytic Methods for Organic Synthesis with Diazo Compounds*, Wiley, New York (1998). b) J. L. Polse, A. W. Kaplan, R. A. Andersen, R. G. Bergman, Synthesis of an η^2-N_2-titanium diazoalkane complex with both imido-like and metal carbene-like reactivity patterns, *J. Am. Chem. Soc.* 120:6316 (1998).

17. a) D. E. Wigley, Organoimido complexes of the transition-metals, *Prog. Inorg. Chem.* 42:239 (1994). b) W. A. Nugent, J. M. Mayer, *Metal-Ligand Multiple Bonds*, Wiley, New York (1988).

18. H. Kwen, V. G. Young, Jr., E. A. Maatta, A diazoalkane derivative of a polyoxometalate. Preparation and structure of $\{Mo_6O_{18}[NNC(C_6H_4OCH_3)CH_3]\}^{2-}$, *Angew. Chem. Int. Ed.* 38:1145 (1999).

19. J. L. Stark, V. G. Young, Jr., E. A. Maatta, A functionalized polyoxometalate bearing a ferrocenylimido ligand. Preparation and structure of $[(FCN)Mo_6O_{18}]^{2-}$, *Angew. Chem. Int. Ed.* 34:2547 (1995).

20. J. L. Stark, A. L. Rheingold, E. A. Maatta, Polyoxometalate clusters as building-blocks. Preparation and structure of bis (hexamolybdate) complexes covalently bridged by organodiimido ligands, *J. Chem. Soc., Chem. Commun.* 1165 (1995).

21. A. R. Moore, H. Kwen, A. M. Beatty, E. A. Maatta, Organoimido-Polyoxometalates as polymer pendants, *Chem. Commun.* 1793 (2000).

22. A. R. Moore, C. G. Hamaker, G. D. Forster, E. A. Maatta, unpublished results and manuscript in preparation.

POLYOXOMETALATE SPECIATION – IONIC MEDIUM DEPENDENCE AND COMPLEXATION TO MEDIUM IONS

Lage Pettersson

Department of Chemistry
Inorganic Chemistry
Umeå University
SE-901 87 Umeå, Sweden

INTRODUCTION

In most aqueous polyoxometalate systems, numerous, often highly negatively charged species, are formed. The speciation is therefore very sensitive to the ionic medium background; and especially to the cations. It is therefore important to use a "constant" ionic medium, where the cation concentration is kept constant and the anion concentration is allowed to vary somewhat. It is of great advantage if the medium ions do not hydrolyse in the pH range studied. Therefore, the most frequently used medium cations/anions are Na^+, K^+ / Cl^-, ClO_4^- and NO_3^-. A commonly used medium is for instance the artificial seawater medium 0.600 M Na(Cl). The formation constants determined in a specific medium are of course only valid in this medium and for the temperature used in the study. Constants obtained in different ionic media, and also for the same ionic medium at different total concentrations, vary substantially. Moreover, in some cases species of different nuclearities are formed (*e.g.* in the hydrolysis of Mo(VI), where a hepta- or an octamolybdate can be formed). The ultimate goal is to determine the ionic strength parameters and/or the extent of medium cation complexation to the polyoxometalate species. Then, the system in question can be modeled at any ionic medium concentration. If the medium contains large bulky organic cations, the complexation to the polyoxometalate species can be expected to be weak and the formation constants obtained at moderate medium concentrations should be similar to those at infinite dilution.

Having studied speciation in polyoxometalate systems for three decades, we found it worthwhile to make a quantitative study of a system to try to determine the extent of medium cation complexation to polyoxometalate anions. We thereby chose the H^+ - V(V) system since ^{51}V is an ideal NMR nucleus, making this system extremely suitable for the powerful EMF-NMR technique. Moreover, the analysis of the experimental data can be performed with the least squares computer program LAKE,[1] capable of simultaneously

Polyoxometalate Chemistry for Nano-Composite Design
Edited by Yamase and Pope, Kluwer Academic/Plenum Publishers, 2002

treating combined EMF-NMR data. As inorganic medium cation, we chose Na^+, and as bulky organic cation, tetrabutylammonium (TBA^+).

Owing to electronic repulsion, medium anions should not be expected to form any complexes with polyoxometalates. They may though have an indirect and competing effect by forming ion pairs with the cations in the medium. The only systems where we so far have found a direct complexation are in peroxomolybdate systems, where Cl^- and HSO_4^- (but not ClO_4^-) form strong complexes with diperoxomolybdates.

This paper will present some earlier findings of the ionic medium dependence in the binary H^+-Mo(VI) and H^+-V(V) systems, and some preliminary results from the ongoing study of the four-component system H^+-HVO_4^{2-}-Na^+-TBA^+ in Cl^- medium. Data can be completely explained with strong Na^+ and weaker TBA^+ complexation to the highly negatively charged vanadate species.

MOLYBDATES

There have been a number of investigations of the equilibrium speciation in the aqueous $pH^+ + qMoO_4^{2-} \rightleftharpoons (H^+)_p(MoO_4^{2-})_q$ system. Most studies have been carried out by means of potentiometric titrations in constant ionic medium. In addition to the monomeric species MoO_4^{2-} (0,1), $HMoO_4^-$ (1,1) and H_2MoO_4 (2,1), Sasaki and Sillén[2] proposed a series of heptamolybdates to predominate in 3.0 M Na(ClO_4) medium, namely $Mo_7O_{24}^{6-}$ (8,7), $HMo_7O_{24}^{5-}$ (9,7), $H_2Mo_7O_{24}^{4-}$ (10,7), $H_3Mo_7O_{24}^{3-}$ (11,7), and in very acidic solutions the high-nuclear species (34,19) and the cation (5,2). The numbers in parentheses refer to the values of p and q in the equilibrium reaction. From an investigation in 0.600 M Na(Cl) media, Yagasaki et al.[3] found that an octameric ion $Mo_8O_{26}^{4-}$ (12,8) is formed instead of (11,7). Figure 1 shows Z(pH) data points obtained from potentiometric titrations on Mo_{tot} = 80 mM from these two media and an unpublished study in 3.0 M Li(ClO_4) medium. The

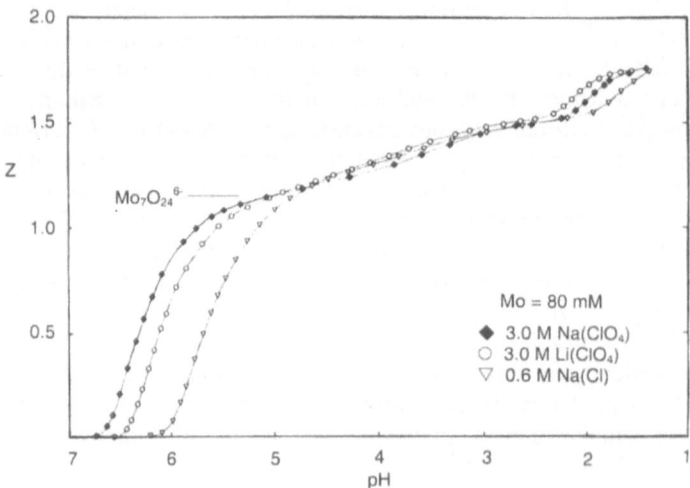

Figure 1. Z (average number of H^+ bound per MoO_4^{2-}) as a function of pH in three ionic media.

experimental points from the 0.600 M Na(Cl) medium are taken from Ref. 3, and those at the 3.0 M Na(ClO_4) medium from our reinvestigation of the system, where the speciation proposed by Sasaki and Sillén was confirmed (only minor adjustments of the formation constants were needed to fit our data). As seen from Figure 1, the medium dependence is considerable. The hydrolysis starts approximately 0.6 pH units earlier when $[Na^+]$ is increased by a factor five. More surprisingly, the hydrolysis differs substantially in the two 3.0 M media. This indicates that the ionic strength is not the only parameter determining the speciation and that Na^+ and Li^+ may form complexes of different strength with the hydrolysis product first being formed, the highly charged $Mo_7O_{24}^{6-}$ anion. This species has a Z value of 8/7 = 1.14 as marked in the figure. Figure 2 shows the molybdate distribution at two different concentrations (Mo_{tot} = 1.25 and 20 mM) in the two Na^+ media. The most striking feature is that an octameric ion (12,8) is formed instead of (11,7), and that the high charged $Mo_7O_{24}^{6-}$(8,7) complex is weaker and the uncharged species H_2MoO_4 (2,1) stronger in the lower $[Na^+]$ medium.

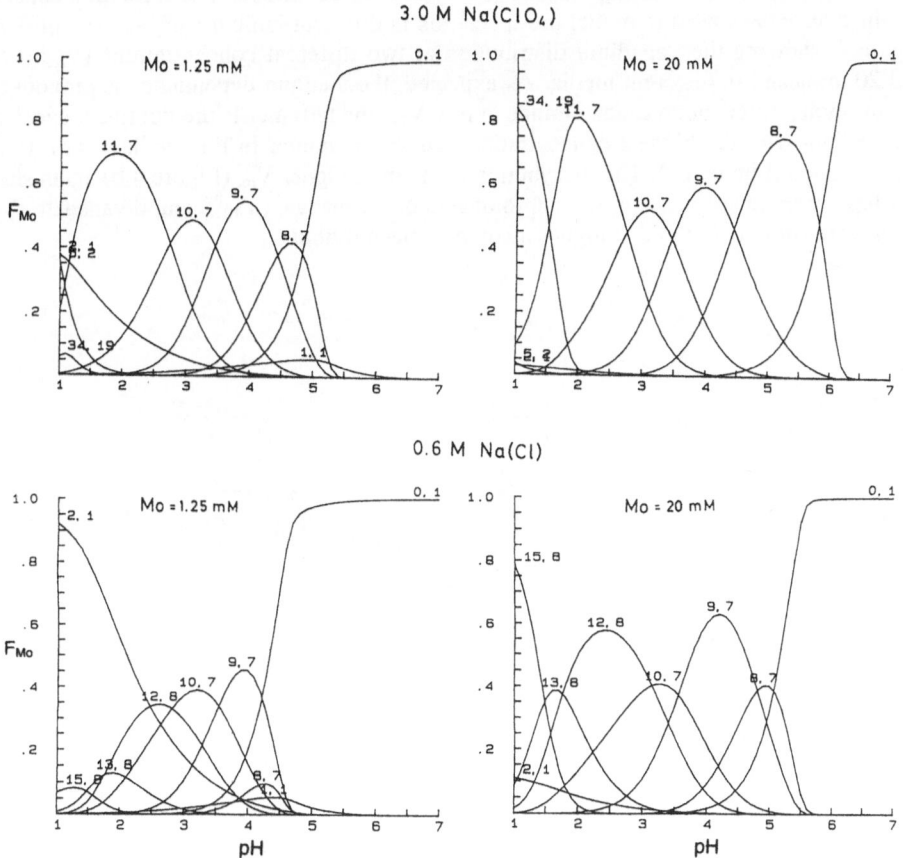

Figure 2. Diagrams showing the distribution of molybdenum, F_{Mo}, as a function of pH in two ionic media.

VANADATES

The hydrolysis of pentavalent vanadium, V(V), is very complex. Besides monomeric species, a variety of polyoxovanadate species with the nuclearities 2-6 and 10 are known to form in equilibrated solutions. The charges vary from +1 to –6. As a consequence the speciation is strongly dependent both on the total concentration of V, V_{tot}, and on the ionic medium. It was not until the early 80's that the full speciation was established for the whole pH range in the same ionic medium.[4] Previously, there was the argument as to whether the polynuclear species in the neutral, the so-called metavanadate range, were trimers, tetramers or both.[5] By combining potentiometric and [51]V NMR data, the co-existence of mono-, di-, tetra-, penta-, and hexanuclear (only at very high V_{tot}) species, all having a charge per vanadium equal to -1, was established.[4,6]

In Figure 3, potentiometric titration data from 0.600 M Na(Cl) are shown.[4] All species found are included in the figure and the line to the left of each formula shows the Z and z values of the species. Later studies in other ionic media showed that the ionic medium dependence is considerable. A compilation of the formation constants obtained at pH > 7 is given in Table 1. It is worth noting that the species formed are identical with the exception that some of the high-nuclear species having high charges can not be detected in low ionic media. This is in contrast to the molybdate system (see above), where different ionic medium background can change the nuclearity of species formed. The formation constants obtained have been used to model the speciation in different ionic media, as exemplified in Figure 4, showing the vanadium distribution at two different concentrations ($V_{tot} = 1.25$ and 20 mM) and in different media. As expected, the medium dependence is pronounced. For example, in the metavanadate range at low V_{tot}, the tetramer is the dominant species in the medium with the highest concentration and the monomer in the medium with lowest concentration (Figure 4 a). The distribution diagram at higher V_{tot} (Figure 4 b) again shows that high-charged species, *e.g.* the unprotonated decavanadate (V_{10}^{6-}) and divanadate (V_2^{4-}) species are strongly favored at higher medium concentrations.

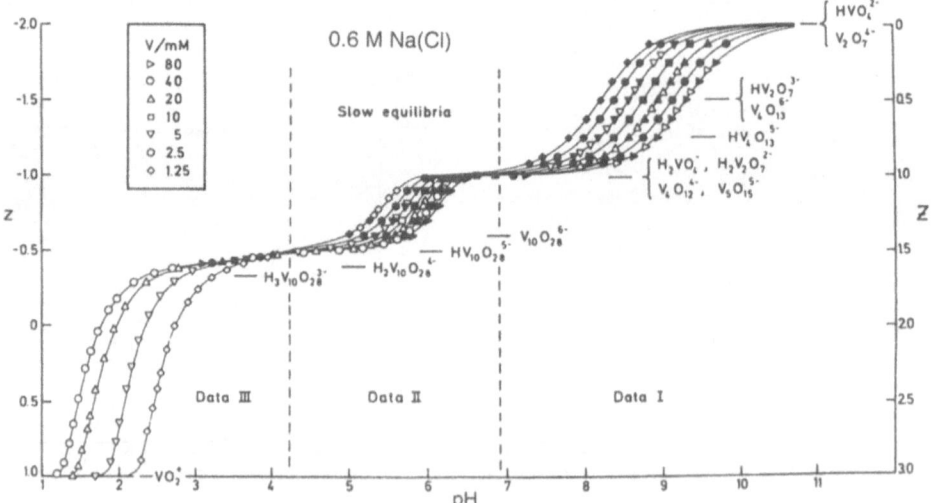

Figure 3. Z (average number of H$^+$ bound per HVO$_4^{2-}$) and z (average charge per vanadium) as a function of pH in 0.600 M Na(Cl) medium.

Table 1. Species and formation constants in different ionic media for the H^+ - HVO_4^{2-} system (pH > 7, 25 °C).

(p,q)	Formula	Notation	log β_{pq}				
			3.0 M Na(ClO₄)	0.6 M Na(Cl)	0.15 M Na(Cl)	0.01 M Na(Cl)	0.05 M TBA(Cl)
-1,1	VO_4^{3-}	V^{3-}	-13.1	-13.36	-	-	-
0,1	HVO_4^{2-}	V^{2-}	0	0	0	0	0
1,1	$H_2VO_4^-$	V^-	8.00	7.946	8.17	8.467	8.430
0,2	$V_2O_7^{4-}$	V_2^{4-}	1.44	0.66	0.15	-	-0.35
1,2	$HV_2O_7^{3-}$	V_2^{3-}	11.12	10.45	10.49	10.48	10.51
2,2	$H_2V_2O_7^{2-}$	V_2^{2-}	18.8	18.68	18.99	19.40	19.47
2,4	$V_4O_{13}^{6-}$	$l\text{-}V_4^{6-}$	24.73	23.19	22.70	-	22.26
3,4	$HV_4O_{13}^{5-}$	$l\text{-}V_4^{5-}$	33.0	31.92	32.05	-	32.02
4,4	$V_4O_{12}^{4-}$	V_4^{4-}	43.24	41.68	41.92	42.25	42.49
5,5	$V_5O_{15}^{5-}$	V_5^{5-}	54.14	51.89	52.02	-	52.38
6,6	$V_6O_{18}^{6-}$	V_6^{6-}	64	61.58	-	-	-

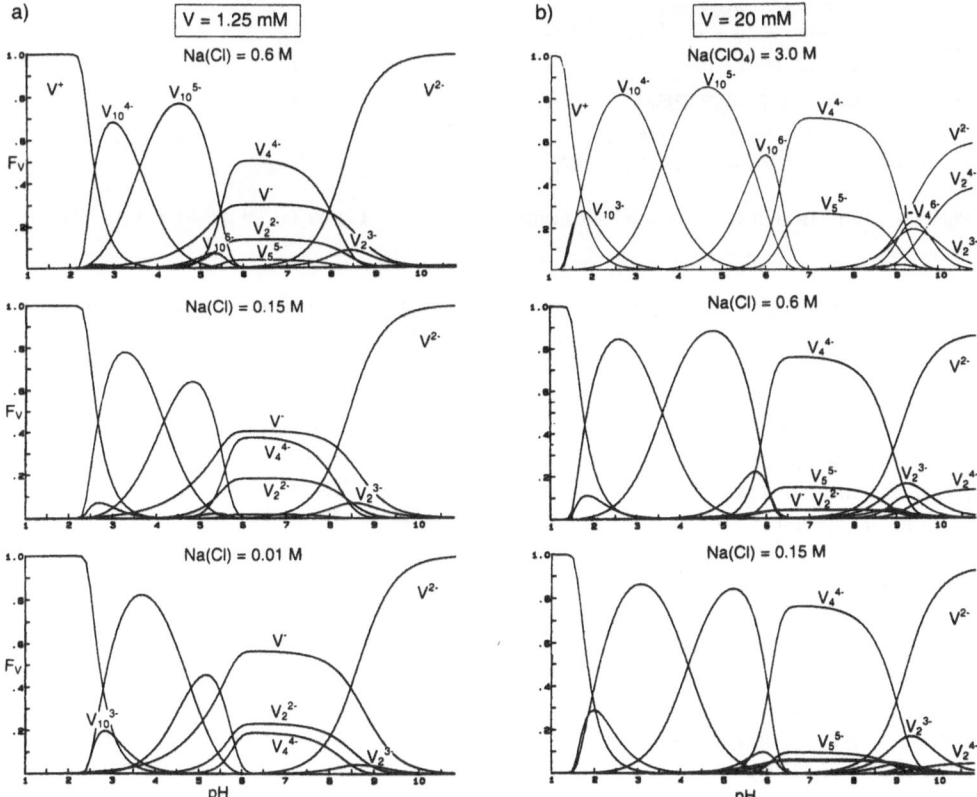

Figure 4. Diagrams showing the distribution of vanadium, F_V, as a function of pH in different ionic media at V_{tot} = 1.25 (a) and 20 mM (b).

The ionic medium dependence on the chemical shift value for highly charged vanadate species strongly indicates interaction with medium cations (Figure 5). In the two lowest constant ionic media studied (50 mM TBA(Cl) and 10 mM Na(Cl), the chemical shifts of the mono- and dimetavanadate species are similar, in contrast to the tetra- and pentameric species V_4^{4-} and V_5^{5-}. Thus, cations in the medium seem to form complexes/ion-pairs with the highly negatively charged vanadate ions.

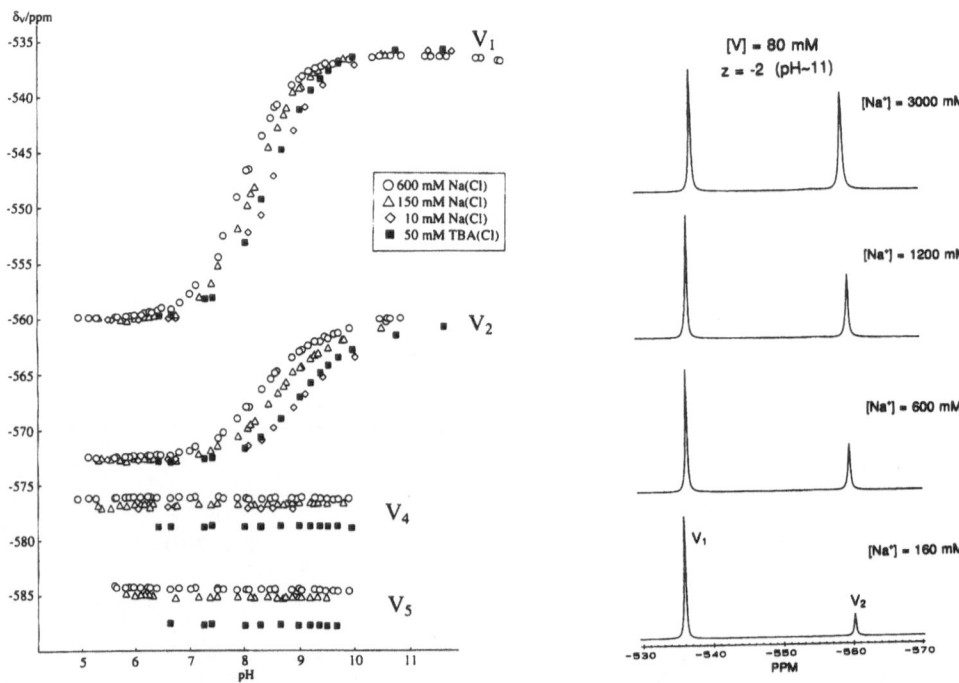

Figure 5. ^{51}V NMR shift vs. pH in four ionic media.   **Figure 6.** ^{51}V spectra at different $[Na^+]$.

To examine this further in a quantitative way, a study was initiated aiming at determining the formation constants of the ion-pairing complexes formed in the $pH^+ + qHVO_4^{2-} + rNa^+ \rightleftharpoons (H^+)_p(HVO_4^{2-})_q(Na^+)_r$ system, by varying all the three components, measuring the pH and recording ^{51}V NMR data of each separately prepared solution. The combined pH and NMR data were then analyzed with the LAKE program. Such an equilibrium analysis should give the medium independent constants (r = 0) at very low $[Na^+]$ and the Na^+ complexes (r > 0) at increased $[Na^+]$. To ascertain the validity of the medium independent constants obtained, an analogous study of the corresponding $pH^+ + qHVO_4^{2-} + rTBA^+ \rightleftharpoons (H^+)_p(HVO_4^{2-})_q(TBA^+)_r$ system has been started as well. As a final check the mixed Na^+/TBA^+ cation system $pH^+ + qHVO_4^{2-} + rNa^+ + sTBA^+ \rightleftharpoons (H^+)_p(HVO_4^{2-})_q(Na^+)_r(TBA^+)_s$ is under study.

In this paper the results obtained so far in the high pH range (pH 10.8-12) will be presented. In this so-called pyrovanadate range, the vanadate system is quite simple (*c.f.* Figure 3). Only two NMR resonances have to be considered, arising from monomeric and dimeric species. Not taking possible cation complexation into account, the dimer is formed according to $2 HVO_4^{2-} \rightleftharpoons V_2O_7^{4-} + H_2O$. This means that high V_{tot} favors the dimer. The formation constant of this (0,2) species varies considerably with the ionic medium as shown in Table 1. In the media for which the constant could be determined, the value for the medium of highest concentration, *i.e.* 3.0 M Na(ClO$_4$), is almost two logarithmic units higher than in 0.05 M TBA(Cl). This strong medium dependence is clearly illustrated in Figure 6, showing [51]V NMR spectra at constant V_{tot} but different $[Na^+]$. Since the medium charge per vanadium is –2, the lowest possible $[Na^+]$ is 2 times V_{tot}. Thus, the bottom spectrum represents a self-medium condition.

In the search for medium complexes we first treated Na^+ data at the lowest possible Na^+ concentrations (self-medium solutions at low V_{tot}) to establish the medium independent constants and obtained $\log\beta_{0,2,0}$ = -0.30. This value was not allowed to vary in subsequent calculations. Then all Na^+ data were treated. As previously found at high V_{tot}, two extremely minor resonances arising from an unprotonated linear trimer $V_3O_{10}^{5-}$ is discernable in such solutions.[7] However, at the present high pH values, the amount is too low to make it possible to determine its formation constant. An approximate value of a monosodium complex, that fairly well explained these resonances, was used in the calculation in the search for all other complexes. All 35 experimental points were perfectly explained with non-sodium and sodium complexes of mono- and divanadates. An attempt to explain the data with extended Debye-Hückel expressions was not successful.

Then the complexation in tetrabutylammonium (TBA^+) solutions was investigated. In this system the NMR integral values of all 21 solutions could be perfectly explained with non-TBA and TBA complexes. However, the TBA^+ complexes were found to be weaker than the sodium complexes. No TBA^+ complex of monovanadate, for instance, was needed to explain the data.

Finally, NMR data were recorded and evaluated from 21 mixed Na^+/TBA^+ solutions and calculations were performed on all Na^+, TBA^+ and Na^+/TBA^+ solutions (77 in total). When including a mixed cation divanadate complex all data were explained within the experimental errors. The result from the final calculation on all data is shown in Table 2. These data comprise the following concentration ranges (in mM): $0 < Na^+_{tot} < 2976$, $0 < TBA^+_{tot} < 600$, and $2 < V_{tot} < 300$. Cation complexation can, without using any Debye-Hückel expressions, explain all vanadate NMR integrals in these vast concentration ranges. It should be noted that the pyrovanadate species are rather highly charged (-2/V) and this may be the reason why cation complexation is the predominant parameter.

Table 2. Species and formation constants in the H^+ - HVO_4^{2-} - Na^+ - TBA^+ system (pH 10.8-12, 25 °C).

(p,q,r,s)	Formula	Notation	logβ (3σ)	pK_{Na}	pK_{TBA}
0,1,0,0	HVO_4^{2-}	V^{2-}	0		
0,1,1,0	HVO_4Na^-	VNa^-	0.02 (07)	0.02	
0,2,0,0	$V_2O_7^{4-}$	V_2^{4-}	-0.30		
0,2,1,0	$V_2O_7Na^{3-}$	V_2Na^{3-}	0.97 (02)	1.27	
0,2,2,0	$V_2O_7Na_2^{2-}$	$V_2Na_2^{2-}$	1.23 (12)	0.26	
0,2,0,1	$V_2O_7TBA^{3-}$	V_2A^{3-}	0.37 (12)		0.67
0,2,0,2	$V_2O_7(TBA)_2^{2-}$	$V_2A_2^{2-}$	0.35 (31)		-0.02
0,2,1,1	$V_2O_7NaTBA^{2-}$	V_2NaA^{2-}	1.14 (39)	0.77	0.17

To illustrate the equilibrium conditions, distribution diagrams have been calculated using the constants from Table 2 and are shown in Figures 7 and 8 for Na^+, TBA^+, and Na^+/TBA^+, respectively. These diagrams are calculated at high V_{tot} to have substantial amounts of the dimer present. As seen from Figure 7, the Na^+ complexation to divanadate starts at lower free concentration of the cation than is the case for TBA^+.

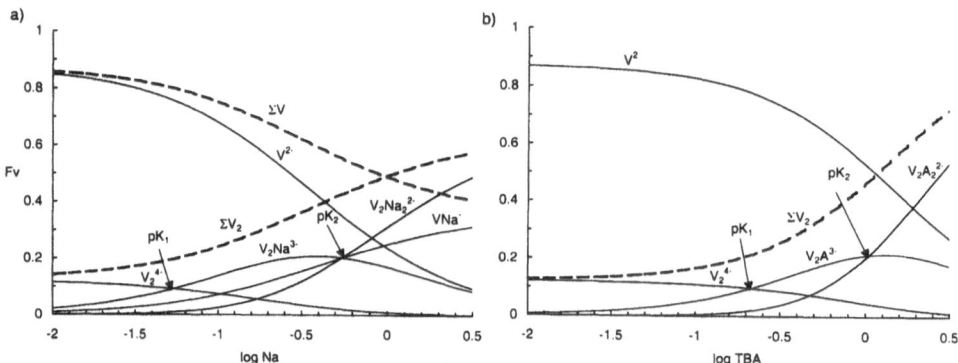

Figure 7. Distribution of vanadium, F_V, at pH = 12 and V_{tot} = 160 mM in a) the HVO_4^{2-} - Na^+ system, b) the HVO_4^{2-} - TBA^+ system, shown as a function of the logarithmic value of the free cation concentration.

Concerning the Na^+ complexation, the conditions in artificial sea water medium, 0.6 M Na(Cl), will be taken as an example. Figure 7(a) shows that at log Na ≈ -0.2, the main species giving rise to the dimeric resonance is $V_2O_7Na^{3-}$ and $V_2O_7Na_2^{2-}$ and that $V_2O_7^{4-}$ is a very minor species. However, for the monomeric resonance, the HVO_4^{2-} species contributes more than HVO_4Na^-. To illustrate the increase of dimeric over monomeric species with increasing cation concentration (*c.f.* Figure 6), the sum of monomeric and dimeric species, respectively, are shown in the figure as well (bold dashed curves). At low cation concentration where no cation complexes are formed, only ca. 12% of vanadium is bound in dimeric species at V_{tot} = 160 mM, but at $[Na^+]$ = 1 M, the monomer and dimer are present in equal amounts.

The TBA^+ formation constants are not as accurately determined as the Na^+ constants. The experimental data are though perfectly explained and the reason for the rather high 3σ values (see Table 2) is probably due to the more restricted TBA^+

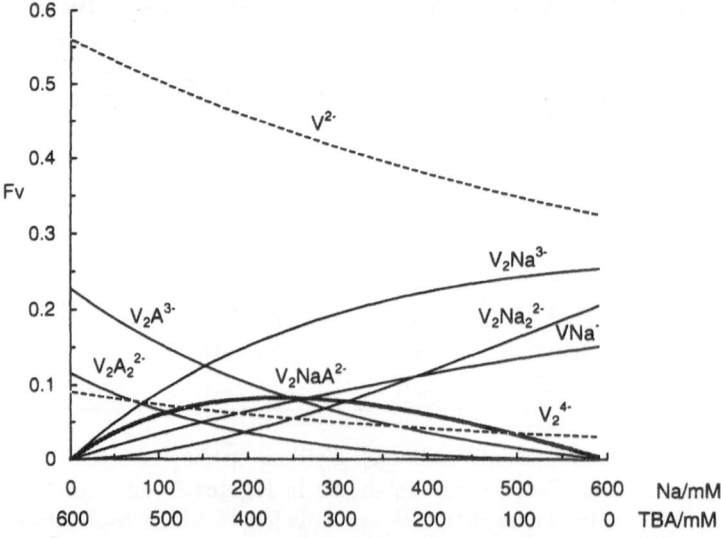

Figure 8. Distribution of vanadium, F_V, at pH = 12 and V_{tot} = 290 mM in the HVO_4^{2-} - Na^+ - TBA^+ system as a function of $[Na^+]$ and $[TBA^+]$ at a total cation concentration of 600 mM.

concentration range studied (0.6 M compared to 3.0 M). The earlier study at constant 0.050 M TBA(Cl) was performed to obtain the vanadate speciation in a medium where complexation to medium cations should be negligible and thus represent conditions at infinite dilution. This assumption seems to be essentially correct. At log TBA = -1.3, only a very minor amount of $V_2O_7 TBA^{3-}$ is formed according to Figure 7 b.

The conditions in mixed cation (Na + TBA) = 600 mM solutions at V_{tot} = 290 mM are illustrated in Figure 8. This diagram represents an approximate self-medium. The distribution curves of non-cation species are marked with dashed lines and the mixed cation dimer $V_2O_7 NaTBA^{2-}$ with a bold line. Since Na^+ complexes are stronger than TBA^+ complexes, the non-cation species decrease with increasing $[Na^+]$. Moreover, the mixed cation species has its optimum in somewhat enriched TBA solutions.

COMPLEXATION WITH MEDIUM ANIONS

One would not expect medium anions to form complexes with polyoxometalates. However, in an ongoing study of peroxomolybdates in 0.300 M $Na_2(SO_4)$ medium, we have found that sulfato species are formed at low pH (pH < 3) having the composition $(H^+)_p(MoO_4^{2-})(H_2O_2)_2(SO_4^{2-})^{(4-p)-}$ with p = 2 and 3. This unexpected finding made us test if the commonly used medium anions Cl^-, ClO_4^- and NO_3^- also form complexes with peroxomolybdates. We found this to be the case for Cl^-, but not for ClO_4^-.

CONCLUDING REMARKS

We have from our study on pyrovanadate equilibria in the pH range 10.8 – 12 at 25 °C found that the medium dependence in Na(Cl), TBA(Cl), and Na,TBA(Cl) media can be explained with medium cation complexation to the vanadate species. Although vast medium concentration ranges have been covered, no Debye-Hückel parameters have to be used. Moreover, since Na^+, TBA^+ and medium independent equilibrium constants have been determined, the pyrovanadate system can be modeled at any V_{tot}, $[Na^+]$, and $[TBA^+]$. An analogous study on the H^+ - HVO_4^{2-} system in the pH range 7 – 12 is in progress and it seems that medium cation complexes can explain all EMF/NMR data.

REFERENCES

1. N. Ingri, I. Andersson, L. Pettersson, A. Yagasaki, L. Andersson, and K. Holmström, LAKE - A program system for equilibrium analytical treatment of multimethod data, especially combined potentiometric and NMR data, *Acta Chem. Scand.* 50:717 (1996).

2. Y. Sasaki, and L.G. Sillén, On equilibria in polymolybdate solutions, *Acta Chem. Scand.* 18:1014 (1964).

3. A. Yagasaki, I. Andersson, and L. Pettersson, Multicomponent polyanions. 41. Potentiometric and ^{31}P NMR study of equilibria in the molybdophenylphosphonate system in 0.6 M Na(Cl) medium, *Inorg. Chem.* 26:3926 (1987).

4. L. Pettersson, B. Hedman, I. Andersson, and N. Ingri, Multicomponent polyanions. 34. A potentiometric and ^{51}V NMR study of equilibria in the H^+-HVO_4^{2-} system in 0.6 M Na(Cl) medium covering the range 1 <≈ -lg$[H^+]$ <≈ 10, *Chem. Scr.* 22: 254 (1983).

5. F. Brito, N. Ingri, and L.G. Sillén, Are aqueous metavanadate species trinuclear, tetranuclear, or both? *Acta Chem. Scand.* 18:1557 (1964).

6. L. Pettersson, I. Andersson, and B. Hedman, Multicomponent polyanions. 37. A potentiometric and ^{51}V NMR study of equilibria in the H^+-HVO_4^{2-} system in 3.0 M Na(ClO$_4$) medium covering the range 1 <≈ -lg$[H^+]$ <≈ 10, *Chem. Scr.* 25: 309 (1985).

7. I. Andersson, L. Pettersson, J.J. Hastings, and O.W. Howarth, Oxygen and vanadium exchange processes in linear vanadate oligomers, *J. Chem. Soc. Dalton Trans.* 3357 (1996).

SOME SMALLER POLYOXOANIONS. THEIR SYNTHESIS AND CHARACTERIZATION IN SOLUTION

Hiroyuki Nakano,[1] Tomoji Ozeki,[2] and Atsushi Yagasaki[1]

[1]Department of Chemistry
 Kwansei Gakuin University
 Nishinomiya 662-8501, Japan
[2]Department of Chemistry and Materials Science
 Tokyo Institute of Technology
 Tokyo 152-8551, Japan

INTRODUCTION

The chemistry of polyvanadates has developed drastically during the last decade. Scores of new species have been reported in recent years.[1] Many of these newer polyvanadates are of mixed-valence and very large. The examples of fully oxidized polyvanadate, on the other hand, are still relatively limited. This is especially true for heteropolyvanadates. Structurally characterized heteropolyvanadates in which all vanadium atoms are in V oxidation state remain to be a rarity. Only a handful of such examples have been reported to date.[2] This apparent lack of information and interest does not necessarily mean that the chemistry of fully oxidized polyvanadates itself is limited and uninteresting. Recent findings from our and other laboratories suggest just the opposite,[3] although they are not as large as mixed-valence polyvanadates. The present article focuses on the synthesis and structural characterization of some smaller polyvanadates that are fully oxidized.

THE $[Se_xV_{4-x}O_{12-x}]$ $(x = 1, 2)$ ANIONS

Synthesis

Reaction of $[(n-C_4H_9)_4N]VO_3$ with an equimolar amount of SeO_2 in acetonitrile at an ambient temperature yields a dark red, hygroscopic compound. This compound gives a single ^{51}V NMR peak at –547 ppm in acetonitrile.* The compound is too hygroscopic for

---

* All ^{51}V NMR chemical shifts are referenced externally against $VOCl_3$.

Polyoxometalate Chemistry for Nano-Composite Design
Edited by Yamase and Pope, Kluwer Academic/Plenum Publishers, 2002

149

further characterization. However, it yields crystalline $[(C_6H_5)_4P]^+$ and $[\{(C_6H_5)_3P\}_2N]^+$ salts when reacted with $[(C_6H_5)_4P]Br$ and $[\{(C_6H_5)_3P\}_2N]Cl$, respectively. These salts give the same single-peak ^{51}V NMR spectra as the $[(n\text{-}C_4H_9)_4N]^+$ compound initially obtained. The elemental analyses of the crystalline materials obtained from acetonitrile/diethyl ether mixed solvent suggested they are salts of a new vanadoselenite anion, $[Se_2V_2O_{10}]^{2-}$.

$$2[(n\text{-}C_4H_9)_4N]VO_3 + 2SeO_2 \rightarrow [(n\text{-}C_4H_9)_4N]_2[Se_2V_2O_{10}]$$

When $[(n\text{-}C_4H_9)_4N]VO_3$ is reacted with SeO_2 in a $3:1$ molar ratio, on the other hand, a different compound is obtained. This compound is colorless and gives two ^{51}V NMR peaks at -556 and -566 ppm in a $2:1$ intensity ratio. It analyzes as $[(n\text{-}C_4H_9)_4N]_3$-$[SeV_3O_{11}]\cdot 0.5H_2O$ after crystallization from acetonitrile/diethyl ether.

$$3[(n\text{-}C_4H_9)_4N]VO_3 + SeO_2 \rightarrow [(n\text{-}C_4H_9)_4N]_3[SeV_3O_{11}]$$

The composition of both $[SeV_3O_{11}]^{3-}$ and $[Se_2V_2O_{10}]^{2-}$ anions implies that they have ring structures. This was confirmed by the x-ray structural analyses.

Solid State Structures

X-ray structural analysis revealed that single crystals of $[\{(C_6H_5)_3P\}_2N]_2[Se_2V_2O_{10}]$ are composed of $[\{(C_6H_5)_3P\}_2N]^+$ cations and discrete $[Se_2V_2O_{10}]^{2-}$ anions having the structure shown in Figure 1.[†] The crystal structure of $[(n\text{-}C_4H_9)_4N]_3[SeV_3O_{11}]\cdot 0.5H_2O$ was also studied by x-ray diffraction. The result revealed that an asymmetric unit of $[(n\text{-}C_4H_9)_4N]_3[SeV_3O_{11}]\cdot 0.5H_2O$ contains six $[(n\text{-}C_4H_9)_4N]^+$ cations, one water molecule and two discrete $[SeV_3O_{11}]^{3-}$ anions having the structure shown in Figure 2.[‡] The water molecule of crystallization is hydrogen-bonded to one of the anions $[O23\cdots O1\ 2.71(8)$ Å, $O23\cdots O6\ 2.86(8)$ Å].

Both $[SeV_3O_{11}]^{3-}$ and $[Se_2V_2O_{10}]^{2-}$ anions are composed of tetrahedral VO_4 and trigonal pyramidal SeO_3 units that share vertices to form ring structures. In $[Se_2V_2O_{10}]^{2-}$ the VO_4 tetrahedra and SeO_3 trigonal pyramids connect alternately in a centrosymmetric manner to complete a $Se_2V_2O_4$ ring. The conformation of this ring can be loosely classified as twisted-chair (TC), although it lacks a two-fold rotation symmetry and a mirror plane necessary for

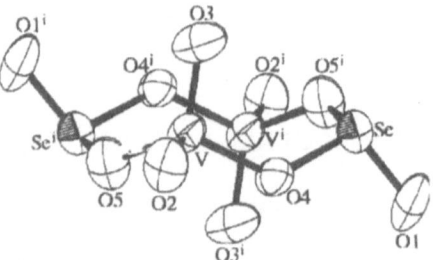

Figure 1. Perspective drawing of the $[Se_2V_2O_{10}]^{2-}$ anion. Displacement ellipsoids are scaled to enclose 50 % probability levels. Atoms labeled with superscripted i are related to those without superscripts by the crystallographic inversion center at $(0, 0, 0)$.

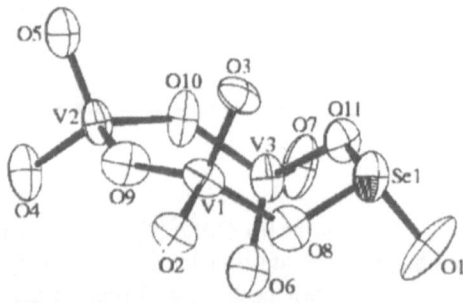

Figure 2. Perspective drawing of the $[SeV_3O_{11}]^{3-}$ anion. Displacement ellipsoids are scaled to enclose 50 % probability levels.

[†] Single crystals of $[\{(C_6H_5)_3P\}_2N]_2[Se_2V_2O_{10}]$ are, at 253 K, monoclinic, space group $P2_1/n$, with $a = 12.2931(3)$ Å, $b = 13.5101(3)$ Å, $c = 20.9793(5)$, $\beta = 106.307(1)°$, and $Z = 4$.

[‡] Single crystals of $[(n\text{-}C_4H_9)_4N]_3[SeV_3O_{11}]\cdot 0.5H_2O$ are, at 93 K, orthorhombic, space group $P2_12_12$, with $a = 22.328(5)$ Å, $b = 44.099(9)$ Å, $c = 12.287(3)$ Å, and $Z = 8$.

an ideal TC.[4] The anion is located on a crystallographic inversion center and thus has a rigorous $\bar{1}$ symmetry. One of the SeO_3 trigonal pyramids in this ring is substituted with a VO_4 tetrahedron in the $[SeV_3O_{11}]^{3-}$ structure. This substitution results in a loss of the center of symmetry and $[SeV_3O_{11}]^{3-}$ has a chiral structure. The less regular structure of the SeV_3O_4 ring in $[SeV_3O_{11}]^{3-}$ can be attributed to the cooperative conformational change associated with the accommodation of a larger VO_4 unit in place of the SeO_3 unit.

Tetrahedral VO_4 units are rare for polyvanadate structures,[3b, 3c, 4c, 5] and the current anions are the first examples of a heteropolyvanadate that is exclusively made up of such units to our knowledge. The $Se_2V_2O_4$ ring found in the $[Se_2V_2O_{10}]^{2-}$ structure has been observed for $VOSeO_3 \cdot H_2O$ [6] and $(VO)_2(SeO_3)_3$ [7] as a building unit of infinite structures. However, the V atoms are in IV oxidation state in these compounds. The compound $CsVSeO_5$ has the same composition as the salts of $[Se_2V_2O_{10}]^{2-}$, [8] but this hydrothermally synthesized compound has an infinite layer structure that has little resemblance to that of $[Se_2V_2O_{10}]^{2-}$. The compound that is most closely related to $[Se_2V_2O_{10}]^{2-}$ from the structural point of view is $S_2V_2O_6(OH_2)_6$, [9] although here the V atoms are in IV oxidation state and octahedral and the ring structure is completed by tetrahedral SO_4 units instead of trigonal pyramidal SeO_3 units. Still $S_2V_2O_6(OH_2)_6$ is molecular and has a discrete structure. It also has a $\bar{1}$ symmetry and its $S_2V_2O_4$ ring has a conformation very similar to that of the $Se_2V_2O_4$ ring in $[Se_2V_2O_{10}]^{2-}$.

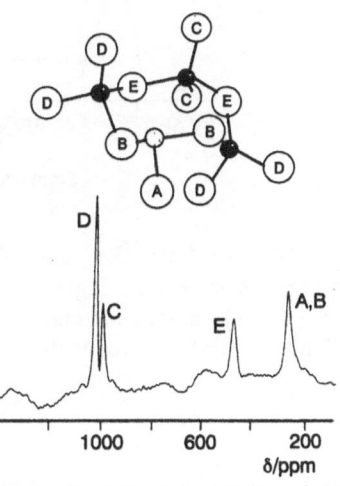

Figure 3. Schematic drawing of the $[SeV_3O_{11}]^{3-}$ anion (top) and its ^{17}O NMR spectrum in acetonitrile (bottom).

Solution Characterization

The $[SeV_3O_{11}]^{3-}$ anion gives two ^{51}V NMR peaks in a 2:1 intensity ratio in solution. These peaks are assignable to the two V atoms that sandwich the SeO_3 unit (–556 ppm) and the unique V atom that is sandwiched between two VO_4 units (–566 ppm). The anion gives a single ^{77}Se NMR peak at 71.9 ppm. Its ^{17}O NMR is given in Figure 3 together with the assignments. [§] The assignments have been made according to the peak intensities and the known correlation between ^{17}O NMR chemical shifts and metal-oxygen bond lengths.[10] Although they are structurally distinct, the oxygens **A** and **B** give a single unresolved peak. Chemical shift difference of the oxygens bound to Se seems to be small. This is in agreement with the general observation that the heavier main group elements yield only a limited range of ^{17}O chemical shifts.[11, 12, 13]

The $[Se_2V_2O_{10}]^{2-}$ anion gives a single peak in both ^{51}V (–547 ppm) and ^{77}Se (66.7 ppm) NMR spectra. Its ^{17}O NMR spectrum is shown in Figure 4. Here again the oxygens bound to Se give an unresolved single peak.

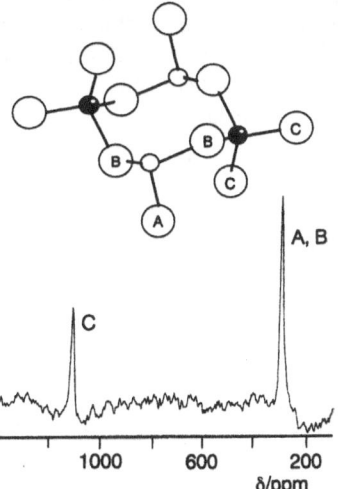

Figure 4. Schematic drawing of the $[Se_2V_2O_{10}]^{2-}$ anion (top) and its ^{17}O NMR spectrum in acetonitrile (bottom).

[§] The ^{77}Se and ^{17}O NMR spectra were referenced externally against SeO_2 saturated in D_2O and neat D_2O, respectively.

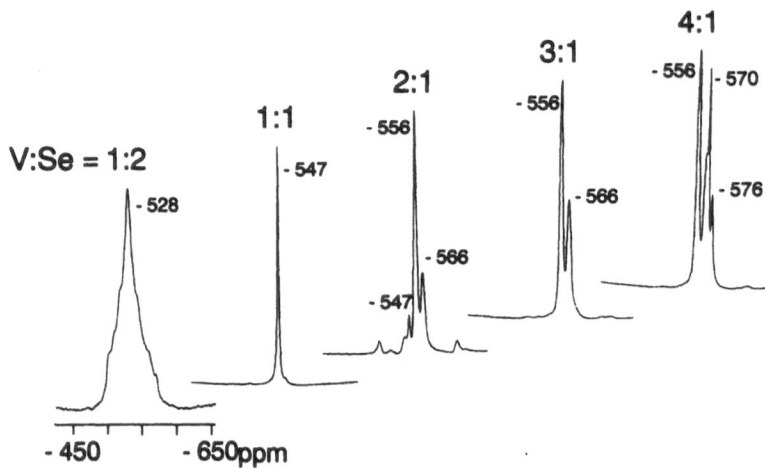

Figure 5 ^{51}V NMR spectra at different V:Se ratios.

All the ^{51}V, ^{77}Se, and ^{17}O NMR spectra described above are consistent with the structures shown in Figures 1 and 2, and strongly suggest that both $[SeV_3O_{11}]^{3-}$ and $[Se_2V_2O_{10}]^{2-}$ anions maintain their respective solid state structures in solution. One more thing worth mentioning is the almost quantitative nature of their formation reaction.

$$2VO_3^- + 2SeO_2 \rightarrow [Se_2V_2O_{10}]^{2-}$$

$$3VO_3^- + SeO_2 \rightarrow [SeV_3O_{11}]^{3-}$$

The chemical equations above indicate that no byproducts would form. The ^{51}V NMR experiments confirmed this (Figure 5). When $[(n\text{-}C_4H_9)_4N]VO_3$ and SeO_2 are mixed in 1:1 molar ratio, the peak for $[Se_2V_2O_{10}]^{2-}$ is the only peak observed in the ^{51}V NMR spectrum of the reaction mixture. When [V]/[Se] = 3, only the $[SeV_3O_{11}]^{3-}$ peak is observed. Moreover, an equimolar mixture of VO_3^- and $[SeV_3O_{11}]^{3-}$ is obtained when [V]/[Se] = 4:1. Basically no peak other than those of VO_3^-, $[Se_2V_2O_{10}]^{2-}$, and $[SeV_3O_{11}]^{3-}$ is observed in the range [V]/[Se] $\geq$ 1.

Something different occurs, however, if the relative amount of Se is increased beyond [V]/[Se] = 1. At [V]/[Se] = 1/3 the dark red mixture gives a single, extremely broad ^{51}V NMR peak at -528 ppm ($\Delta\nu_{1/2}$ = 2200 Hz). The compound in this mixture has so far evaded our effort for further characterization.

THE $[V_4O_{12}]^{4-}$ ANION, OR H_2O SUPPORTED ON MOLECULAR OXIDE

Since its first preparation, $[(n\text{-}C_4H_9)_4N]VO_3$ has been widely used as starting material for polyvanadates that are soluble in aprotic solvents.[3b, 4c, 5c, 14, 15] However, $[(n\text{-}C_4H_9)_4N]$-VO_3 itself has not been fully characterized to date. It is assumed to be a tetrameric compound with a ring structure, $[(n\text{-}C_4H_9)_4N]_4[V_4O_{12}]$, because it readily gives organometallic compounds that have a V_4O_{12} ring. But this assumptions has never been confirmed by an x-ray structure analysis. What makes the things complicated is its ^{51}V NMR spectra. The compound gives two ^{51}V peaks in solution, not a single-peak spectrum expected for a $[V_4O_{12}]^{4-}$ anion that has a ring structure. We found some clue to this mystery in the solid state structure and solution behavior of its tetraethylammonium analogue, $[(C_2H_5)_4N]VO_3$.[**]

Preparation and Crystal Structure

The compound is readily obtained by reacting V_2O_5 with an appropriate amount of aqueous $[(C_2H_5)_4N]OH$.

$$V_2O_5 + 2[(C_2H_5)_4N]OH \rightarrow 2[(C_2H_5)_4N]VO_3 + H_2O$$

This hygroscopic powder can be crystallized from acetonitrile/*m*-xylene mixed solvent. X-ray structural analysis revealed that these crystals are composed of $[(C_2H_5)_4N]^+$ cations, water molecules of crystallization, and discrete $[V_4O_{12}]^{4-}$ anions having the structure shown in Figure 6.[†] Much to our surprise, the water molecules are not just filling the lattice, but they are bound on the surface of the $[V_4O_{12}]^{4-}$ anion through hydrogen bonds (O2···O7 2.792 Å, O4···O7 2.853 Å). The water molecules are bound tightly. The compound does not lose these water molecules even if it is dried under vacuum over P_2O_5 for more than 12 hours. The overall structure of the $[V_4O_{12}(H_2O)_2]^{4-}$ anion closely resembles that of $[\{(\eta^3\text{-}C_4H_7)_2Rh\}_2(V_4O_{12})]^{2-}$,[4c] a vanadate-supported organorhodium compound. In other words, the current compound can be described as two water molecules supported on the surface of $[V_4O_{12}]^{4-}$.

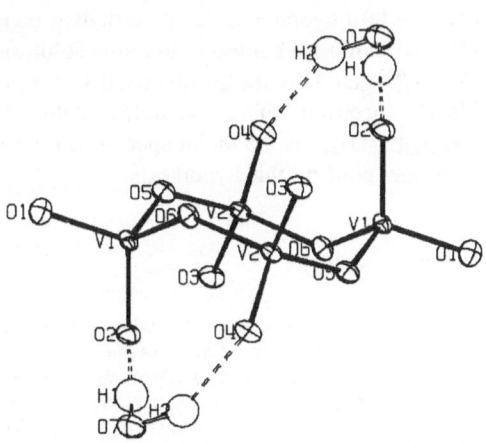

Figure 6 Perspective drawing of the $[V_4O_{12}(H_2O)_2]^{4-}$ anion. Displacement ellipsoids are scaled to enclose 50 % probability levels.

Solution Behavior

As mentioned above, $[(n\text{-}C_4H_9)_4N]VO_3$ gives two ^{51}V NMR peaks in acetonitrile solution: one at −571 ppm and the other much smaller peak at −578 ppm. Relative intensities of these peaks depend on the water content of the sample. The peak at the higher field is small when the sample is dry. Its intensity increases when the samples is wet or water is added to the solution. Similar phenomenon is observed for the tetra*ethyl*ammonium analogue, $[(C_2H_5)_4N]VO_3$. The tetraethylammonium analogue also gives two ^{51}V NMR peaks in acetonitrile; −569 ppm and −574 ppm. Here again the intensity of the higher field peak increases when the sample is wet.

[**] Details will be published elsewhere.

[†] Single crystals of $[(C_2H_5)_4N]_4[V_4O_{12}]\cdot2H_2O$ are, at 293 K, monoclinic, space group $P2_1/n$, with a = 14.0187(3) Å, b = 10.5729(2) Å, c = 16.0212(4), β = 103.836(1)°, and Z = 2.

How can we explain this? Formation of another species by hydrolysis is one possibility. However, this explanation is not consistent with the concentration dependence of the ^{51}V NMR spectra of $[(C_2H_5)_4N]VO_3$. The two peaks observed for the solutions of $[(C_2H_5)_4N]VO_3$ change their relative intensities with concentration. At lower concentrations, the peak at the lower field increases. The higher field peak grows when the concentration is increased. If the higher field peak arouse from hydrolysis, the reverse concentration dependence should have been observed.

Our tentative explanation for this behavior is hydration, not hydrolysis, of the $[V_4O_{12}]^{4-}$ in solution. In other words, formation of vanadate-supported water molecules. The water molecules shown in Figure 6 is bound on the surface of $[V_4O_{12}]^{4-}$ with some strength. It is plausible that some of them remain trapped on the vanadate in solution. The ^{51}V NMR peak at the higher field can be assigned to this hydrated species. The lower field peak then can be assigned to the unhydrated species. The hydrated species would naturally increase when water is added to the solution. It would decrease under more dry conditions. Solvent molecules would compete with the water molecules, and as a result we would observe less hydrates at lower vanadate concentrations. Hydration seems to stabilize the $[V_4O_{12}]^{4-}$ anion in solution. In dilute solutions, where the unhydrated anion is predominant, $[V_4O_{12}]^{4-}$ gradually transforms itself into a pentamer, $[V_5O_{14}]^{3-}$. On the other hand, the ^{51}V NMR spectrum of a saturated solution of $[(C_2H_5)_4N]VO_3$, where the hydrated $[V_4O_{12}(H_2O)_2]^{4-}$ is the major species, does not change after two days. Efforts are underway to further confirm this hypothesis.

REFERENCES

1. (a) A. Müller, H. Reuter, and S. Dillinger, Supramolecular inorganic chemistry: small guests in small and large hosts, *Angew. Chem. Int. Ed. Engl.* 34:2328 (1995). (b) A. Müller, F. Peters, M. T. Pope, and D. Gatteschi, Polyoxometalates: very large clusters—nanoscale magnets, *Chem. Rev.* 98:239 (1998). (c) Y. Zhang, R. C. Haushalter, and J. Zubieta, A new form of decavanadate: hydrothermal synthesis and structural characterization of $Ba_{5.33}(H_3O)_{0.33}[V_{10}O_{30}]$, *Inorg. Chim. Acta* 277:263 (1998) and the references cited therein.

2. (a) A. Kobayashi and Y. Sasaki, Crystal structure of $K_7[NiV_{13}O_{38}] \cdot 18H_2O$, *Chem. Lett.* 1123 (1975). (b) A. Durif and M. T. Averbuch-Pouchot, Structure d'un arséniato vanadate d'ammonium: $(NH_4)_4H_6(As_6V_4O_{30}) \cdot 4H_2O$, *Acta Crystallogr. Sect. B* B35:1441 (1969). (c) R. Kato, A. Kobayashi, and Y. Sasaki, 1:14 Heteropolyvanadate of phosphorus: preparation and structure, *J. Am. Chem. Soc.* 102:6571 (1980). (d) R. Kato, A. Kobayashi, and Y. Sasaki, The heteropolyvanadate of phosphorus. Crystallographic and NMR studies, *Inorg. Chem.* 21:240 (1982). (e) Y. Michiue, H. Ichida, and Y. Sasaki, Structure of hexasodium dihydrogendivanadodiperiodate decahydrate *Acta Crystallogr. Sect. C* C43:175 (1987). (f) T. Ozeki, H. Ichida, and Y. Sasaki, Structure of heptaammonium hydrogendecavanadotetraselenite nonahydrate, *Acta Crystallogr. Sect. C* C43:1662 (1987). (g) H. Ichida, K. Nagai, Y. Sasaki, and M. T. Pope, Heteropolyvanadates containing two and three manganese(IV) ions: unusual structural features of $Mn_2V_{22}O_{64}{}^{10-}$ and $Mn_3V_{12}O_{40}H_3{}^{5-}$, *J. Am. Chem. Soc.* 111:586 (1989). (h) G.-Q. Huang, S.-W. Zhang, Y.-G. Wei, and M.-C. Shao, Synthesis and crystal structure of the heteropolyvanadate anion $[AsV_{12}O_{40}(VO)_2]^{9-}$, *Polyhedron* 12:1483(1993). (i) Q. Chen and J. Zubieta, Polyoxovanadium-organophosphonates — properties and structure of an unusual pentavanadate species with a V=O group directed toward the interior of the molecular cavity, *Angew. Chem. Int. Ed. Engl.* 32:261 (1993). (j) K. Nagai, H. Ichida and Y. Sasaki, The structure of heptapotassium tridecavanadomanganate(IV) octadecahydrate, $K_7[MnV_{13}O_{38}] \cdot 18H_2O$, *Chem. Lett.* 1267 (1986)

3. (a) N. Kawanami, T. Ozeki, and A. Yagasaki, NO^- anion trapped in a molecular oxide bowl, *J. Am. Chem. Soc.* 122:1239 (2000). (b) M. Abe, K. Isobe, K. Kida, and A. Yagasaki, Crystallographic and dynamic NMR evidence for organometallic fragments pivoting on a molecular oxide surface, *Inorg. Chem.* 35:5114 (1996). (c) D. Hou, K. S. Hagen, and C. L. Hill, Pentadecavanadate, $V_{15}O_{42}{}^{9-}$, a new highly condensed fully oxidized isopolyvanadate with kinetic stability in water, *J. Chem. Soc., Chem. Commun.* 426 (1993). (d) D. Hou, K. S. Hagen, and C. L. Hill, Tridecavanadate $[V_{13}O_{34}]^{3-}$, a new

high-potential isopolyvanadate, *J. Am. Chem. Soc.* 114:5864 (1992). (e) Y. Hayashi, Y. Ozawa, and K. Isobe, Site-selective oxygen exchange and substitution of organometallic groups in an amphiphilic quadruple-cubane-type cluster. Synthesis and molecular structure of $[(MCp^*)_4V_6O_{19}]$ (M = Rh, Ir), *Inorg. Chem.* 30:1025 (1991). (f) W. G. Klemperer T. A. Marquart, and O. M. Yaghi, Shape-selective binding of nitriles to the inorganic cavitand, $V_{12}O_{32}^{4-}$ *Mater. Chem. Phys.* 29:97 (1991). (g) V. W. Day, W. G. Klemperer, and O. M. Yaghi, Synthesis and characterization of a soluble oxide inclusion complex, $[CH_3CNC(V_{12}O_{32}^{4-})]$, *J. Am. Chem. Soc.* 111:4518 (1989). (h) M. Abe, H. Akashi, K. Isobe, A. Nakanishi, and A. Yagasaki, Effects of a vanadate-support on the reactivity of the *bis*(methallyl)rhodium complex $[\{(\eta^3\text{-}C_4H_7)_2Rh\}_2(V_4O_{12})]^{2-}$. A comparison with the acetylacetonato complex, *J. Cluster Sci.* 7:103 (1996).

4. (a) J. B. Hendrikson, Molecular geometry. VI. Methyl-substituted cycloalkanes, *J. Am. Chem. Soc.* 89:7043 (1967). (b) E. L. Elile and S. H. Wilen, *Stereochemistry of Organic Compounds*, John Wiley & Sons, New York, p765 (1994). (c) H. Akashi, K. Isobe, Y. Ozawa, and A. Yagasaki, β-Methallylrhodium(III) supported on a vanadium oxide cluster: synthesis, structure and reactivity, *J. Cluster Sci.* 2:291 (1991).

5. (a) J. Fuchs and J. Pickardt, Structure of the "true" metavanadate ion, *Angew. Chem. Int. Ed. Engl.* 15:374 (1976). (b) V. W. Day, W. G. Klemperer, and O. M. Yaghi, A new structure type in polyoxoanion chemistry: synthesis and structure of the $V_5O_{14}^{3-}$ anion, *J. Am. Chem. Soc.* 111:4518 (1989). (c) V. W. Day, W. G. Klemperer, and A. Yagasaki, Synthesis and structure of the new organometallic polyoxovanadates, $\{[(\eta\text{-}C_8H_{12})Ir]_2(V_4O_{12})\}^{2-}$ and $[(\eta\text{-}C_8H_{12})Ir(V_4O_{12})]^{3-}$, *Chem. Lett.* 1267 (1990). (d) Román, A. S. José, A. Luque, and J. M. Gutiérrez-Zorrilla, Observation of a novel cyclic tetrametavanadate anion isolated from aqueous solution, *Inorg. Chem.* 32:775 (1993).

6. G. Huan, J. W. Johnson, A. J. Jacobson, D. P. Goshorn, and J. S. Merola, Hydrothermal synthesis, single-crystal structure, and magnetic properties of $VOSeO_3 \cdot H_2O$, *Chem. Mater.* 3:539 (1991).

7. P. S. Halasyamani and D. O'Hare, A new three-dimensional vanadium selenite, $(VO)_2(SeO_3)_3$, with isolated and edge-shared VO_6 octahedra, *Inorg. Chem.* 36:6409 (1997).

8. Y.-U. Kwon,; K.-S. Lee, and Y. H. Kim, $AVSeO_5$ (A = Rb, Cs) and $AV_3Se_2O_{12}$ (A = K, Rb, Cs, NH$_4$): hydrothermal synthesis in the V_2O_5 SeO$_2$ AOH system and crystal structure of Cs $VSeO_5$, *Inorg. Chem.* 35:1161 (1996).

9. P. F. Théobald, and J. Galy, Structure cristalline de $VOSO_4 \cdot 3H_2O$, *Acta Crystallogr. Sect. B* B29:2732 (1973).

10. W. G. Klemperer, ^{17}O-NMR spectroscopy as a structural probe, *Angew. Chem. Int. Ed. Engl.* 17:246 (1978).

11. J.-P. Kintzinger, *NMR Basic Principles & Progress* 17:1 (1981).

12. W. G. Klemperer, Application of ^{17}O NMR spectroscopy to structural problems, in: *The Multinuclear Approach to NMR Spectroscopy*; J. B. Lambert and F. G. Riddell, eds. D. Reidel Publishing Company, New York, pp 245-260 (1983).

13. W. McFarlane, and H. C. E. McFarlane, Oxygen, in: *Multinuclear NMR*, J. Mason, ed., Plenum, New York, pp 403-416 (1987).

14. K. Kida, Synthesis, structure, and dynamic behavior of polyoxovanadate $V_4O_{12}^{4-}$ supported organorhodium(I) and -iridium(I) complexes, Ph. D. Thesis, Kwansei Gakuin University (1996).

15. D. Attanasio, F. Bachechi, and L. Suber, Polyanion-supported, atomically dispersed cyclooctadieneruthenium(II) complexes, *J. Chem. Soc., Dalton Trans.* 2373 (1993).

POLYOXOMETALATES: FROM MAGNETIC MODELS TO MULTIFUNCTIONAL MATERIALS

Juan M. Clemente-Juan, Miguel Clemente-León, Eugenio Coronado, Alicia Forment, Alejandro Gaita, Carlos J. Gómez-García, Eugenia Martínez-Ferrero

Instituto de Ciencia Molecular.Univ. of Valencia. Dr. Moliner 50, E-46100 Burjasot (Spain) E-mail: eugenio.coronado@uv.es

INTRODUCTION

In the past few years polyoxometalate (POM) molecular science has grown vastly and major developments have occurred in chemistry, physics and applications. Thus, in chemistry a strong effort has been devoted to rationalize and control the formation of these high-nuclearity metal-oxygen cluster anions in terms of self-assembly processes, as well as to develop the supramolecular chemistry of these species. Efforts in this direction nucleated with the development of a coordination chemistry of POMs and with the synthesis of organic and organometallic derivatives. The understanding of the electronic structure of these compounds has also been an active focus of research in molecular physics. In this context, POMs have shown to be ideal chemical objects to study metal oxide conductivity, intramolecular and intermolecular electron transfer in mixed-valence systems, magnetic interactions in large clusters, and theory of multinuclear NMR chemical shifts and electron spin couplings. On the applications front, POMs have proven to be enormously valuable industrial catalysts (especially for "green" chemistry), potent antitumor and antiviral agents, and, more recently, useful inorganic components for novel molecular materials with magnetic, photophysical and electrical properties.[1]

During the past ten years our group has been involved in two aspects of this development namely the study of the magnetic properties of POMs, and the use of these molecular anions as inorganic building blocks for the construction of organic/inorganic molecular materials with magnetic or/and electrical properties. In molecular magnetism we have shown that these compounds provide ideal examples of magnetic and mixed-valence clusters of increasing nuclearity, wherein magnetic exchange interactions as well as electron transfer processes can be studied at the molecular level.[2] In connection with the molecular materials, we have shown that POMs can be combined with organic π-electron donors of the tetrathiafulvalene (TTF) type to form radical salts with unprecedented organic packings and coexistence of localized magnetic moments with itinerant electrons.[3] Furthermore, we

Polyoxometalate Chemistry for Nano-Composite Design
Edited by Yamase and Pope, Kluwer Academic/Plenum Publishers, 2002

have shown that they can also be incorporated into Langmuir-Blodgett films to produce organic-inorganic hybrids containing well-organized monolayers of polyoxometalates.[4]

In this paper we illustrate with novel examples the considerable impact of POMs in molecular magnetism and molecular materials. In the first part we show that ideal examples of large magnetic clusters can be designed, wherein magnetic exchange interactions can be chemically tuned and physically studied at the molecular level. In the second part we report the synthesis of highly conductive molecular materials based upon polyoxometalates and perylene. In the last section we show the possibilities offered by POMs to obtain multifunctional materials.

MAGNETIC CLUSTERS

Polyoxometalates provide excellent examples of magnetic clusters. Recent reviews accounting for the state-of-the art in this area can be found in references 2 and 5. The ability of tungstates, and to a less extent of molybdates, for acting as ligands toward 3d-transition metal ions leads to the encapsulation by the polyoxometalate framework of a variety of magnetic clusters possessing different spins and showing either ferromagnetic or antiferromagnetic exchange couplings. The bulky nonmagnetic polyoxometalate framework guarantees an effective magnetic isolation of the cluster, imposing at the same time its geometry. Furthermore, its chemistry allows the assembly of stable anion fragments into larger clusters. A chemical control of the magnetic nuclearity is therefore possible. The above characteristics make these complexes ideal candidates for modeling the magnetic exchange interactions in clusters of increasing nuclearities and definite topologies. In this section we will present two examples that illustrate the possibilities offered by polyoxometalate chemistry to produce well-insulated magnetic clusters with predictable magnetic properties for which a detailed information on the nature of the magnetic exchange interactions can be extracted.

Compounds

The first polyoxometalate salt, $Na_{12}[Co_3W(H_2O)_2(CoW_9O_{34})_2]\cdot46H_2O$ contains a pentanuclear Co_5 magnetic cluster (Figure 1).[6] This polyanion consists of two ligands $(CoW_9O_{34})^{12-}$ encapsulating a rhomb-like Co_3WO_{16} unit of edge-sharing MO_6 octahedra where the W occupies one of the two short diagonal sites. Each tetrahedral CoO_4 unit of the $(CoW_9O_{34})^{12-}$ core is connected through oxo groups to two of the CoO_6 octahedra in the central Co_3 trimer.

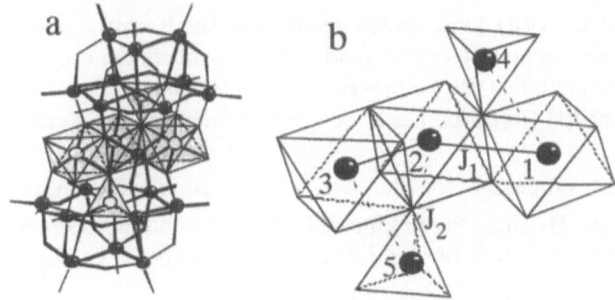

Figure 1. (a) Structure of the $[Co_3W(D_2O)_2(CoW_9O_{34})_2]^{12-}$ polyanion and (b) polyhedral representation of the Co_5 cluster showing the different exchange pathways.

The salt $K_3Na_{11}\{[Co_7(H_2O)_2(OH)_2P_2W_{25}O_{94}]Co(H_2O)_4\}\cdot15H_2O$ contains an heptanuclear Co_7 cluster (Figure 2). Its structure is unprecedented and is formed by two reconstituted Keggin units $\{PW_9Co_3\}$. These two units are linked together by a central unit formed by a tetrahedral Co site and seven edge-sharing WO_6 octahedra. The magnetic cluster can be viewed as formed by two triangular Co_3O_{13} units of edge-sharing CoO_6 octahedra bridged by a central tetrahedral CoO_4 site sharing vertices with these two units. Note that the related Co(II) POMs $[Co_2(H_2O)(W_{11}O_{39})]^{8-}$,[7] $[Co_4(H_2O)_2(PW_9O_{34})_2]^{10-}$,[8] and $[Co_9(OH)_3(H_2O)_6(HPO_4)_2(PW_9O_{34})_3]^{16-}$,[9] with magnetic nuclearities 2, 4 and 9, are also known.

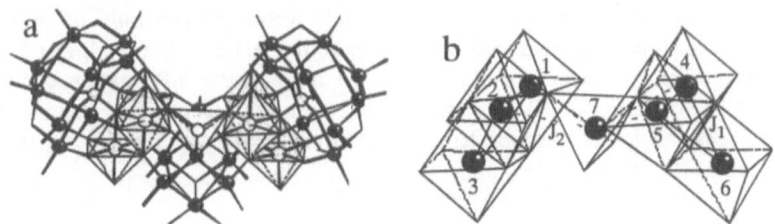

Figure 2. (a) Structure of the $[Co_7(H_2O)_2(OH)_2P_2W_{25}O_{94}]^{16-}$ polyanion and (b) polyhedral representation of the Co_7 cluster showing the different exchange pathways.

Magnetic Properties

In this type of polyoxometalates the exchange interaction between edge-sharing CoO_6 octahedra has been proved to be ferromagnetic. In contrast, the study of the $[Co_2(H_2O)(W_{11}O_{39})]^{8-}$ anion indicates an antiferromagnetic exchange between the octahedral CoO_6 sites and the tetrahedral CoO_4 one. Taking into account these previous results, we should expect in the two magnetic clusters reported here the coexistence of ferromagnetic exchange interactions between the edge-sharing octahedral Co(II) ions, with antiferromagnetic interactions between tetrahedral and octahedral Co(II) pairs. In order to understand the magnetic properties of these clusters the anisotropy of the exchange interaction must also be considered. The anisotropy arises from the $^4T_1$ ground state of the high-spin octahedral Co(II) ions, which splits into six Kramers doublets by spin-orbit coupling and the low symmetry-crystal field. At low temperature only the lowest Kramers doublet is populated and this can be described by an anisotropic S = 1/2 spin state. On the contrary, the tetrahedral Co(II) ion has a $^4A_2$ ground state which is described as a spin only S = 3/2.

The magnetic properties for the Co_5 cluster are in full agreement with these expectations (Figure 3). Thus, the product χT shows a continuous decrease upon cooling which indicates the presence of dominant antiferromagnetic $Co(O_h)$-$Co(T_d)$ exchange interactions.

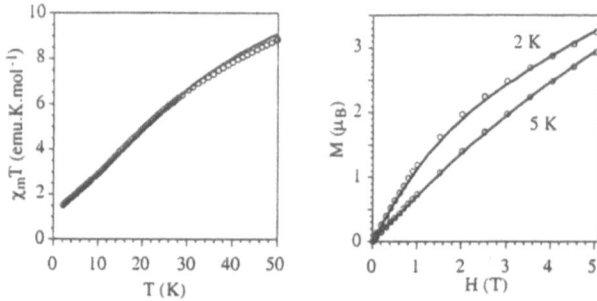

Figure 3. $\chi_m T$ vs T at low temperature and magnetization curves for the Co_5 cluster. Solid lines are the best fit to the model (see text).

However, the information provided by the magnetic data is insufficient for the determination of the magnetic parameters since this technique can only provide reliable information on the sign of the two types of exchange interactions, but is almost insensitive to the presence of an exchange anisotropy. To get more direct information on this point Inelastic Neutron Scattering (INS) spectroscopy has been used.[10] The INS spectra shows up to three cold peaks which correspond to magnetic transitions from the ground state to the first excited states (Figure 4).

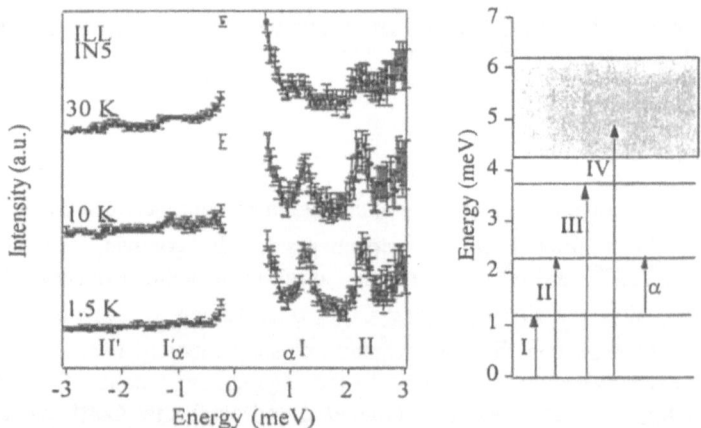

Figure 4. (a) INS spectra of the deuterated salt of the $[Co_3W(D_2O)_2(CoW_9O_{34})_2]^{12-}$ polyanion mesured on the IN5 spectrometer with an incident neutron scattering of 4.1 Å at 1.5, 10 and 30 K. (b) Energy level diagram of the Co_5 cluster derived from the INS experiments.

The energy of these peaks directly gives the energy splitting pattern due to the exchange interactions. It can be satisfactorily reproduced by assuming an anisotropic exchange Hamiltonian (1) and making use of the parameter information obtained from the more simple related clusters (Co_2 and Co_4).

$$H = -2 \sum_{i=x,y,z} J_{1i} \left(S_{1i}S_{2i} + S_{2i}S_{3i} \right) + J_{2i} \left(S_{1i}S_{4i} + S_{2i}S_{4i} + S_{2i}S_{5i} + S_{3i}S_{5i} \right) \quad (1)$$

The best fit gives the following set of parameters: $J_{x1} = 5.60$ cm^{-1}, $J_{y1} = 3.44$ cm^{-1}, $J_{z1} = 12.08$ cm^{-1}, $J_{x2} = -9.92$ cm^{-1}, $J_{y2} = -4.24$ cm^{-1}, $J_{z2} = -11.52$ cm^{-1}. This result demonstrates that the two kinds of exchange interactions have different sign and that they are anisotropic.

The magnetic properties of Co_7 are shown in Figure 5. In this case the presence of a maximum in χT at low temperatures indicates that the ferromagnetic interactions are the dominant ones. As expected, the magnetic coupling within the two Co_3O_{13} units is ferromagnetic. As the exchange interaction between tetrahedral Co and octahedral Co is expected to be antiferromagnetic, the two ferromagnetic Co_3O_{13} units must have parallel magnetic moments. Accordingly with the exchange pathway topology of the cluster, the following Hamiltonian (2) has been proposed:

$$H = -2 \sum_{i=x,y,z} J_{1i} \left(S_{1i}S_{2i} + S_{1i}S_{3i} + S_{2i}S_{3i} + S_{4i}S_{5i} + S_{4i}S_{6i} + S_{5i}S_{6i} \right)$$
$$+ J_{2i} \left(S_{1i}S_{7i} + S_{2i}S_{7i} + S_{4i}S_{7i} + S_{5i}S_{7i} \right) \quad (2)$$

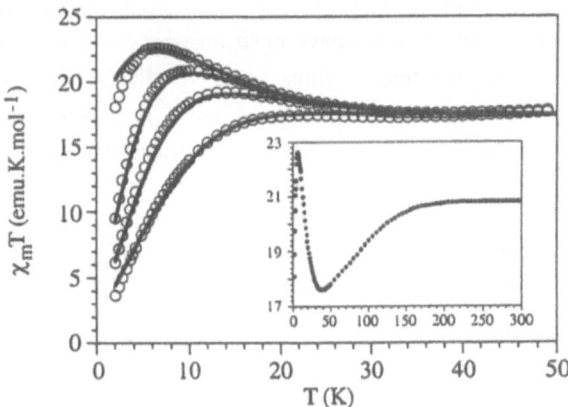

Figure 5. $\chi_m T$ vs T at low temperature for the Co_7 cluster at different magnetic field (0.1, 1.0, 2.0 and 4.0 T from top to bottom). Solid lines are the best fit to the model (see text). Inset: full range of tempearure for the product $\chi_m T$ at 0.1 T.

From a simultaneous fit of the magnetic susceptibility curves at different fields we have obtained the following set of parameters: $J_{z1} = 9.54$ cm^{-1}, $J_{z2} = -6.09$ cm^{-1},

$J_{xyi}/J_{zi} = 0.67$, $g(T_d) = 2.00$, $g_\parallel = 5.25$ and $g_\perp = 4.69$. These parameters are qualitatively in agreement with the parameters obtained for the other related clusters.

To conclude this part it is important to emphasize that this kind of study has been possible thanks to the great versatility of polyoxometalate chemistry in providing examples of largely insulated magnetic clusters of increasing nuclearities, definite topologies and high symmetries, which can be easily deuterated in big amounts to perform INS studies. The INS study of these clusters in combination with other magnetic techniques should provide a unique opportunity to progress in the understanding of the magnetic exchange interaction phenomenon in polynuclear metal complexes, as well as in its parametrization using effective Hamiltonians. Finally, the unique possibility provided by polyoxometalate chemistry to obtain magnetic clusters of various nuclearities having similar nearest neighbor geometries, has enabled to use the small magnetic clusters as fragments to understand the properties of higher nuclearity magnetic clusters.

CONDUCTIVE MOLECULAR MATERIALS

In view of their unique molecular characteristics, polyoxometalate complexes have been extensively used to obtain organic/inorganic solids formed by stacks of partially oxidized π-electron organic donor molecules.[3] Owing to their specific shapes and large sizes and charges, these molecular metal-oxides have shown to stabilize many unprecedented structural packings in the organic component. In most of them the electrons are strongly localized and show semiconducting behaviors, but in a few cases large electronic conductivities and metallic behaviors are also observed.[11] So far however only the TTF-type donors have been used to construct these hybrid molecular materials. Here we show that polyoxometalates can also be combined with the perylene molecule (Figure 6) to afford conducting molecular solids.

In fact, perylene has been used to produce many cation-radical salts with simple inorganic monoanions (Br$^-$,[12] I$_3^-$,[13] ClO$_4^-$,[14] or PF$_6^-$ and AsF$_6^-$),[15] and charge-transfer salts with organic acceptors such as TCNQ[16] and per-fluoroanil.[17] Electrical conductivities at room temperature up to 1400 S.cm^{-1} have been measured in some cases. On the other hand, perylene salts with magnetic anions such as FeX$_4^-$ (X$^-$ = Cl, Br) [18] and M(mnt)$^-$ anions (M = Au, Co, Cu, Fe, Ni, Pd, Pt; mnt = maleonitriledithiolate)[19,20] have also been reported. The last series constitutes a unique example of coexistence of conducting organic chains with magnetic inorganic chains.

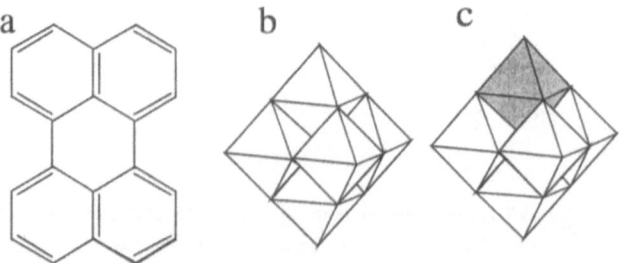

Figure 6. (a) The perylene molecule, (b) the Lindqvist anions [M$_6$O$_{19}$]$^{2-}$ (M = Mo and W) and (c) [VW$_5$O$_{19}$]$^{3-}$.

A general feature of the perylene salts is its one-dimensional electronic character due to the absence of atoms on the perylene molecule that could allow in-plane interactions. Thus, the transport properties of the perylene-containing solids are mostly determined by the packing pattern of the organic molecules and their degree of partial oxidation. While the former depends on the geometry and volume of the anion, the charge distribution within the organic stacks varies with the number of charge-compensating anions and its charge. Taking advantage of this fact, we have combined polyanions with the same size and shape but with two different charges with the aim of studying the differences induced by the variation of this charge on the transport and magnetic properties of the perylene salt. Three polyoxometalates of the Lindquist type, namely $[Mo_6O_{19}]^{2-}$, $[W_6O_{19}]^{2-}$, and $[VW_5O_{19}]^{3-}$ (see Figure 6) have been used to obtain single crystals of the compounds $per_5[Mo_6O_{19}]$, $per_5[W_6O_{19}]$, and $per_5[VW_5O_{19}]$. The compounds are semiconductors with room temperature conductivities as high as 45 S.cm^{-1}. These values represent the highest conductivities so far reported in radical-cation salts containing polyoxometalates.

The overall structure of these compounds consists of slightly interpenetrated layers of polyoxometalates and perylene molecules in the ac plane alternating in the b direction (Figure 7). The organic layer has 3 types of crystallographically independent perylene molecules, named A, B, and C. These layers are formed by chains of parallel A and B type molecules running along the c axis. These chains are separated by perpendicular C type molecules (Figure 7). In the chains the two kinds of dimers (AA and BB) are alternating and display a criss-cross type arrangement. As far as we know, this is a new organic packing mode. From the electronic point of view the different charge of the polyanions has shown to have important consequences on the electronic band filling of the perylene sublattice and hence on the transport and magnetic properties. Thus, for the salts containing the dianions $[Mo_6O_{19}]^{2-}$ and $[W_6O_{19}]^{2-}$, the charge carriers responsible for conductivity have been found to be positive, while for that containing the trianion $[VW_5O_{19}]^{3-}$ they have been found to be negative. At the same time, the magnetic properties have changed from diamagnetic in the former salts to paramagnetic in the latter. Other polyoxometalates having larger nuclearities, higher charges and/or more interesting electronic properties can now be used to prepare novel hybrid salts based upon the perylene molecule.

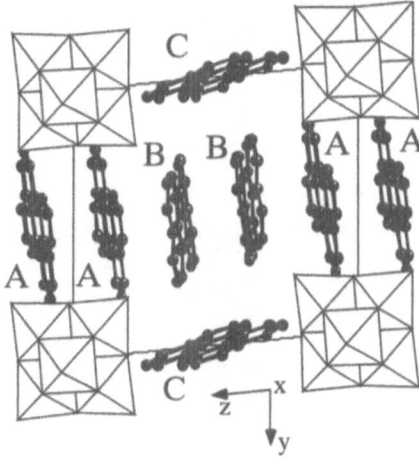

Figure 7. Structure of the radical salts $(per)_5[M_6O_{19}]$ (M = Mo and W) and $(per)_5[VW_5O_{19}]$, showing the perylene chains along the c axis.

MULTIFUNCTIONAL MOLECULAR MATERIALS

The search for bi-functionality in molecular materials constitutes a contemporary challenge in materials science. Using the two-network approach described in the previous section it has been possible to construct hybrid crystalline solids formed by a partially oxidized π-electron donor network that support electronic conductivity or even superconductivity, and transition metal complexes containing magnetic moments. Some relevant examples worth to mention are the paramagnetic conductors [21] and superconductors[22] obtained by combining the discrete paramagnetic anions $[M(ox)_3]^{3-}$ (M = Fe^{III} and Cr^{III}; ox = oxalate) and $[MX_4]^{n-}$ (M = Cu^{II}, Fe^{III} and X = Cl⁻, Br⁻) with the organic donor bis(ethylenedithio)tetrathiafulvalene (BEDT-TTF) or its selenium derivative bis(ethylenedithio)tetraselenafulvalene (BETS), and the ferromagnetic conductor formed by the polymeric magnetic anion $[Mn^{II}Cr^{III}(ox)_3]^-$ and BEDT-TTF,[23] which represents the first example wherein two useful solid-state properties (metal-conductivity and ferromagnetism) coexist in a single molecular crystal.

Since polyoxometalates are electron acceptors and can incorporate one or more paramagnetic centers at specific sites of the polyoxoanion structure, they have enabled the formation of materials with coexistence of delocalized electrons in both sublattices,[24] or with coexistence of localized magnetic moments and delocalized electrons.[25] Here we report a different approach to prepare hybrid materials formed by two functional molecular networks. The approach is based on the use of the Langmuir-Blodgett technique. In the past we have shown that this technique allows to obtain well-oganized LB films wherein monolayers of POM's are alternating with bi-layers of the organic surfactant dimethyldioctadecyl-ammonium cation (DODA⁺). This method is general and has been used to insert in between the organic layers a variety of POMs of increasing nuclearities.[26] Thus, monolayers of magnetic POMs of the type $[Co_4(H_2O)_2(PW_9O_{34})_2]^{10-}$ and $[Co_4(H_2O)_2(P_2W_{15}O_{62})_2]^{16-}$ that encapsulate a Co_4O_{16} ferromagnetic cluster,[27] (Figure 8), as well as the giant POM $[M_9(OH)_3(H_2O)_6(HPO_4)_2(PW_9O_{34})_3]^{16-}$, that encapsulates a nonanuclear M_9O_{36} magnetic cluster (M = Co and Ni), have been constructed.

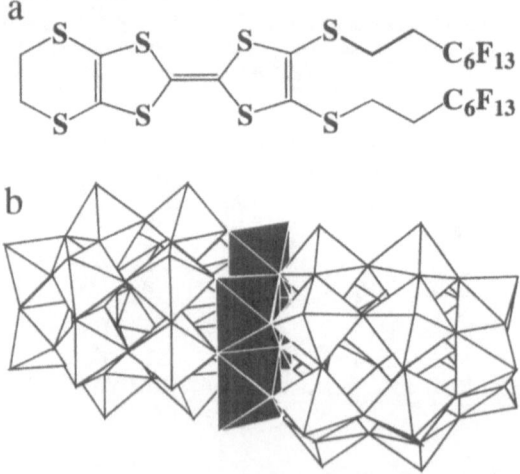

Figure 8. Molecules incorporated in the LB films: **(a)** The SF-EDT molecule and **(b)** the POM $[Co_4(H_2O)_2(P_2W_{15}O_{62})_2]^{16-}$ with the Co_4O_{16} cluster (in black).

So far however, the electronic properties incorporated to the film are those coming from the POMs, as the organic amphiphilic cation only plays a structural role. A step forward is to incorporate as organic component a TTF derivative which can introduce into the LB film an electron delocalization. With this aim, we have used the amphiphilic organic molecule SF-EDT (Figure 8), as it has already shown a good ability to form stable monolayers in the air-water interface that can be transferred to a solid substrate to produce well-defined lamellar structures. As POM we have used the polyanion $[Co_4(H_2O)_2(P_2W_{15}O_{56})_2]^{16-}$ that contains the ferromagnetic cluster Co_4O_{16} (Figure 8). Using the LB technique, an alternating LB film has been constructed which is formed by successive monolayers of POM/DODA/SF-EDT (Figure 9).

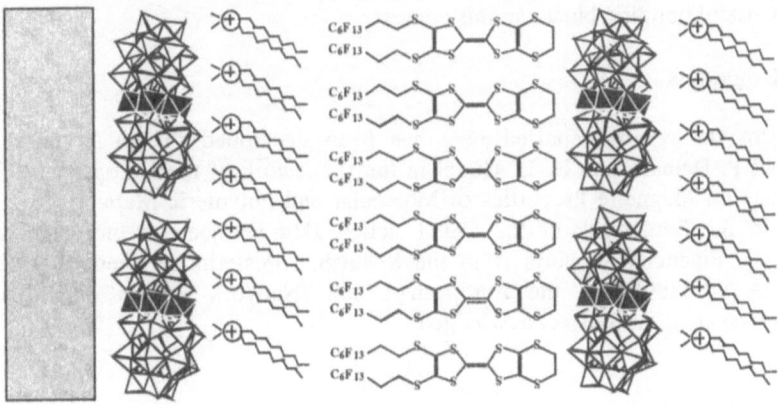

Figure 9. The lamellar structure of the LB film prepared with SF-EDT and the $[Co_4(H_2O)_2(P_2W_{15}O_{56})_2]^{16-}$ polyoxometalate.

The lamellar structure so obtained has been demonstrated by infrared (IR) linear dichroism and X-ray diffraction experiments. In order to introduce electron delocalization into the SF-EDT layers, the LB film has been oxidized with iodine vapors. A charge transfer band centered at 2850 cm^{-1} has been observed in the IR spectrum of the oxidized film, which is a clear evidence of the presence of a mixed-valence state in the SF-EDT layer. This feature demonstrates the presence of an electron delocalization within the organic monolayer at the local level. On the other hand, EPR spectroscopy on films deposited on quartz substrates has shown the presence of the ferromagnetic Co_4 cluster. This hybrid film represents the first example of organized LB films formed by alternating monolayers of polyoxometalate clusters that possess high magnetic moments, and SF-EDT π-electron donor molecules that support an electron delocalization. It demonstrates that the LB technique can be a useful approach to create multilayered structures comprising two different types of functional molecular units, which is the critical step towards the construction of multifunctional molecular materials exhibiting magnetic, electrical and optical properties.

CONCLUSIONS

In this article we have presented some recent achievements of the polyoxometalates in the fields of molecular magnetism and functional molecular materials. In the context of the molecular magnetism we have shown that POMs provide ideal examples of magnetic clusters with coexisting ferro and antiferromagnetic exchange interactions which can be investigated using a variety of magnetic techniques, including the Inelastic Neutron Scattering spectroscopy, to get a thorough characterization of the magnetic levels in these large clusters, and to test the validity of the spin hamiltonians commonly used in magnetism.

In the context of the molecular materials we have shown that these cluster anions can be combined with perylene organic donors to form crystalline radical salts with unprecedented organic packings and large electrical conductivities. On the other hand, we have shown that nanostructured hybrid films formed by monolayers of ferromagnetic polyoxometalates alternating with of mixed-valence TTF-type monolayers can be constructed by using the Langmuir-Blodgett technique. More and more complex systems are expected to be obtained in the future starting from molecular units. POMs have shown to be very useful building blocks in this context.

Acknowledgments

Part of the results reported here has been developed in collaboration with C. Mingotaud, P. Delhaes and H. U. Güdel in the framework of the European COST action 518 (Project on Magnetic Properties of Molecular and Polymeric Materials) and with M. Almeida in the framework of the COST action D14 (Project on Inorganic Molecular Conductors). Financial supports from the Spanish Ministerio de Ciencia y Tecnología (Grant MAT98-0880) and the European Union (Network Project on Molecules as Nanomagnets) are gratefully acknowledged.

REFERENCES

1. For a review presenting the recent developments occurred in this field see for example the special issue devoted to polyoxometalates in *Chem. Rev.* 98 (1998).
2. J. M. Clemente-Juan and E. Coronado Magnetic Clusters from Polyoxometalate Complexes *Coord. Chem. Rev.* 193-195:361 (1999).
3. E. Coronado and C. J. Gómez-García Polyoxometalate-Based Molecular Materials *Chem. Rev.* 98:273 (1998).
4. E. Coronado and C. Mingotaud ybrid Organic/Inorganic Langmuir-Blodgett-Films - A Supramolecular Approach to Ultrathin Magnetic-Films *Adv. Mater.* 11:869 (1999).
5. A. Müller, F. Peters, M. T. Pope and D. Gatteschi Polyoxometalates - Very Large Clusters - Nanoscale Magnets *Chem. Rev.* 98:239 (1998).
6. C. M. Tourné, G. F. Tourné and F. Zonnevijlle Chiral Polytungstometalates $[WM_3(H_2O)_2(XW_9O_{34})_2]^{12-}$ (X = M = Zn or Co^{ii}) and Their M-Substituted Derivatives - Syntheses, Chemical, Structural and Spectroscopic Study of Some D,L Sodium and Potassium-Salts *J. Chem. Soc. Dalton Trans.* 143 (1991).
7. (a) L. C. W. Baker and T. P. McCutcheon, Heteropoly Salts Containing Cobalt and Hexavalent Tungsten in the Anion, *J. Am. Chem. Soc.* 78:4503 (1956); (b) L. C. W. Baker, S. W. Baker, K. Eriks, M. T. Pope, M. Shibata, O. W. Rollins, J. H. Fang and L. L. Koh, A New General Structural Category of Heteropolyelectrolytes. Unusual Magnetic and Thermal Contraction Phenomena, *J. Am. Chem. Soc.* 88:2329 (1966).
8. (a) N. Casañ-Pastor, J. Bas, E. Coronado, G. Pourroy and L. C. W. Baker 1st Ferromagnetic Interaction in a Heteropoly Complex - $[Co^{ii}_4O_{14}(H_2O)_2(PW_9O_{27})_2]^{10-}$ - Experiment and Theory for Intramolecular Anisotropic Exchange Involving the 4 Co(II) Atoms *J. Am. Chem. Soc.* 114:10380 (1992); (b) C. J. Gómez-García, E. Coronado and J. J. Borrás-Almenar Magnetic Characterization of Tetranuclear Copper(II) and Cobalt(II) Exchange-Coupled Clusters Encapsulated in Heteropolyoxotungstate Complexes - Study of the Nature of the Ground-States *Inorg. Chem.* 31:1667 (1992); (c) H. Andres, J. M. Clemente-Juan, M. Aebersold, H. U. Güdel, E. Coronado, H. Büttner, G. Kearly, J. Melero and R. Burriel Magnetic Excitations in Polyoxometalate Clusters Observed by Inelastic Neutron-Scattering - Evidence for Anisotropic Ferromagnetic Exchange Interactions in the Tetrameric Cobalt(II) Cluster $[Co-_4(H_2O)_2(PW_9O_{34})_2]^{10-}$ - Comparison with the Magnetic and Specific-Heat Properties *J. Am. Chem. Soc.* 121:10028 (1999).

9. (a) T. J. R. Weakley, The Identification and X-Ray Structure of the Diphosphatotris(nonatungstophosphonato)nonacobaltate(II) Heteropolyanion, *J. Chem. Soc. Chem. Commun.* 1406 (1984); (b) C. J. Gómez-García, E. Coronado and J. R. Galán-Mascarós *Adv. Mater.* 6:221 (1994).

10. H. Andres, J. M. Clemente-Juan, R. Basler, M. Aebersold, H. U. Güdel, J. J. Borrás-Almenar, A. Gaita, E. Coronado, H. Büttner and S. Janssen, Magnetic Polyoxometalates: Anisotropic Antiferro- and Ferromagnetic Exchange Interactions in the Pentameric Cobalt(II) Cluster $[Co_3W(D_2O)_2(CoW_9O_{34})_2]^{12-}$. A Magnetic and Inelastic Neutron Scattering Study, *Inorg. Chem.* 40:1943 (2001).

11. (a) S. Triki, L. Ouahab, D. Grandjean, R. Canet, C. Garrigou-Lagrange and P. Delhaes New Conducting Radical-Cation Salt Based on BEDT-TTF and Substituted Polyoxometalate Anions - (BEDT-TTF)$_5$(VW$_5$O$_{19}$) 6H$_2$O *Synth. Met.* 55-57:2028 (1993); (b) E. Coronado, J. R. Galán-Mascarós, C. Gimenez-Saiz, C. J. Gómez-García and V. N. Laukhin The First Radical Salt of the Polyoxometalate Cluster (P$_2$W$_{18}$O$_{62}$)(6-Theta) with bis(Ethylenedithio)Tetrathiafulvalene (Et) - Et$_{11}$(P$_2$W$_{18}$O$_{62}$)·3H$_2$O *Adv. Mater.* 8:801 (1996).

12. H. Akamatu, H. Inokuchi and Y. Matsunaga, Organic Semiconductors with High Conductivity. I. Complexes between Polycyclic Aromatic Hydrocarbons and Halogens, *Bull. Chem. Soc. Jpn.* 29:213 (1956).

13. H. C. I. Kao, M. Jones and M. M. Labes, Metallic Conductivity in a Perilene·4I$_2$ Complex, *J. Chem. Soc. Chem. Commun.* 329 (1979).

14. D. Schweitzer, I. Henning, K. Bender, H. Endres and H. J. Keller, Organic Metals form Simple Aromatic Hydrocarbons: Perylene Radical Salts, *Mol. Crys. Liq. Cryst.* 120:213 (1985).

15. (a) H. J. Keller, D. Nöthe, H. Pritzkov, D. Wehe, M. Werner, P. Koch and D. Schweitzer, Electrochemically Generated Peryleniumylhexafluorophosphate and Hexafluoroarsenate, *Mol. Cryst. Liq. Cryst.* 62:181 (1980); (b) W. Brütting and W. Riess Peierls Instability and Charge-Density-Wave Transport in Fluoranthene and Perylene Radical-Cation Salts *Acta Phys. Pol. A* 87:785 (1995).

16. I. J. Tickle and C. K. Prout, Molecular Complexes. Part XVII. Crystal and Molecular Structure of Perylene-7,7,8,8-Tetracyanoquinodimethane Molecular Complex, *J. Chem. Soc. Perkin Trans.* 2:720 (1973).

17. H. Kokado, K. Hasewaga and W. G. Schneider, Photoconduction and Semiconduction in Monocrystals of the Charge-transfer Complex, Perilene-fluoranil, *Can. J. Chem.* 42:1084 (1964).

18. J. A. Ayllón, I. C. Santos, R. T. Henriques, M. Almeida, E. B. Lopes, J. Morgado, L. Alcácer, L. F. Veiros and M. T. Duarte Perylene Salts with Tetrahalogenoferrate(III) Anions - Synthesis, Crystal-Structure of [(C$_{20}$H$_{12}$)$_3$](FeCl$_4$) and Characterization *J. Chem. Soc. Dalton Trans.* 3543 (1995).

19. (a) V. Gama, M. Almeida, R. T. Henriques, I. C. Santos, A. Domingos, S. Ravy and J. P. Pouget Low-Dimensional Molecular Conductors (per)$_2$M(MNT)$_2$, M = Cu and Ni - Low- Conductivity and High-Conductivity Phases *J. Phys. Chem.* 95:4263 (1991); (b) V. Gama, R. T. Henriques, G. Bonfait, L. C. Pereira, J. C. Waerenborgh, I. C. Santos, M. T. Duarte, J. M. P. Cabral and M. Almeida Low-Dimensional Molecular-Metals (per)$_2$M(MNT)$_2$ (M = Fe and Co) *Inorg. Chem.* 31:2598 (1992).

20. M. Almeida and R. T. Henriques in *Handbook of Organic Conductive Molecules and Polymers*, vol. 1, H. S. Nalwa (ed.), John Wiley, Chinchester, p. 87 (1997).

21. P. Day, M. Kurmoo, T. Mallah, I. R. Marsden, R. H. Friend, F. L. Pratt, W. Hayes, D. Chasseau, J. Gaultier, G. Bravic and L. Ducasse Structure and Properties of Tris(bis(Ethylenedithio)Tetrathiafulvalenium)Tetrachlorocopper(II) Hydrate, (BEDT-TTF)$_3$CuCl$_4$.H$_2$O - 1st Evidence for Coexistence of Localized and Conduction Electrons in a Metallic Charge-Transfer Salt *J.Am. Chem. Soc.* 114:10722 (1992).

22. (a) M. Kurmoo, A. W. Graham, P. Day, S. J. Coles, M. B. Hursthouse, J. L. Caulfield, J. Singleton, F. L Pratt, W. Hayes, L. Ducasse and P. Guionneau Superconducting and Semiconducting Magnetic Charge-Transfer Salts - (BEDT- TTF)$_4$AFe(C$_2$O$_4$)$_3$·C$_6$H$_5$CN (A=H$_2$O, K, NH$_4$) *J. Am. Chem. Soc.* 117:12209 (1995); (b) H. Kobayashi, H. Tomita, T. Naito, A. Kobayashi, F. Sakai, T. Watanabe and P. Cassoux New Bets Conductors with Magnetic Anions (Bets-Equivalent-to-bis(Ethylenedithio)Tetraselenafulvalene) *J. Am. Chem. Soc.* 118:368 (1996).

23. E. Coronado. J. R. Galán-mascarós, C. J. Gómez-García and V. N. Laukhin Coexistence of Ferromagnetism and Metallic Conductivity in a Molecule- Based Layered Compound *Nature* 408:447 (2000).

24. L. Ouahab, M. Bencharif, A. Mhanni, D. Pelloquin, J. F. Halet, O. Peña, J. Padiou, D. Grandjean, C. Garrigou-Lagrange, J. Amiell and P. Delhaes Preparations, X-Ray Crystal-Structures, Eh Band Calculations, and Physical-Properties of (TTF)$_6$H(XM$_{12}$O$_{40}$)(Et$_4$N) (M = W, Mo X = P, Si) - Evidence of Electron-Transfer Between Organic Donors and Polyoxometalates *Chem. Mater.* 4:666 (1992).

25. (a) C. J. Gómez-García, C. Giménez-Saiz, S. Triki, E. Coronado, P. Le Magueres, L. Ouahab, L. Ducasse, C. Sourisseau and P. Delhaes Coexistence of Magnetic and Delocalized Electrons in Hybrid Molecular Materials - The Series of Organic-Inorganic Radical Salts (BEDT-TTF)$_8$(XW$_{12}$O$_{40}$)(Solv)$_n$(X=2H$^+$, B^{III}, Si^{IV}, Cu^{II}, Co^{II}, Fe^{III} Solv=H$_2$O, CH$_3$CN) *Inorg. Chem.* 34:4139 (1995); (b) E. Coronado, J. R. Galán-Mascarós, C. Giménez-Saiz, C. J. Gómez-García and S. Triki Hybrid Molecular Materials Based upon Magnetic Polyoxometalates and Organic Pi-Electron Donors - Syntheses, Structures, and Properties of bis(Ethylenedithio)Tetrathiafulvalene Radical Salts with Monosubstituted Keggin Polyoxoanions *J. Am. Chem. Soc.* 120:4671 (1998).
26. M. Clemente-León, C. Mingotaud, C. J. Gómez-García, E. Coronado and P. Delhaes Polyoxometalates in Langmuir-Blodgett-Films - Toward New Magnetic- Materials *Thin solid films* 327-329:439 (1998).
27. E. Coronado and C. J. Gómez-García, Polyoxometalates: from Magnetic Clusters to molecular Materials, *Comments Inorg. Chem.* 17:255 (1995).

MAGNETIC EXCHANGE COUPLING AND POTENT ANTIVIRAL ACTIVITY OF [(VO)$_3$(SbW$_9$O$_{33}$)$_2$]$^{12-}$

Toshihiro Yamase*[1], Bogdan Botar,[1] Eri Ishikawa,[1] Keisuke Fukaya,[1] and Shiroh Shigeta[2]

[1]*Chemical Resources Laboratory, Tokyo Institute of Technology, 4259 Nagatsuta, Midori-ku, Yokohama 226-8503*
[2]*Department of Microbiology, Fukushima Medical College, Fukushima 960-12, Japan*

INTRODUCTION

Reaction of α-B [SbW$_9$O$_{33}$]$^{9-}$ with Mn^{2+} gives [{Mn(H$_2$O)}$_3$(SbW$_9$O$_{33}$)$_2$]$^{12-}$,[1] similar to the reaction of α-B [AsW$_9$O$_{33}$]$^{9-}$ with Cu^{2+} to yield [Cu$_3$(H$_2$O)$_2$(AsW$_9$O$_{33}$)$_2$]$^{12-}$.[2] A crystal structure determination of Na$_{11}$(NH$_4$)[{Mn(H$_2$O)}$_3$(SbW$_9$O$_{33}$)$_2$]•45H$_2$O shows that the structure of [{Mn(H$_2$O)}$_3$(SbW$_9$O$_{33}$)$_2$]$^{12-}$ consists of two α-B [SbW$_9$O$_{33}$]$^{9-}$ anions linked via three equatorial Mn atoms with square-pyramidal coordination by exterior H$_2$O as unshared ligands and four oxygen atoms belonging to [SbW$_9$O$_{33}$]$^{9-}$. On the other hand, the structure of the related Cu^{2+} complex [Cu$_3$(H$_2$O)$_2$(AsW$_9$O$_{33}$)$_2$]$^{12-}$ shows that one Cu atom is in square-planar coordination through four oxo groups and two Cu atoms are in exterior sites (of exterior H$_2$O) with square-pyramidal coordination. The three equatorial Mn/Cu atoms are well-separated in 5.255-5.297 Å/4.669-4.707 Å. We report here the synthesis and structural characterization of isostructural tris(vanadyl)-substituted tungsto-antimonate(III) [(VO)$_3$(SbW$_9$O$_{33}$)$_2$]$^{12-}$ and -bismutate(III) [(VO)$_3$(BiW$_9$O$_{33}$)$_2$]$^{12-}$ with large V^{IV}••V^{IV} separation of 5.3-5.5 Å, which reveal the weak spin-exchange interaction between V^{IV} centers linked by μ$_2$-OWO(μ)WO bridges and provide simple models for understanding intramolecular spin-spin interactions between V^{IV} centers of the magnetically intriguing high-nuclearity-polyoxovanadates.[3,4] So far, the presence of the weak intramolecular coupling (|J|<10 cm^{-1}) of vanadyl cluster fragment in the polyoxovanadates has been revealed for μ$_2$-OAO bridges (A=AsO$_2$ and PO$_2$) with V^{IV}••V^{IV} distances of 4.5-5.7 Å.[3] Although in K$_6$[V^{IV}$_{15}$As$_6$O$_{42}$(H$_2$O)]•8H$_2$O the presence of the magnetically isolated V^{IV}-triangle (with V^{IV}••V^{IV} distances of 5.290(5)-5.603(4) Å) layer sandwiched by two antiferromagnetic hexagonal V^{IV} layers at 20 K<T<100 K has been proposed, there was no evidence of the intramolecular spin-exchange interaction in the vanadyl triangle.[5] We also describe potent antiviral activities of the tris(vanadyl)tungstoantimonate(III) anions against a wide variety of the enveloped viruses which infect high-risk individuals, such as infants born with prematurity, cardiovascular failure, pulmonary dysplasia, and immunodeficiency. This finding indicates

Polyoxometalate Chemistry for Nano-Composite Design
Edited by Yamase and Pope, Kluwer Academic/Plenum Publishers, 2002

169

that the tris(vanadyl)tungstoantimonate(III) anion derivative is important as a candidate for a new antiviral drug.

EXPERIMENTAL

The $[(VO)_3(SbW_9O_{33})_2]^{12-}$ anion has been prepared as crystalline of potassium salt $K_{11}H[(VO)_3(SbW_9O_{33})_2] \cdot 27H_2O$ (1): $Na_9[SbW_9O_{33}] \cdot 19.5H_2O$ used as a starting material was obtained according to the literature[6] and identified by IR spectrum. $Na_9[SbW_9O_{33}] \cdot 19.5H_2O$ (11.3 g, 4 mmol) was added to a solution containing $VOSO_4$ $5H_2O$ (3.1 g, 12 mmol) in CH_3CO_2Na/CH_3CO_2H buffer (pH 4.8, 70 ml) at room temperature. The dark-brown solution was heated at 60-70°C for 1 h and filtered off. The filtrate was cooled to room temperature. Addition of finely ground KCl solids (12 g, 0.16 mol) to the filtrate led to formation of brown precipitates which were recrystallized from hot water. Dark-brown plate crystals were produced within 2-3 days with yield of 8.0 g (70 % based on W). Anal. Calc. for $H_{55}O_{96}K_{11}V_3Sb_2W_{18}$: K, 7.51; V, 2.67; Sb, 4.25; W, 57.78. Found: K, 7.7; V, 2.7; Sb, 4.2; W, 56.2 %. IR (metal oxygen stretching region, KBr disk, cm^{-1}): 988(m), 943(s), 901(m), 857(s), 740(vs), 692(s). Electronic spectrum in H_2O, λ_{max}, nm (ε_{max}, in $mol \cdot dm^{-3}$ cm^{-1}): 255 (6.90 x 10^4), 550(sh, 900), 759 (525).

The $[(VO)_3(BiW_9O_{33})_2]^{12-}$ anion has been prepared as crystalline of potassium salt $K_{12}[(VO)_3(BiW_9O_{33})_2] \cdot 29H_2O$ (2): $Na_9[BiW_9O_{33}] \cdot 16H_2O$ used as a starting material was obtained according to the literature[7] and identified by IR spectrum. $Na_9[BiW_9O_{33}] \cdot 16H_2O$ (12.2 g, 4.2 mmol) was added to a solution containing $VOSO_4$ $5H_2O$ (3.1 g, 12 mmol) in CH_3CO_2Na/CH_3CO_2H buffer (pH 4.8, 150 ml) at room temperature. The dark-brown solutiob was heated to 40-45°C for 3-4 h and then filtered off. Addition of finely ground KCl solids (20 g, 0.16 mol) to the filtrate led to formation of brown precipitates which were recrystallized from warm water (at 45°C). Dark-brown plate crystals of the desired product were obtained within 1 weak with yield of 5.5 g (45 % based on W). Anal. Calc. for $H_{58}O_{98}K_{12}V_3Bi_2W_{18}$: K, 7.85; V, 2.56; Bi, 7.005; W, 57.78. Found: K, 7.6; V, 2.7; Bi, 7.4; W, 57.0 %. IR (metal oxygen stretching region, KBr disk, cm^{-1}): 990(m), 938(s), 897(m), 847(s), 727(vs), 659(m, sh).

IR and UV-vis spectra were recorded on Jasco FT-IR 5000 and Jasco V-570 UV/VIS/NIR spectrometers, respectively. The contents of K, V, Sb, W, and Bi were determined by X-ray fluorescence analysis on a JEOL JSX-3200 spectrometer. The water content was measured by thermogravimetric method on an ULVAC-TGD9600/MTS9000 instrument. ESR measurement was carried out on a JEOL X-band ESR spectrometer (JES-RE3X) equipped with a temperature-controlling unit (ES-DVT3). The magnetic susceptibility in the range of 4-300 K was measured under the magnetic field of 1.0 T with a Quantum Design, MPMS-5S SQUID magnetometer and the experimental data were corrected for diamagnetic contribution using standard Pascal constants.

Single crystals of 1 and 2 with 0.12 x 0.10 x 0.03 and 0.2 x 0.1 x 0.05 mm sizes were collected on a Rigaku-RAXIS-RAPID imaging-plate diffractometer employing the monochromatized Mo-K_α radiation (λ=0.71069 Å) at 0 and -100°C, respectively. All of the calculations were performed using the TEXSAN software package for the structure determination.

Crystal data for $H_{55}O_{96}K_{11}V_3Sb_2W_{18}$ (1): M=5727.08, triclinic, space group P$\bar{1}$ (No.2), a=12.725(1), b=18.203(1), c=20.670(2) Å, α=87.719(3)°, β=87.121(4)°, γ=75.583(5)°, V=4629.5(7) Å^3, Z=2, D_c=4.108 g·cm^{-3}, μ=237.58 cm^{-1}. The structure was solved by direct methods (MULTAN 88) and refined using a full-matrix least-squares refinement procedure; 19552 unique reflections (R_{int}=0.077), 10246 observed reflections with I>3$\sigma(I)$, R=0.067, R_w=0.078, 680 refined parameters, and 2θ_{max}=55°. Lorentz polarization and absorption correction (ABSCOR) were applied to the intensity data. The transmission factors range from 0.302-0.691.   K9-K12 and O atoms were refined isotropically, and all other non-hydrogen atoms were refined anisotropically.   The site occupancies of K11 and K12 atoms are fixed at 0.5 throughout the refinements, since distance (2.87(5) Å) between these (symmetry-related) atoms is very short, which may be brought about by the disordered structure of these atoms.   This let us formulate **1** as $K_{11}H[(VO)_3(SbW_9O_{33})_2] \cdot 27H_2O$, to maintain electrical neutrality.

Crystal data for $H_{58}O_{98}K_{12}V_3Bi_2W_{18}$ (**2**): M=5975.66, triclinic, space group P$\bar{1}$ (No.2), a=12.6455(5), b=18.1169(8), c=20.7898(7) Å, α=88.261(2)°, β=87.197(1)°, γ=74.698(2)°, V=4587.8(3) Å^3, Z=2, D_c=4.32 g·cm^{-3}, μ=272.6 cm^{-1}.   The structure was solved by direct methods (SIR 92) and refined using a full-matrix least-squares refinement procedure; 19347 unique reflections (R_{int}=0.082), 12426 observed reflections with $I>2\sigma(I)$, R=0.047, R$_w$=0.1166, 707 refined parameters, and $2\theta_{max}$=52°.   Lorentz polarization and a numerical absorption correction (Numas and Shape) were applied to the intensity data.   The transmission factors range from 0.143-0.345.   K11, K13, and O atoms were refined isotropically, and all other non-hydrogen atoms were refined anisotropically.   The site occupancies of K12 and K13 atoms are fixed at 0.5 throughout the refinements, since distance (3.11(1) Å) between these atoms is very short, which may be brought about by the disordered structure of these atoms.

The cells for DFV (dengue virus), FluV-A (influenza virus), RSV (respiratory syncytial virus), PfluV (parainfluenza virus), CDV (distemper virus), HIV (human immunodeficiency virus), and HSV (herpes simplex virus) are CV-1, MDCK, HEp-2, HMV-2, Vero, MT-4 (or Hela CD4), and RPM18226 cells, respectively.   The cells were suspended in the culture medium at 1×10^5 cells/ml and infected with corresponding virus. After 4-6 days of incubation at 37°C, the cell viability was determined by the 3-(4,5-dimethylthiazol-2-yl)-2,5-diphenyltetrazolium bromide (MTT) method.[8]   Antiviral activity was estimated by the reduction of a 50% tissue-culture-infective dose (TCID$_{50}$) under the treatment of **1** and other polyoxometalates.

CRYSTAL STRUCTURES

The anion (Fig. 1a) of **1** consists of two α-B $[SbW_9O_{33}]^{9-}$ ligands linked by three exterior VO^{2+} groups into an assembly of virtual D_{3h} symmetry.[9]   Each V atom is coordinated to four oxygen atoms (with V-O distances 1.92(2)-1.99(3) Å) and an exterior vanadyl O atom (with V-O distances 1.56(3)-1.59(3) Å) in a square pyramidal geometry. The V atoms are on average 0.54(1) Å above the plane of the four oxygen atoms belonging to α-B $[SbW_9O_{33}]^{9-}$ ligands.   V atoms in the equatorial vanadyl triangle (with V-V-V angles of 59.4(1)-60.4(1)°) are 5.411(8)-5.464(8) Å apart.   This well-separated long distance is significantly longer than X$\cdots$X (X=Mn^{2+} and Cu^{2+}) distances for isostructural anions.[1,2]   The structure of anion is stabilized by three equatorial K$^+$ ions linking three VO$_5$ square-pyramids in alternate positions to produce the 6-member ring with K$\cdots$K distances of 7.64(1)-7.80(1) Å and K-K-K angles of 59.2(1)-61.2(1)° (Fig. 1b).   The least-square plane containing vanadyl- and K-triangles (V1,V2,V3,O39,O42,O45 and K1,K2,K3) is positioned at symmetrical distance from the two Sb atoms Sb1 and Sb2 in Sb$\cdots$Sb distance of 4.780(3) Å.   The incorporation of K triangle into the the $[(VO)_3(SbW_9O_{33})_2]^{12-}$ anion is reminiscent of a recent structure of the tris(phenyltin)-substituted complex $Cs_6[(PhSn)_3Na_3(SbW_9O_{33})_2] \cdot 20H_2O$, in which the $[(PhSn)_3(SbW_9O_{33})_2]^{9-}$ anion is stabilized by the incorporation of three Na$^+$ ions.[10]   The crystal structure of $K_{12}[(VO)_3(BiW_9O_{33})_2] \cdot 29H_2O$ (**2**) also exhibited the $[(VO)_3(BiW_9O_{33})_2]^{12-}$ anion to be stabilized by three equatorial K$^+$-cations with the same linkage.   Thus, it is possible to say that the $[X_3(YW_9O_{33})_2]^{12-}$ (X=Cu^{2+}, Mn^{2+}, PhSn^{3+}, VO^{2+}; X=Sb^{3+}, Bi^{3+}) anion is commonly stabilized by a capping Na$^+$ and K$^+$ to form a

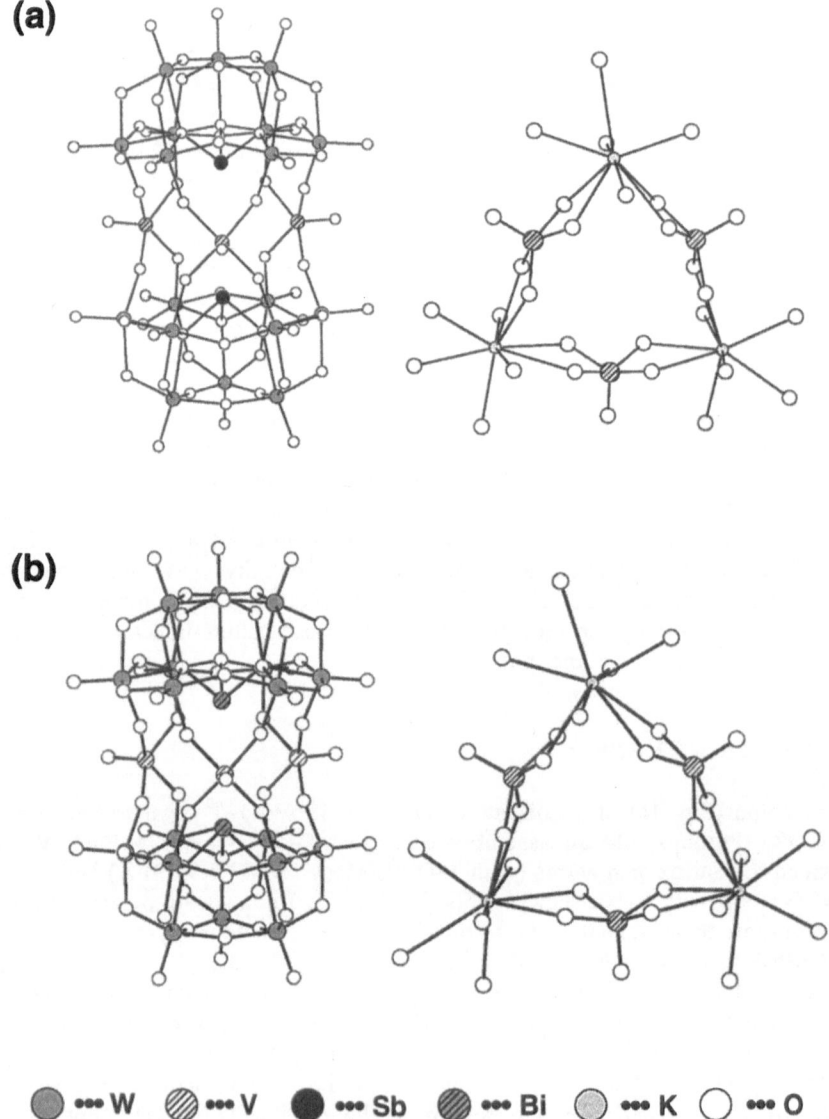

Figure 1. Ball-and-stick structural representations of $[(VO)_3(SbW_9O_{33})_2]^{12-}$ (a) and $[(VO)_3(BiW_9O_{33})_2]^{12-}$ (b) with coordination geometries of the equatorial V and K atoms (three for each atom).

horizontal plane of the six-member ring at the belt. The overall structure of anion (Fig. 1b) for **2** obtained under essentially the same experimental conditions, is almost the same as for **1**: the central vanadyl V atoms (in distances of 1.59(1)-1.61(1) Å with exterior O atoms) are 5.376(4)-5.474(4) Å apart and on average 0.591(7) Å above the square pyramidal plane of the four O atoms (with V-O distances 1.94(1)-1.98(1) Å). A short Bi•••Bi distance (4.5035(9) Å) compared to the Sb•••Sb distance in **1** arises from longer Bi-O distances (2.09(1)-2.11(1) Å) compared to Sb-O distances (1.98(1)-2.00(1) Å) in **1**, as reflected by the fact that the O-Bi-O angles (85.7(4)-87.6(4)º) of the BiO_3 trigonal pyramid are smaller than the O-Sb-O angles (90.5(4)-91.5(5)º).

ELECTROCHEMISTRY

The neutral ligand tungstoantimonate(III) as a sodium salt $Na_9[SbW_9O_{33}] \cdot 19.5H_2O$ was electrochemically inactive at the potential range of 0.3-0.8 V vs. Ag/AgCl (Fig. 2a). The cyclic voltammogram of **1** showing a rest potential at 0.13 V in aqueous solutions at natural pH level of 5.8 exhibits three anodic peaks at 0.25, 0.35, and 0.44 V (vs. Ag/AgCl) which are accompanied by a composite cathodic peaks at 0.19, 0.30, and 0.39 V as three quasi-reversible one-electron oxidations corresponding to $V^{IV}_2V^V/V^{IV}_3$, $V^{IV}V^V_2/V^{IV}_2V^V$, and $V^V_3/V^{IV}V^V_2$ (Fig. 2b). Three successive sets of the redox peaks were observed for the aqueous solutions of **1** at 3<pH<11 without significant change in the peak potential. The stepwise coalescence of the anodic/cathodic peak couples occurred at pH<3: the 0.35/0.26 V peak couple as a result of the coalescence of the first and second ones was observed at pH=2.4 (Fig. 2c) and the 0.46/0.35 V peak couple as a result of coalescence of the three peak couples at pH 1.5 (Fig. 2d). The neutralization of the highly acidic samples at pH<3 brought about a restoration of the pattern consisting of three redox peak couples. The redox currents were decreased at pH>11. Such results of the pH dependence of the cyclic voltammograms of **1** indicate that the [(VO)₃(SbW₉O₃₃)₂]$^{12-}$ framework is maintained at pH<11. Similarly, [(VO)₃(BiW₉O₃₃)₂]$^{12-}$ is stable at pH<11. The cyclic voltammograms of **2** (1 mM) at 3<pH<11 exhibit the three composite couples around 0.26/0.23, 0.37/0.44, and 0.47/0.44 V due to three sets of quasi-reversible one-electron redox reactions. In the case of **2**, however, other redox peaks as well as the coalescence of the redox peak couples at pH< 3 appeared probably due to the difference in the redox potentials between the protonated and unprotonated species. Fig. 2e shows the cyclic voltammogram of **2** at pH=1.4. The cyclic voltammogram of the [{Mn(H₂O)}₃(SbW₉O₃₃)₂]$^{12-}$ anion of $Na_{11}(NH_4)[\{Mn(H_2O)\}_3(SbW_9O_{33})_2]•45H_2O$ showing a coalescent pattern of three composite redox peak couples around 0.50 V in 1,2 dichloroethane[1] showed also three quasi-reversible redox peaks around 0.30/0.24、 0.44/0.38, and 0.56/0.50 V for the aqueous solution at pH=6.5, and the electrochemical formation of Mn^{IV} at more than 0.7 V led to the decomposition of the anion.

MAGNETIC INTERACTION

The magnetic exchange that may be expected in V^{IV} with a ^2D free ion ground term, d¹ electron structure, and spin S=1/2 is the isotropic Heisenberg spin Hamiltonian ($\mathcal{H}=-2JS_iS_j$). The magnetic spin-exchange interaction within the vanadyl coplanar triangle will enable three spins (S=1/2 for each) to be coupled into a quartet (S=3/2) and two doublets (S=1/2). The ESR spectrum (Fig. 3a) of **1** (11.3 mM) dissolved in water at 300 K is the broad structured multi-line band around g=1.94 (ΔH_{pp}=34 mT at 300 K) that seems to include an isotropic signal with a 22-line pattern assignable to hyperfine interactions with

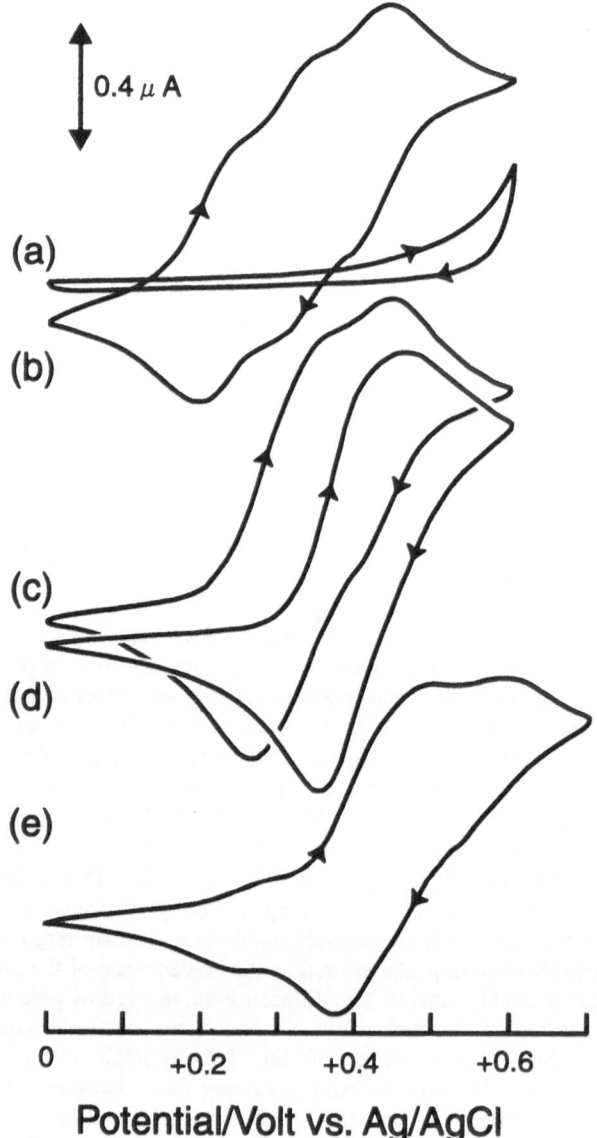

Figure 2. Cyclic voltammograms (with a sweep rate of 0.1 mV/s) of aqueous solutions of Na$_9$[SbW$_9$O$_{33}$]•19.5H$_2$O (2 mM), **1** (1mM), and **2** (1 mM) in 0.1 M KCl at pH<11 under the nitrogen atmosphere. (a): Na$_9$[SbW$_9$O$_{33}$]•19.5H$_2$O at pH=6.2, (b): **1** at pH=5.8, (c): **1** at pH=2.4, (d): **1** at pH=1.5, and (e): **2** at pH=1.4. Glassy carbon rod (φ=1 mm) electrode, Pt wire auxiliary electrode, and Ag/AgCl reference electrode.

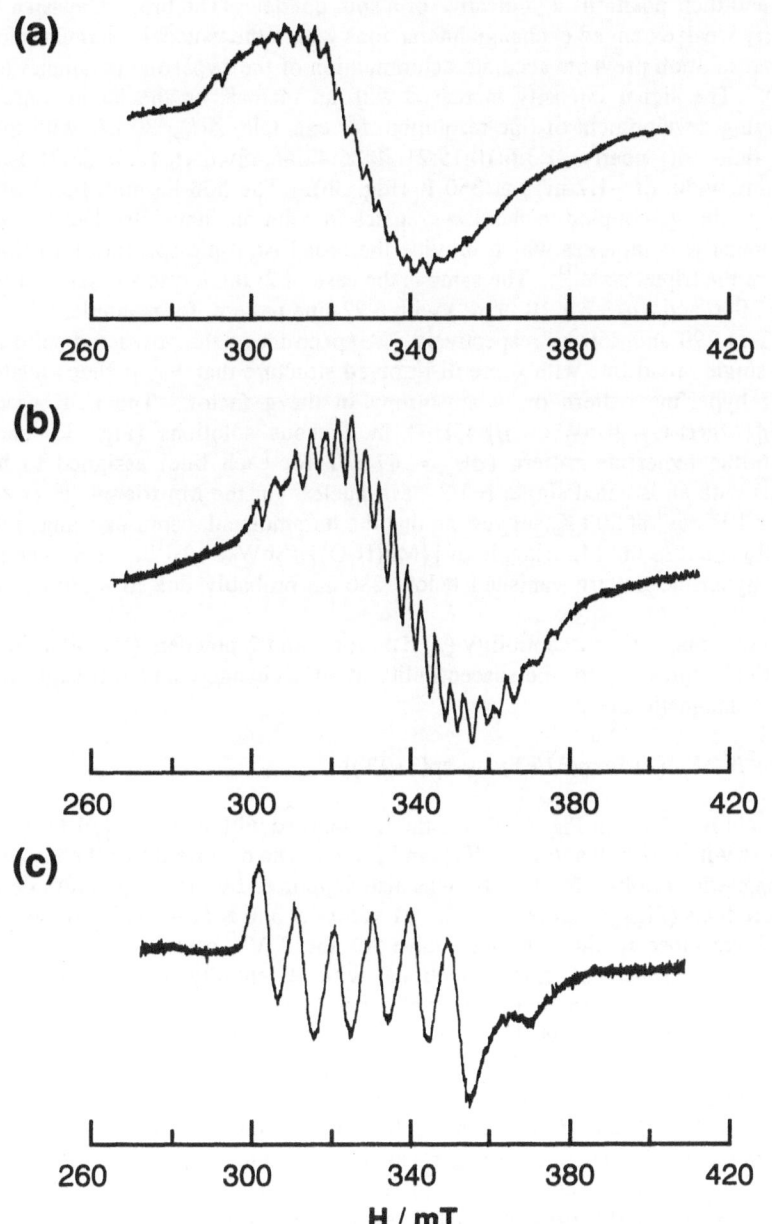

Figure 3. ESR spectra of **1** (11.3 mM) in H_2O at 300 K (a) and 350 K (b). ESR spectrum (c) of $Na_{11}(NH_4)[\{Mn(H_2O)\}_3(SbW_9O_{33})_2]\cdot45H_2O$ (10 mM) in H_2O at 300 K is represented for comparison.

three equivalent $I=7/2$ ^{51}V nuclei.[9] The number of features that are present in the spectrum and their positions is indicative of a spin quartet. The broad Gaussian ESR line reflects very weak extended exchange interactions within the vanadyl triangular cluster, and poor peak resolution prevents accurate determination of the hyperfine parameter ($A_V \approx 3.2$ x 10^{-3} cm^{-1}). The signal intensity increased with an increase in the temperature with an accompanying development of line-resolution, for example, ΔH_{pp}=30 mT with the 22-line intensity ratio of nearly 1:3:6:10:15:21:28:36:42:46:48:48:46:42:36:28:21:15:10:6:3:1 (average line width of ~1.7 mT) at 350 K (Fig. 3b). The 300-K multi-line ESR spectra due to the exchange-coupled vanadyl aggregates in solutions have also been observed for binuclear vanadyl complexes which provide the broad isotropic spectra of 15-line pattern arising from the triplet state.[11] The same is the case of **2**: there was a broad isotropic ESR signal (g=1.946 and $A_V \approx 3.2$ x 10^{-3} cm^{-1}) with a 22-line pattern, for example, with ΔH_{pp}=40 and 30 mT at 290 and 350 K, respectively. A spectrum of the powdered solid at 300 K showed a single broad line with some ill-resolved structure that was perhaps related to the underlying hyperfine pattern or to anisotropy in the g factor. The ESR spectrum of Na$_{11}$(NH$_4$)[{Mn(H$_2$O)}$_3$(SbW$_9$O$_{33}$)$_2$]•45H$_2$O in aqueous solutions (Fig. 3c) showed an isotropic 6-line hyperfine pattern (ΔH_{pp}=4.47 mT for each line) assigned to hyperfine interactions with an isolated single $I=5/2$ ^{55}Mn nuclear for the Mn triangle at g=2.006 and A_{Mn}=9.00 x 10^{-3} cm^{-1} at 300 K, suggesting that the intramolecular spin-exchange interaction is negligibly small in the Mn triangle of [{Mn(H$_2$O)}$_3$(SbW$_9$O$_{33}$)$_2$]$^{12-}$. The spectrum with the 6-line hyperfine pattern vanished below 280 K, probably due to a strong zero-field splitting.

Molecular magnetic susceptibility (χ) data for **1** and **2** powders (Fig. 4) were fitted to the theoretical expression for the susceptibility of an exchange-coupled triangle with three identical V^{IV} magnetic sites,

$$\chi=(Ng^2\beta^2/4kT)[1+5\exp(3J/kT)/1+\exp(3J/kT)].^{12}$$

The solid line for **1** in Fig. 4a shows the fit obtained with g=1.779 and J=-1.4(1) cm^{-1}, and similarly with g=1.849 and J=-1.4(1) cm^{-1} for **2**. The occurrence of the intramolecular antiferromagnetic coupling for **1** and **2** was also supported by the temperature dependence of the χT products (Figs. 4a and 4b). The χT values at 300 K (1.05 and 1.16 emu.mol^{-1} K) of **1** and **2** are close to the spin-only value for the 3 V^{IV} centers (1.13 emu.mol^{-1} K), indicating the presence of three electrons with essentially uncoupled spin. The intramolecular antiferromagnetic coupling of **1** and **2** implies that the lowest multiplets of the vanadyl triangle are the two S=1/2 states (four magnetic levels) and are about 2 cm^{-1} below the S=3/2 state (as affirmed by the 300-K 22-line ESR spectrum) for [(VO)$_3$(YW$_9$O$_{33}$)$_2$]$^{12-}$ (Y=Sb^{3+} and Bi^{3+}) with the V•••V separation of 5.43-5.44 Å. Such formation of the S=1/2 states (corresponding to χT=0.37 emu.mol^{-1} K) results from the spin frustration in the vanadyl coplanar triangle which occurs at temperatures lower than 10 K.

In contrast, as can be seen in Fig. 4c, the χT plot against temperature for the powder of Na$_{11}$(NH$_4$)[{Mn(H$_2$O)}$_3$(SbW$_9$O$_{33}$)$_2$]•45H$_2$O exhibits essentially the same plateau value (12.4 emu.mol^{-1} K which corresponds to μ_{eff}=9.96 μ_B) at T>~10 K. This reveals that the [{Mn(H$_2$O)}$_3$(SbW$_9$O$_{33}$)$_2$]$^{12-}$ anion have a S=15/2 ground state (with a spin-only value of 9.95 μ_B) due to the presence of three unpaired Mn^{II} sites (S=5/2 for each) substantially uncoupled and exhibits no significant intramolecular antiferromagnetic interaction between the Mn^{II} sites at T>~10 K irrespective of slightly short Mn•••Mn separation of 5.255-5.297 Å compared with the V•••V separation of 5.4-5.5 Å. Thus, the tendency to lose the spin-exchange over long distances 5.3-5.5 Å seems to be due to a strong zero-field splitting in the S=5/2 ground state for each Mn^{II} site, where unpaired electrons will be separately positioned in each of five d orbitals.

K$_{10}$Na[(VO)$_3$(SbW$_9$O$_{33}$)$_2$]•26H$_2$O (**1'**), the mixed-valence triangular vanadyl

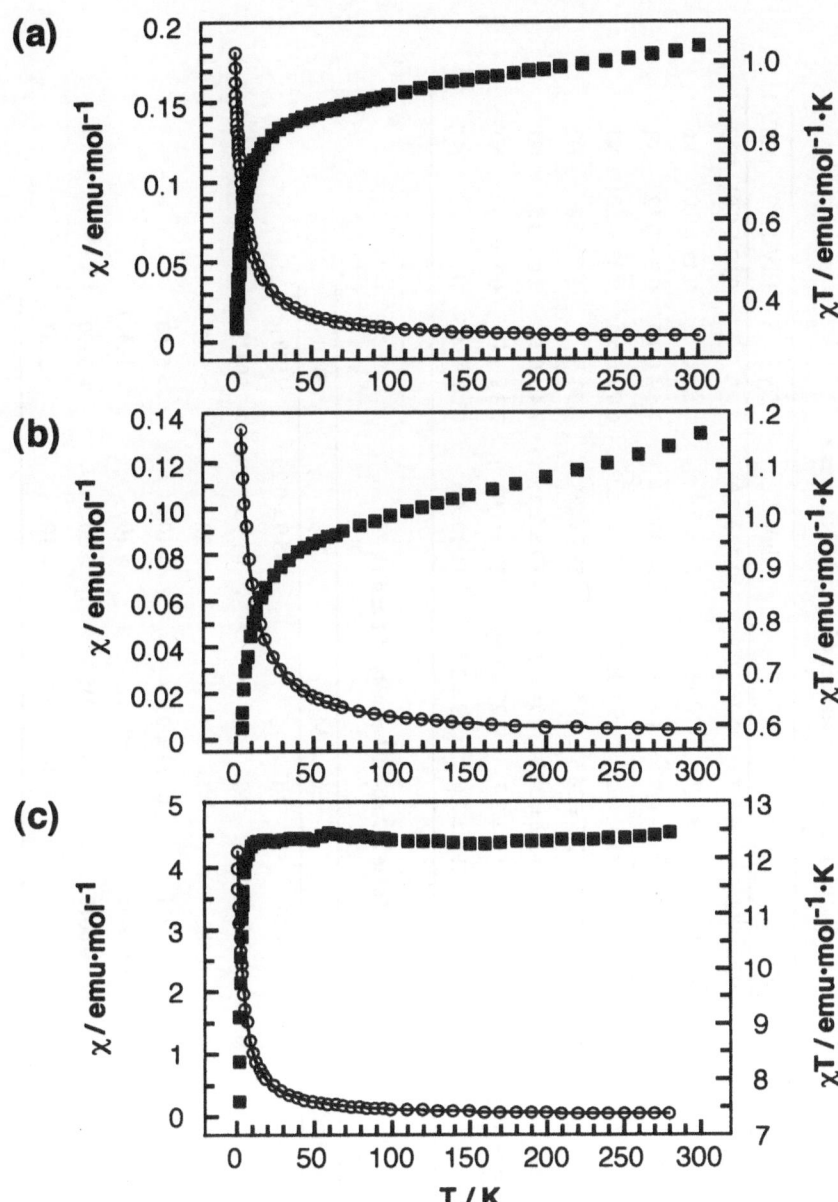

Figure 4. Temperature (T) variations of magnetic susceptibility (○) and χT (■) for **1**(a), **2**(b), and $Na_{11}(NH_4)[\{Mn(H_2O)\}_3(SbW_9O_{33})_2]\cdot45H_2O$ (c) powders. The solid line corresponds to the calculated curve by the theoretical equation as described in the text.

Table 1. $EC_{50}(\mu M)$ of polyoxometarates of several RNA viruses.

Compounds	DFV	FluV-A	RSV	PfluV-2	CDV	HIV-1	HSV-1
1	1.95 ± 1.42	4.6 ± 1.7	0.75 ± 0.05	0.75 ± 0.05		0.03 ± 0.01	0.3
1'	0.45 ± 0.35	5.3 ± 0.7	< 0.16	1.1 ± 0.9		0.14 ± 0.07	1.62
$K_5[SiVW_{11}O_{40}]$	10.7 ± 6.7	8.4 ± 6.5	1.6	> 100	7.5 ± 0.95	0.3 ± 0.12	NT
$K_7[BVW_{11}O_{40}]$	10.5 ± 6.9	11.5 ± 0.6	29 ± 15.7	67.1 ± 32.9	6.0 ± 0.59	0.03 ± 0.01	NT
$[Et_2NH_2]_7[PTi_2W_{10}O_{40}]$	36.8 ± 9.5	62.3 ± 26.5	25.6 ± 8.1	53.2 ± 39.2	> 50	2.0 ± 0.8	NT
$[^iPr_2NH_2]_5[PTiW_{11}O_{40}]$	11.1 ± 7.1	45.2 ± 25.8	0.74 ± 0.58	3.2 ± 2.4	7.4 ± 0.39	2.0 ± 0.5	NT
$[^iPrNH_3]_6H[PTi_2W_{10}O_{38}(O_2)_2]\cdot H_2O$	> 61.5	6.5 ± 3.7	1.27 ± 0.46	2.5 ± 1.0	7.3 ± 1.12	0.3 ± 0.07	NT
Ribavirin	> 100	9.6 ± 2.6	3.9 ± 3.1	14.0 ± 4.8	73.6 ± 34.5	NT	NT

Table 2. Anti-HIV activity of 1 and 1'.

Cell	Compound	$EC_{50}(\mu M)$	$EC_{90}(\mu M)$	CC_{50}	SI
MT-4	1	0.0300 ± 0.010	ND	45.9 ± 0.29	1530
	1'	0.138 ± 0.067	ND	41.9 ± 2.88	304
	DS5000	0.652 ± 0.286	ND	> 20	> 30.7
HeLa CD4	1	0.0177 ± 0.0071	0.112 ± 0.054	> 100	> 5650
	1'	0.0096 ± 0.0056	0.11 ± 0.057	> 100	> 10417
	DS5000	0.0059 ± 0.0018	0.127 ± 0.055	> 20	> 3390
	AZT	0.037 ± 0.023	0.416 ± 0.327	> 100	> 2703

Aggregate ($V^V/V^{IV}=1/2$ with disorder) was prepared by the reaction of $Na_9[SbW_9O_{33}] \cdot 19.5H_2O$ (12 g) with $NaVO_3$ (1.5g) under the condition (for 3 h at 70°C) similar to the case of **1**. The ESR spectrum of **1'** in aqueous solutions at 300 K showed also the broad signal comprising unresolved 22 lines, implying that the spin-exchange coupling of **1'** reflects the disorder of three sets of the V^{IV}-V^{IV} interaction within the $(V^VO)(V^{IV}O)_2$ triangle.

ANTIVIRAL ACTIVITY

1, **1'**, and other polyoxometalates were examined for antiviral activities against DFV, FluV-A, RSV, PfluV-2, CDV, HIV, and HSV-1. **1** and **1'** showed activities against all viruses examined. EC_{50} values of **1** and **1'** were less than 1 μM against RSV, PfluV-2 and HIV, and lower than those of ribavirin. Like ribavirin, **1** and **1'** exhibited a relatively high EC_{50} against FluV-A (Tab. 1). As predicted, none of the compounds evaluated were markedly cytotoxic in stationary cells (Tab. 2). Anti-HIV effects of **1** and **1'** were further examined in the MT-4 and HeLa CD4 cell lines infected with 100 $TCID_{50}$, together with the effect of DS5000 and AZT as positive controls. Both **1** and **1'** exhibited high selective index (SI) compared with positive controls DS5000 and AZT (Tab. 2). The $[(VO)_3(SbW_9O_{33})_2]^{11-}$ anion of **1'** as a mixed-valence species for **1** also exhibits strong potency for the same variety of viruses, especially RSV and HIV. The results in Tabs. 1 and 2 indicate that both **1** and **1'** have a broad spectrum against the enveloped viruses, implying that $[(VO)_3(YW_9O_{33})_2]^{n-}$ (n=10-12, $Y=Sb^{3+}$ and Bi^{3+}) with the tri(vanadyl)-ring sandwiched by two α-B $[YW_9O_{33}]^{9-}$ ligands, inhibits either fusion between the virus envelope and the cellular membrane or adsorption of the virus onto the cell membrane.[13]

ACKNOWLEDGEMENT

One (T.Y.) of us acknowledges Grants-in-Aid for Scientific Research, Nos. 09354009 and 10304055, from the Ministry of Education, Science, Sports, and Culture and for Research project No. 99P01201 from RFTF/JSPS for support of this work.

REFERENCES

1. M. Bösing, A. Nöh, I. Loose, and B. Krebs, Highly efficient catalysts in directed oxygen-transfer process. Synthesis, structures of novel manganese containing heteropolyznions, and applications in regioselective epoxidation of dienes with hydrogen-peroxide, *J. Amer. Chem. Soc.*, 120:252 (1998).
2. F. Robert, M. Leyríe, and G. Herve, Structure of potassium diaguatricuprooctadecatungstodiarsenate(III)(12-)undecahydrate, *Acta Crystallogr., Sect. B*, B38:358 (1982).
3. a) A. Müller, F. Peter, M. T. Pope, and D. Gatteschi, Polyoxometalates. Very large clusters. nanoscale magnets, *Chem. Rev.*, 98:239 (1998) and b) T. Yamase and H. Makino, Photochemistry of polyoxovanadates. Part4. Crystal and electronic-structures and magentic-susceptibility of the photochmically prepared layered vanadyl phosphate $Na[VO]_2[PO_4]_2 \cdot 4H_2O$, *J. Chem. Soc. Dalton Trans.*, 1143 (2000).
4. a) G. B. Karet, Z. Sun, D. D. Heinrich, J. K. McCusker, K. Folting, W. E. Streib, J. C. Huffmann, D. N. Hendrickson, and G. Christou, Tetranuclear and pentanuclear vanadium(IV/V) carboxylate complexes. $[V_4O_8(NO_3)(O_2Cr)_4]^{2-}$ and $[V_5O_9X(O_2Cr)_4]^{2-}$ (X=Cl⁻, Br⁻) salts, *Inorg. Chem.*, 35:6450 (1996) and b) E. Coronado and C. J. Gómez-Garcia, Polyoxometalate-based molecular materials, *Chem. Rev.*, 98:273 (1998).

5. A.-L. Barra, D. Gatteschi, L. Pardi, A. Müller, and J. Döring, Magnetic-properties of high-nuclearity spin clusters. 14-Oxovanadium(IV) and 15-oxovanadium(IV) clusters, *J. Am. Chem. Soc.*, 114:8509 (1992) and A. Müller, and J. Döring, Reduced clusters with remarkable topological and electronic-properties of the type of $[V_{18}O_{42}(X)]^{n-}$ (X=SO_4, VO_4) with T_d-symmetry and related clusters $[V_{(18-P)}As_{2p}O_{42}(X)]^{m-}$ (X= SO_3, SO_4, H_2O; p=3,4), *Z. anorg. Allg. Chem.*, 595:251 (1991).

6. M. Bösing, I. Loose, H. Pohlman, and B. Krebs, New strategies for the generation of large heteropolymetalate clusters: The β-B-SbW_9 fragment as a multifunctional unit, *Eur. J. Chem.*, 3:1232 (1997).

7. B. Botar, T. Yamase, and E. Ishikawa, A highly nuclear vanadium-containing tungstobismutate. Synthesis and crystal-structure of $K_{11}H[(BiW_9O_{33})_3Bi_6(OH)_3(H_2O)_3V_4O_{10}]\cdot25H_2O$, *Inorg. Chem. Commun.*, 3:579 (2000).

8. R. Pauwels, J. Balzarini, M. Baba, R. Snoeck, D. Schols, P. Herdewijn, J. Desmyter, and E. De Clercq, Rapid and automated tefrazalium-based calorimetric assay for the detection of anti-HIV compounds, *J. Vitrol. Meth.*, 20:309 (1988).

9. T. Yamase, B. Bogdan, E. Ishikawa, and K. Fukaya, Chemical structure and intramolecular spin-exchange interaction of $[(VO)_3(SbW_9O_{33})_2]^{12-}$, *Chem. Lett.*, 1:56 (2001).

10. G. Sazani, M. H. Dickman, and M. T. Pope, Organotin derivatives of α-$[X(III)W_9O_{33}]^{9-}$ (X=As, Sb) heteropolytungstates. Solution state and solid state characterization of $[((C_6H_5Sn)_2O)_2H(\alpha-AsW_9O_{33})_2]^{9-}$ and $[(C_6H_5Sn)_3Na_3(\alpha-SbW_9O_{33})_2]^{6-}$, *Inorg. Chem.*, 39:939 (2000).

11. a) M. R. Bond, L. M. Mokry, T. Otieno, J. Thompson, and C. J. Carrano, Three new polynuclear bis(μ-phosphato)vanadyl clusters. $[HB(pz)_3VO(\mu-(C_6H_5O)_2PO_2)]_2$, $[HB(3,5-Me_2pz)_3VO(\mu-(C_6H_5O)_2PO_2)]\cdot C_7H_8$, and $H_2O\subset[t-Bupz(\mu-C_6H_5OPO_3)VO]_6(H_2O)_2\cdot2CH_3CH_2OH$. Adaptability of the cyclic $(OV)(OPO)_2(VO)$ bridging unit, *Inorg. Chem.*, 34:1894 (1995); b) C. W. Hahn, P. G. Rasmussen, and C. Bayón, A dinuclear vanadyl(IV) complex of 3,5-dicarboxypyrazole. Synthesis, crystal structure, and electron spin resonance spectra, *Inorg. Chem.*, 31:1963 (1992); c) V. Thanabad and V. Krishnan, Cation-induced crown porphyrin dimers of oxovanadium(IV), *Inorg. Chem.*, 21:3606 (1982).;d) P. D. W. Boyd,and T. D. Smith, An electron spin resonance study of the metal ion separations in dimeric copper(II) and vanadyl chelates of 4,4',",4'''-tetrasulphophthalocyanine, *J. Chem. Soc. Dalton Trans.*, 839 (1972); e) B. L. Belford, N. D. Chasteen, H. So, and R. E. Tapscott, Triplet state of vanadyl tartrate binuclear complexes and electron paramagnetic resonance spectra of the vanadyl α-hydroxycarboxylates, *J. Amer. Chem. Soc.*, 91:4675 (1969).

12. C. J. O'Connor, Magnetochemistry-Advances in theory and Experimentation, *Prog. Inorg. Chem.*, 29:203 (1982).

13. S. Shigeta, S. Mori, J. Watanabe, S. Soeda, K. T9akahashi, ad T. Yamase, Synergistic antiinfluenza virus-A (H1N1) activities of PM-523 (polyoxometalate) and ribavirin in-Vitro and in-Vivo, *Antimicrob. Agents Chemother.*, 41:1423 (1997) and S. Shigeta, S. Mori, J. Watanabe, T. Yamase, and R. Schinazi, In-vitro anti myxovirus activity and mechanism of anti-influenzavirus activity of polyoxometalates PM-504 and PM-523, *Antiviral Chem. Chemother.*, 7:346 (1996).

TETRAVANADATE, DECAVANADATE, KEGGIN AND DAWSON OXOTUNGSTATES INHIBIT GROWTH OF *S. cerevisiae*

Debbie C. Crans,[1]* Harvinder S. Bedi,[3] Sai Li,[3] Boyan Zhang,[1]
Kenji Nomiya,[2]* Noriko C. Kasuga,[2] Yukihiro Nemoto,[2]
Keiichi Nomura,[2] Kei Hashino,[2] Yoshitaka Sakai,[2] Yosief Tekeste,[3]
Gary Sebel,[3] Lori-Ann E. Minasi,[3] Jason J. Smee,[1] and Gail R. Willsky[3]*

[1]Department of Chemistry, Colorado State University
 Fort Collins, Colorado 80523-1872
[2]Department of Materials Science, Kanagawa University
 Hiratsuka, Kanagawa 259-1293, Japan
[3]Department of Biochemistry, SUNY at Buffalo School of Medicine and
 Biomedical Sciences, 140 Farber Hall, Buffalo, New York 14214
*To whom correspondence should be addressed.

INTRODUCTION

Interactions of oxometalates with proteins has been of considerable interest since the initial discovery of the anti-HIV activity of 9-antimonio(III)-21-tungsten(VI)-sodate (HPA-23)[1] and a series of polyoxometalates.[2, 3] These studies were followed by a series of reports concerning the beneficial effects of this class of complex anions. The ribosomal structure represents a high point in these studies.[4-11] In these crystallo-graphic studies the polyoxometalate anions play a critical role as seen through MIR and MAD phasing. The early low-resolution phase sets established by the use of oxometalate derivatives have been dramatically extended by the use of other heavy atom derivatives to give 2.4 Å resolution for the larger ribosomal subunit.[5,6,9] However, details of the bonding interactions of these anions with the macromolecular structures typically have not been reported since the oxometalates were only used in connection with the low-resolution data.[5,11] Few structural model studies have been carried out with simple organic ligands; this work documents some of the anticipated molecular types of interactions.[12-14]

Studies with simple oxometalates have also been carried out, and even the smaller anions of this class of compounds are found to have affinities for proteins.[2] Specifically, in mixtures of oxovanadates, the tetramer (V_4) and the decamer (V_{10}) are found to be better inhibitors than the other oxovanadates in solution for many enzymes (for representative publications see Ref. 15-18). Since many of these enzymes are proteins which lack any significant positively charged surface regions, the affinity between oxometalate and protein appears to be governed by factors other than charge.[2] Several studies have also been

Polyoxometalate Chemistry for Nano-Composite Design
Edited by Yamase and Pope, Kluwer Academic/Plenum Publishers, 2002

181

carried out exploring additional types of protein-oxometalate interactions.[19-21]   Our studies have led us to propose that two types of proteins exist, one in which Coulombic interactions are most important,[21] and a second type in which efficacy is based on dispersion forces, hydrogen bonding and oxometalate-protein complementarity.[2]

Given the applications of oxometalates in crystallographic studies and their potential pharmacological use, information regarding the interaction of oxometalates with proteins and how the oxometalates enter cells is needed.   Such investigations require selection of an appropriate biological assay to report on the problem under examination.   We have previously examined the interaction of Keggin and Dawson compounds with isolated soluble enzymes using spectrophotometric and NMR assays.[22]   However, the examination of membrane-bound enzymes requires a different approach.   The effect of oxometalates on growth can be used as an initial screen for specific oxometalate interaction with the cell surface.[23]   At this time we do not know the mechanism of such interaction and whether protein-oxometalate interactions are involved.   However, it is important to document the interaction and investigate the compounds' efficacy before undertaking further mechanistic studies.   Herein we report a screen of the effect of simple oxovanadate, decavanadate, Keggin and Dawson oxotungstates on the growth of yeast *S. cerevisiae*.

EXPERIMENTAL

Materials

The chemicals used for preparation of the oxometalates were of reagent grade and obtained from several sources.   The chemicals used for the growth experiments were of analytical grade and obtained from Sigma Chemical Company.   The yeast strain used was the auxotrophic *S. cerevisiae* strain LL20 (his⁻, leu⁻) which was obtained from Joel Huberman.[23,24]   The rich media used was Yeast Extract-Peptone- 2% dextrose (YPD).   The minimal medium used was Yeast Nitrogen Base with amino acids (YNB) with 2% dextrose.   Yeast Extract, BactoPeptone and Yeast Nitrogen Base were obtained from Difco Laboratories, Detroit, Michigan.   The media in some experiments were buffered with 100 mM tris-succinate in order to maintain the same pH during the entire experiment;[24] see Growth Measurements section below.

Preparation of oxometalates and determination of purity

The oxometalates ($Na_6[V_{10}O_{28}] \cdot 18H_2O$ (V_{10}),[25] $Na_3[PW_{12}O_{40}] \cdot 11H_2O$ (PW_{12}), $K_4[PV(V)W_{11}O_{40}] \cdot 4H_2O$ ($PV(V)W_{11}$),[26] $K_5[PV(IV)W_{11}O_{40}] \cdot 7H_2O$ ($PV(IV)W_{11}$),[27] $K_6[1,2,3$-$PV_3W_9O_{40}] \cdot 5H_2O$ ($1,2,3$-PV_3W_9),[26,28] $K_6[1,4,9$-$PV_3W_9O_{40}] \cdot 5H_2O$ ($1,4,9$-PV_3W_9),[28] $K_6[P_2W_{18}O_{62}] \cdot 10H_2O$ (P_2W_{18}),[29] $K_7[P_2V(V)W_{17}O_{62}] \cdot 10H_2O$ ($P_2V(V)W_{17}$),[30] $K_8[P_2V(IV)W_{17}O_{62}] \cdot xH_2O$ (where x = 10-11) ($P_2V(IV)W_{17}$)[30] and $K_8[HP_2V_3W_{15}O_{62}] \cdot 9H_2O$ ($1,2,3$-$P_2V_3W_{15}$)[31]) were prepared according to literature methods.   The yields were reasonable, and their compositions and purities were determined by using elemental analysis, FT-IR, TG/DTA, ^{31}P and ^{51}V and when appropriate ^{183}W NMR spectroscopy.[32] Elemental analyses were performed at Mikroanalytisches Laboratorie Pascher, Germany and at Desert Analytics, Tucson, Arizona.   TG/DTA curves were measured on a Seiko SSC 5000 TG-DTA 300 thermometer with a temperature ramp of 4 °C per min and the temperature range scanned was 20 °C to 550 °C under air atmosphere unless otherwise noted.   The absorption spectra were recorded on a Lambda 4B spectrophotometer and the IR spectra were recorded on a JASCO 300 FT-IR spectrometer as KBr disks at ambient temperature.   The NMR spectra were recorded on an ACP-300 MHz Bruker or a JEOL EX-400 FT NMR spectrometer.   The ^{31}P NMR spectra were recorded at 121 or 162 MHz using 80% H_3PO_4 as an external reference.   The ^{51}V NMR spectra were recorded at 79 or 105 MHz using $VOCl_3$ as an external reference.[33,34] ^{183}W NMR spectra were recorded at 4.22 MHz.

Oxometalate Properties

Absorption spectra are obtained for each compound to determine which optical density (OD) to use for the growth experiment, and at what pH to prepare the solutions and the growth media.   These studies also defined what pH value to use for the reference sample (control).   The solubility of the oxometalates were first determined in distilled water and then in growth media.   With a few exceptions, all the compounds used in this study were soluble and stable at 5.0 mM in distilled water.   However, the solubility was significantly less for compounds PW_{12} (0.50 mM) and $PV(V)W_{11}$ (1.3 mM) in the growth media.   The stability of the complexes were determined in the stock solutions prepared in distilled water and then in media.   The polyoxoanion solution in media is prepared without buffer to increase the solubility and to assure that the pH in media does not overpower the pH resulting from the oxometalate.

Growth Measurements

Yeast growth at 30 °C was monitored as changes in light scattering as seen in the changes in optical density (OD) with time; the OD measurements were made after 0, 6, 12 and at 24 hours.   Commonly the light scattering is measured at 600 nm (OD_{600}).[23,24]   The OD_{800} was used in this work if the oxometalate absorbs at 600 nm.   (The media itself does not show any significant absorbance at either 600 or 800 nm.)   If absorption was observed in the oxometalate-media mixture at both wavelengths, the wavelength with the smaller absorption was chosen for observation.   Cell stocks and growth experiments concerning vanadate were incubated in 10-20 ml volumes in Erlenmeyer flasks and aerated using a G76 Gyrotory water bath (New Brunswick Scientific, Edison, NY).   Other growth experiments were done in 2.5 or 5.0 ml volumes in 18 x 150 mm capped culture tubes aerated on a TC-7 Roller Drum from New Brunswick.

Yeast strains acidify the pH with growth and yeast growth typically results in a pH drop of 0.5 unit for freely growing yeast in buffered media after 24 hours.[23]   In unbuffered media the pH drop is greater.   In most experiments containing the poly-oxoanions no buffer was added to the media.   The pH of the medium was monitored at each time-point when growth was monitored.   A buffered control at the same pH as the original media containing the polyoxometalate was used as a control to evaluate the inhibition exhibited by the oxometalate.

The yeast strain was maintained on yeast extract-peptone-2% dextrose (YPD) plates at 0 °C, which were good for two months.   Single colonies were re-suspended into 5 mL of YPD liquid media for overnight growth.   A 1:20 dilution was made into minimal medium (yeast nitrogen base with amino acids and 2% dextrose containing 5 mg/mL histidine and lysine buffered to pH 6.5 with 100 mM tris-succinate).   This stock was good for 2-3 weeks at 0°C.   The evening before an experiment the stock was diluted into the buffered or unbuffered medium to generate cells growing in the log phase the following day (1:20 to 1:100 dilution depending on the cell concentration of the stock and the incubation time).   For each experiment, cells were diluted into a culture containing tenfold more cells than desired for the experiment.   The experiments were started by diluting this culture 1:10 into fresh media to obtain an OD_{600} of about 0.1.   The growth stocks used to generate the yeast samples contained tris-succinate buffer, and the overnight incubations of yeast before each experiment contained either this buffer[23,24] or no buffer if the experiment involved the addition of polyoxometalates to the samples.   Each experimental point represents the mean ± the standard deviation of the OD of three experimental cell cultures containing media with or without the polyoxoanions. The final growth OD is the original sample value less the OD of the media without cells.[23,24] This subtraction is needed to correct for any changes in absorption of the polyoxometalate containing media in the presence of cells. The pH of each growth point is also determined. Each time point involves taking 0.8 to 1.0 mL of sample. The three time points for each sample are pooled and the pH determined of the pooled sample.

RESULTS

Defining Growth Experiments for Comparing Effects of Compounds

In a study where a wide range of compounds are to be screened, developing an assay that can be investigated in the absence of buffer would be preferable. Yeast is an organism that grows under a wide range of conditions, and this flexibility may be well suited for screening the effects of a wide range of compounds. In such an assay, yeast cells are grown overnight to obtain cells in the early log phase. These cells are re-suspended in growth media containing the oxometalate under examination. The parameters important to yeast growth are temperature, nutrients, salts, pH of the medium and having cells that are in exponential growth. The optimum growth temperature for yeast is 30 °C and this temperature was used throughout these studies. The yeast will grow in a minimal medium in the presence of a variety of nutrients (including $(NH_4)_2SO_4$) and often a buffer (to maintain the pH of the solution during growth). Since buffers are likely to react with some of the more reactive oxometalates, attempts were made to investigate the possibility that the experiments could be carried out in the absence of buffer. Studies were conducted at a series of pH 2 - 9 in which the growth (OD) and the pH of the medium were monitored (data not shown). We found that the yeast grew with similar rates in the pH range 3 to 7.2. At pH 2, 8 and 9 no growth was observed and at pH 2.5 the growth was slightly slower. These results suggest that the effects of the various oxometalates on the yeast growth can be compared at different pH. Thus, the screen of compounds may be carried out at the respective pH where oxometalates are most stable.

The pH of the medium is decreased during the growth of the yeast. For yeast growing freely in 100 mM tris-succinate buffer from pH 3 to 7.2 the pH of the medium decreased slightly. The decrease was greatest in the pH 7.2 samples, where the pH had dropped one unit after 24 hours of growth, but in general a decrease of 0.5 pH units were observed in the samples. Although changes were greater in the absence of buffer, the pH never went below 2.5. Since the changes in the pH of the medium during the incubation may affect the oxometalate composition it was necessary to measure the pH at each time-point.

Given the high charges of some of the oxometalates, the possibility that ionic strength would affect growth was investigated. In the present study up to 5.0 mM oxometalate would be added and compared with similar levels of electrolyte. No observable effect of 250 mM NaCl and KCl on growth was found; the effects of 1, 1.5 and 2 M NaCl was also tested and at these significantly higher levels growth was inhibited. These studies show that should inhibition be observed at 5.0 mM oxometalate, the effect is due to the oxometalate and is not merely an ionic strength effect.

In summary, we have shown that the effects of oxometalate on yeast growth can be compared from pH 3 - 7. Determining yeast growth in this pH range would be a viable assay. Using this growth assay we are well placed to examine the effects of a series of Keggin and Dawson oxometalates.

Inhibition of *S. cerevisiae* Growth by Vanadate: V_4 is the Inhibiting Species

Vanadate is known to inhibit yeast growth.[23]   There is no inhibition in 1.0 mM vanadate but at 5.0 mM vanadate the inhibition is complete and no growth is observed at pH 6.5.   Based on the concentration dependence of this inhibition and on spectroscopic characterizations by the Willsky group,[23] it seemed possible that the inhibitory species was one of the higher oligomeric forms of vanadate.[35]   Specifically it was found that the cell-associated fraction from cells growing normally did not contain high concentrations of V_1 or any oligomeric vanadate.   The cell-associated fraction from cells that were inhibited by vanadate did contain high concentrations of V_1 and some oligomeric vanadate could be observed.[23]   These data, combined with understanding the speciation chemistry of oxovanadates led us to recognize that inhibition occurred when small concentrations of tetrameric vanadate was found in the spectra of the cell-associated fraction.   Thus, *we hypothesized that the inhibition was due to V_4, and designed a set of experiments to examine this hypothesis.*[35]   Yeast cells were grown overnight and a series of cultures were prepared in which from 0 to 5.0 mM vanadate was added and the growth monitored at 0, 4, 8 and 20 hrs.   The yeast growth is shown in semi-log plots showing OD_{600} as a function of time (Figure 1).   The yeast incubated in the presence of less than 1.5 mM vanadium(V) did not show any inhibition of growth, however, in all the yeast cultures containing 2.0 mM or more vanadium(V) growth, inhibition was observed.

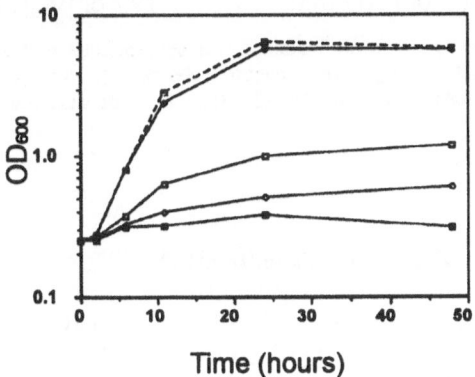

Fig. 1.   Growth of the yeast *S. cerevisiae* in the presence of vanadate.   Control (no additions) (----), Vanadate added as: 1.0 mM ($\blacklozenge$), 2.0 mM ($\square$), 3.5 mM (O) and 5.0 mM ($\blacksquare$).   Vanadate added as V_1 from sterile 100 mM sodium orthovanadate stock.[35]

Studies were conducted in which the growth was monitored as a function of time, and at the same time [51]V NMR spectra of the original growth media were recorded.[17,33,35] These spectra allowed the calculation of the concentration of species of the vanadium(V) that existed in the growth medium.[17]   Figures 2A and 2B show inhibition of cell growth plotted as a function of the concentration of vanadium species obtained by [51]V NMR spectroscopy.   Inhibition, plotted as a percentage, is calculated as 100% × (observed growth) / (growth of uninhibited cells).   The concentration of oxovanadate is calculated from the mole fraction of the species that is obtained from the [51]V NMR spectra.   If one species were the major inhibiting form, the observed inhibition would be expected to be linearly proportional with the concentration of the inhibiting species.   Neither the concentration of V_1 nor that of V_2 shows a linear relationship with the growth inhibition. Although a linear relationship is observed between inhibition and [V_5], the fact is that the yeast is severely inhibited before a significant concentration of V_5 has accumulated, this species cannot alone be responsible for inhibition of the yeast. Stated differently, the intercept of the line does not go through the zero point as needed, for a compound that can explain all the observed inhibition. In contrast, the concentration of V_4 is found to be linearly proportional with growth inhibition, and this species can alone account for all of the observed inhibition. We conclude that the V_4 is the most inhibiting simple oxovanadate.[35] Further support for this conclusion comes from experiments in which 5.0 mM of the more stable molybdenum tetramer inhibited yeast cell growth, while 5.0 mM of the monomeric form of molybdenum did not inhibit cell growth.[35]

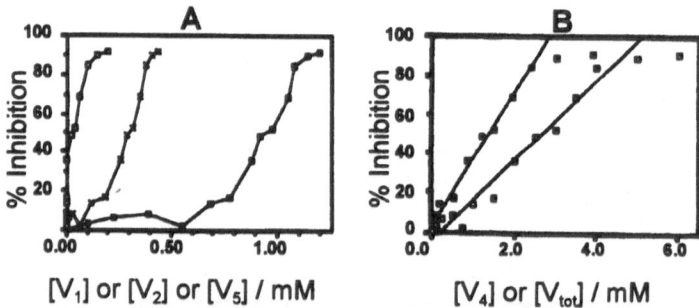

Fig. 2. The relationships between growth inhibition and concentration of vanadium species. Growth inhibition was calculated as (observed growth of control - observed growth)/(observed growth of control). A. The vanadium species: V_1 (■), V_2 (×), and V_5 (□) [35]. B. The vanadium species: V_4 (■) and V_{total} (□).[35]

Inhibition of *S. cerevisiae* Growth by Vanadium(IV)

The effect of vanadium(IV) on yeast growth was examined since some oxometalates investigated contain vanadium(IV). Yeast growing at pH 3.0 in the presence of added $VOSO_4$ were inhibited (Figure 3). In contrast, no inhibition was observed for yeast growing at pH 6.5 in the presence of added $VOSO_4$ (Figure 3). The source of vanadium(IV) used in these studies is $VOSO_4$, however, given the complex chemistry of aqueous vanadium(IV) in the neutral pH range, little vanadium(IV) remains in the form of vanadyl cation ($[VO(H_2O)_5]^{2+}$ or $[VO(OH)(H_2O)_4]^+$). Studies with vanadium(IV) are complicated by the oligomerization and polymerization reactions of the vanadyl cation in the neutral pH range.[36,37] Thus, the addition of $VOSO_4$ to a media solution with a pH > 4 will result in the conversion of some of the $VO(H_2O)_5^{2+}$ and $VO(OH)(H_2O)_4^+$ to EPR silent oligomeric and polymeric forms of vanadium(IV). Thermodynamically, less than 10^{-7} mM vanadyl cation is present at pH 7 regardless of how much $VOSO_4$ has been added to the solution.[36] Unfortunately, the kinetics of these conversions is poorly understood, and under some conditions the conversions have been found to be fast and under other conditions slow.[36,37] As seen in Figure 3, yeast growing at pH 3.0 was inhibited in the presence of added $VOSO_4$ that is mostly in the form of $VO(H_2O)_5^{2+}$. Yeast growing at pH 6.5 was not inhibited when the added $VOSO_4$ was converted into EPR silent oligomeric and polymeric forms. These observations suggest that the formation of oligomeric and polymeric species that are generated from vanadyl cation at pH 6.5, in effect, protects the yeast from the vanadyl cation.

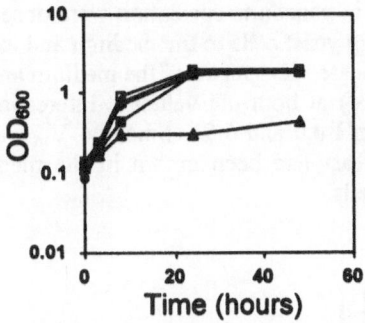

Fig. 3. Effect of VOSO₄ on the growth of the yeast *S. cerevisiae*. The symbols represent: control at pH 3.0 (♦), 1.0 mM VOSO₄ at pH 3.0 (■), 5.0 mM VOSO₄ at pH 3.0 (▲); control at pH 6.5 (□), 1.0 mM VOSO₄ at pH 6.5 (△), 5.0 mM VOSO₄ at pH 6.5 (○).

Inhibition by V_{10} of *S. cerevisiae* Growth and Demonstration of Cell Associated V_{10}

Studies carried out using 5.0 mM vanadate at pH 4.0 where all the vanadium(V) in solution is converted to V_{10} did not show any inhibition compared to a control culture containing no vanadate (Figure 4). However, recognizing that 5.0 mM vanadate corresponded to 0.50 mM V_{10}, the studies were repeated at the 50 mM vanadate level, and at this concentration inhibition was observed (Figure 4). At 5.0 mM vanadate did inhibit in a pH range where V_4 formed (Figure 3) and further illustrates the higher potency of this simple oxovanadate over V_{10}.

In order to obtain information on the cellular responses to the added vanadium(V) spectroscopic studies were carried out (Figure 5). Centrifugation of growth samples led to a media and cell fraction that were investigated by ^{51}V NMR spectroscopy. In the absence of yeast the 5.0 mM vanadate had formed mainly V_{10} with a small component of V_1 in the media at pH 4.0; V_2 and V_4 were predominant in the media at pH 6.5 as observed in the ^{51}V

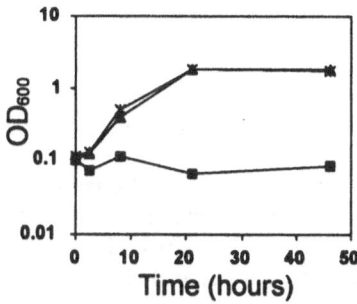

Fig. 4. Effect of decavanadate on the growth of yeast. Control (no addition) (*), 0.50 mM V_{10} (▲) and 5.0 mM V_{10} (■).

NMR spectra.   No changes in vanadium speciation were observed after 20 hours for these media samples.   After adding yeast cells to the medium and exposing them for 15 hours to the presence of 5.0 mM vanadate, the spectra of the medium and the cells show evidence of several forms of vanadium(V) at both pH values.   Interestingly, V_{10} was found in both media of samples grown at pH 4.0 and 6.5.   Since no V_{10} was found in media at pH 6.5 after 15 hours unless the yeast had been grown in the media, the formation of V_{10} is apparently due to the yeast cells.

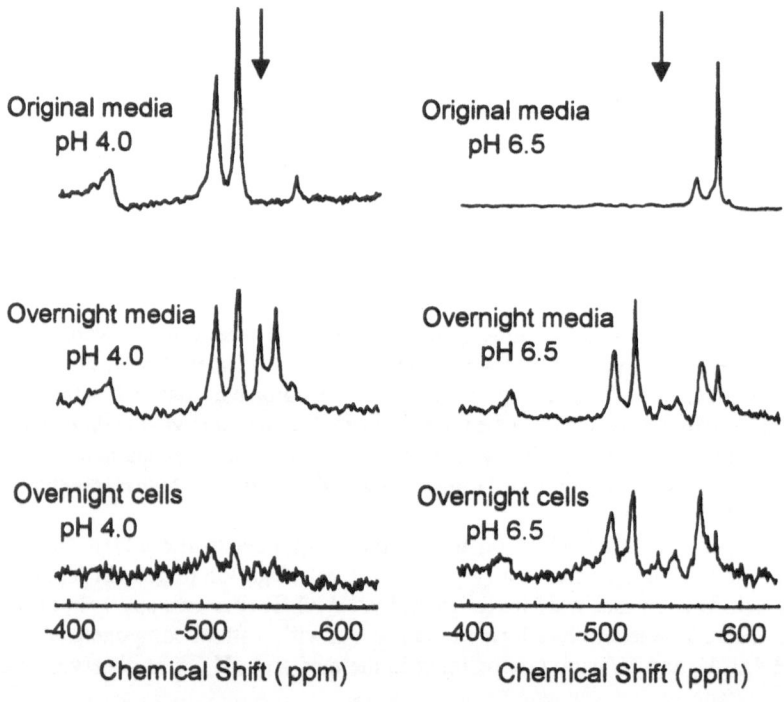

Fig. 5. ^{51}V NMR spectra of media and cell-associated fractions of the yeast *S. cerevisiae* exposed to 5.0 mM vanadate at pH 4.0 (left) and pH 6.5 (right).   The arrow indicates the chemical shift of the orthovanadate resonance at -541 ppm.   Reproduced from [38] with permission from the American Chemical Society.

Examination of the spectra of the cell-associated fraction, showed a large concentration of V_{10} in the yeast grown at pH 6.5 and only a small amount was detected in the yeast grown at pH 4.0.   Since the observed vanadium(V) can be attributed to vanadium(V) in media trapped between cells in the cell-associated fraction, it is possible that no V_{10} can enter or associate with the cells treated with 5.0 mM vanadium(V) (or 0.50 mM V_{10}) and grown at pH 4.0.   The cells treated with the labile oligovanadates at pH 6.5, show a significant concentration of cell-associated vanadium(V).   In addition to V_{10}, the medium and cell-associated samples grown at pH 6.5 also contains V_1, V_2, V_4 and a peak suggestive of some additional vanadium(V) complex.   The V_1 species is presumed to enter cells through anion transporters.[39]   The media used for these experiments contains high concentrations of phosphate (10 mM) and the phosphate specific transport system, which will accept vanadate as an alternative substrate, are not induced and will not serve as a transport system for V_1.[40]   Since the labile oligovanadates rapidly generate V_1,[34] the observed V_{10} can readily form from any V_1 that became cell-associated.   Although we cannot rule out the possibility that V_4 enters the cell, and there is evidence that oxometalates can enter cells,[19,20,41,42] the data shown in Figures 1 and 5 do not require that V_4 enter cells.

Since V_{10} is found in the media spectra obtained from cells exposed to both pH 6.5 and pH 4.0 media, after formation of the cell-associated V_{10} it is most likely that V_{10} has been extruded into the medium. After exposure to cells overnight, the pH 6.5 medium did not drop below pH 6.2, making it unlikely that the V_{10} formed spontaneously. Since the formation of V_{10} requires an acidic environment and the cytoplasmic pH values are nearly neutral, it is likely that V_{10} is formed in acidic vesicles prior to extrusion. There is precedence for the concentration of vanadium from the ocean by ascidians.[43] These invertebrate animals are believed to concentrate the vanadium in vesicles that are acidic and/or contain the appropriate chelating agents.[43] During the uptake/concentration the vanadium(V) is reduced to vanadium(IV) and/or vanadium(III) in the ascidians. It is possible that our observations in yeast may use an analogous mode of action.

Cells grown at pH 4.0 in the presence of V_{10} (and little V_4) showed little cell-associated V_{10}, and what is observed is readily attributed to trapped media. The likelihood that V_{10} cannot enter the cells is based upon EPR spectroscopic studies that were carried out (data not shown). The EPR spectra of the media showed no presence of vanadium(IV) in the absence of cells. However, the spectra of media in which cells were present for 15 hours contains vanadium(IV) presumably in the form of vanadyl succinate at pH 4.0 and at pH 6.5 (data not shown). Given the chemistry of vanadium(IV) in aqueous solution at pH 6.5, less vanadyl cation would be observed in the EPR spectra (due to the oligomerization and polymerization reactions) unless the vanadium(IV) was complexed by a ligand. Although these spectra show less vanadium(IV) in the yeast grown at higher pH, the fact that significantly less vanadium was observed in both the media samples compared to the cellular samples supports the interpretation that vanadium(IV) forms inside the cells and is then extruded. These EPR studies suggest that vanadium(V) gets into the cell even at pH 4 where most of the vanadium(V) is in the form of V_{10}. These experiments do not establish that V_{10} gets into the cells, since alternative interpretations exist. It is possible that somehow the cell surface interacts with the V_{10} and breaks it down. However, such an interpretation is less likely based on what is known about the chemical stability[44] and the interactions of vanadium with ligands of biological relevance.[12-14] Once inside the cell and regardless of the form by which it entered, the vanadium(V) is converted to vanadium(IV). Willsky previously pro-posed that yeast cells respond to vanadate treatment by reducing the vanadium(V) and extruding the vanadium(IV) into the medium.[45] Based on these studies it is possible that the system that carries out the vanadium(V) reduction may be able to process vanadium(V) in oxometalate structures.

In summary, these studies show that V_{10} weakly inhibits cell growth. Regardless of what pH the yeast are incubated, we show that cell-associated vanadium(IV) is formed. Furthermore at pH 6.5 cell-associated V_{10} is definitely seen in cells exposed to V_4 in the growth medium. We propose that V_{10} is extruded into the medium. To the best of our knowledge, this is the first experimentally documented observation of the formation of an oxometalate in a biological system.

Inhibition of *S. cerevisiae* Growth by Keggin Oxometalates

The initial screen of oxometalates was designed to identify compounds with a greater effect than V_{10}, using V_{10} as a reference compound and 5.0 mM as the initial target concentration. The effect on yeast growth by 5.0 mM of a series of Keggin structures was first investigated. The compounds were selected in order to begin to develop some structure and activity information with regard to biological effects of oxometalates with the Keggin base-unit. The oxotungstates under investigation are: $Na_3[PW_{12}O_{40}] \cdot 11H_2O$

(PW$_{12}$), K$_4$[PV(V)W$_{11}$O$_{40}$]•4H$_2$O (PV(V)W$_{11}$), K$_5$[PV(IV)W$_{11}$O$_{40}$]•7H$_2$O (PV(IV)W$_{11}$), K$_6$[1,2,3-PV$_3$W$_9$O$_{40}$]•5H$_2$O (1,2,3-PV$_3$W$_9$), and K$_6$[1,4,9-PV$_3$W$_9$O$_{40}$]•5H$_2$O (1,4,9-PV$_3$W$_9$). Previous studies with this series of oxometalates have documented the suitability of this series of compounds in identifying whether Coulombic charges are governing the effects of these oxometalates.[22]   Before the studies were conducted the solubilities and stabilities of compounds needed to be determined.

Solubility and Stability of Keggin Oxometalates.   The solubilities of the compounds were first determined in distilled water, and a 5.0 mM solution was readily found to be prepared from solid materials of all the oxometalates listed above.   The white PW$_{12}$ decomposed when dissolved in distilled water, but when dissolved in 0.01 M sulfuric acid at pH 2.0 the compound was stable for days.   Since the yeast grew poorly at this pH, yeast growth studies with this oxometalate were carried out at slightly higher pH (3.5). The yellow PV(V)W$_{11}$ formed a yellow solution at pH 3.0 and the black PV(IV)W$_{11}$ formed a purple solution at pH 5.2 when dissolved directly in distilled water.   The orange 1,2,3-PV$_3$W$_9$ and 1,4,9-PV$_3$W$_9$ formed orange solutions at pH 3.0 and 3.5, respectively.   When dissolving these compounds directly in growth media, compounds PW$_{12}$ and PV(V)W$_{11}$ did not dissolve at the 5.0 mM level.   In these cases saturated solutions were prepared at the concentrations of 0.50 and 1.4 mM, respectively.

The stability of the oxometalates was monitored under a variety of conditions by following the NMR and/or absorption spectra as a function of time.   The stabilities of the compounds were investigated at a wide range of pH values and in the presence of a variety of buffers.   Specifically, a series of buffers including imidazole, cacodylic acid, hepes, mes, tris and maleic acid, were investigated at pH values 4 to 8.   The initial studies of the Keggin structures showed that the conditions for optimum stabilities varied significantly, and as more oxometalates were included the number of acceptable buffers decreased. Initial studies were carried out using hepes as a buffer, however later studies showed that maleic acid was preferable for studying this series of compounds.

Inhibition of *S. cerevisiae* by Keggin Oxometalates.   Absorbances were measured at 0, 4, 8 and 24 hours after the oxometalates (PW$_{12}$, PV(V)W$_{11}$, PV(IV)W$_{11}$, 1,2,3-PV$_3$W$_9$, 1,4,9-PV$_3$W) were added to the media.   Control pH matched growth experiments were done in buffered media in the absence of oxometalate.   The data is shown in Figure 6.   In addition, the possibility that hydrolyzed oxotungstate could be responsible for any observed inhibition was investigated.   No inhibition was observed at the 5.0 mM Na$_2$WO$_4$ level (Figure 6) suggesting that should compound hydrolysis products form, they would not significantly contribute to the inhibition of growth.   The values shown are mean values of three determinations, and the SD in most cases is smaller than the size of the symbol.   The two most inhibitory growth curves were observed with the Keggin ions 1,2,3-PV$_3$W$_9$ and 1,4,9-PV$_3$W$_9$ which both carry a -6 charge.   The Keggin ions PV(V)W$_{11}$ and PW$_{12}$ were not soluble at the 5.0 mM level in the growth medium and the growth curves for these compounds showed some growth (i.e. less than full inhibition).   The growth curve for PW$_{12}$ shows initial growth followed by inhibition.   In this case solid compound added to the media was found to slowly dissolve, which might explain the unusual curve observed for this compound.   (True growth followed by inhibition would not result in a drop in the OD, as observed in this case.)   The measured curves suggest the following trend in growth inhibition: 1,2,3-PV$_3$W$_9$ = 1,4,9-PV$_3$W$_9$ > PV(IV)W$_{11}$ > PV(V)W$_{11}$ > PW$_{12}$.   Given the differences in oxometalate concentrations measured, it is difficult to accurately evaluate the order of effectiveness for these compounds without further experimentation.   At this time it appears that the preliminary screen suggests that Keggin ions, with greater charge, are more effective inhibitors.   An expanded study in which lower concentrations of Keggin compounds are investigated will show the validity of this hypothesis.

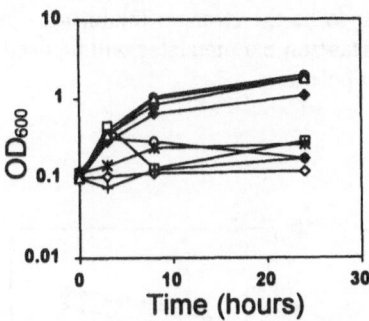

Fig. 6. Incubation of the yeast *S. cerevisiae* in the presence of Keggin oxometalates. The symbol, original pH and concentration of the oxometalate in the medium are as indicated in parenthesis: PW_{12} (□, pH 3.5, 0.50 mM), $PV(V)W_{11}$ (O, pH 3.0, 1.4 mM), $PV(IV)W_{11}$ (*, pH 5.2, 5.0 mM), $1,2,3\text{-}PV_3W_9$ (◇, pH 3.0, 5.0 mM), $1,4,9\text{-}PV_3W_9$ (+, pH 3.5, 5.0 mM), Na_2WO_4 (△, pH 6.7, 5.0 mM). Buffered controls with no oxometalate addition were carried out in the entire pH range 2.5–5.5: pH 2.5 (♦), pH 3.0 (■), and pH 5.5 (●) are shown.

Inhibition of *S. cerevisiae* Growth by Dawson Oxometalates

The initial screen of oxometalates was chosen to be at the 5.0 mM level which was necessary to observe inhibition by V_{10}. The effect on yeast growth by 5.0 mM of a series of Dawson oxometalates was determined and compared to the effects exhibited by the Keggin oxometalates examined above. The compounds were selected in order to begin to develop some structure and activity information with regard to biological effects of oxometalates with the Dawson base-unit. The oxotungstates under investigation are: $K_6[P_2W_{18}O_{62}]\cdot10H_2O$ (P_2W_{18}); $K_7[P_2V(V)W_{17}O_{62}]\cdot10H_2O$ $(P_2V(V)W_{17})$; $K_8[P_2V(IV)W_{17}O_{62}]\cdot xH_2O$ (x = 10 - 11) $(P_2V(IV)W_{17})$; and $K_8[1,2,3\text{-}HP_2V_3W_{15}O_{62}]\cdot9H_2O$ $(P_2V_3W_{15})$.

Solubility and Stability of Dawson Oxometalates. The solubilities of the compounds were first determined in distilled water, and a 5.0 mM solution was readily found to be prepared from all the Dawson oxometalates listed above. The pale yellow-green P_2W_{18} formed a greenish solution in distilled water at pH 4.0, and readily dissolved in growth media. The greenish-yellow $P_2V(V)W_{17}$ formed a greenish solution at pH 5.0 and the dark green $P_2V(IV)W_{17}$ formed a green solution at pH 4.0 when dissolved directly in distilled water. The red-orange $1,2,3\text{-}P_2V_3W_{15}$ formed an orange-red solution at pH 4.0. The stability of the oxometalates was monitored under a variety of conditions by following the NMR and/or absorption spectra as a function of time. The maximum stability for these compounds was assumed to be that obtained when dissolving the compound in distilled water.

Inhibition of *S. cerevisiae* Growth by Dawson Oxometalates. The absorbances were measured at 0, 4, 8, and 24 hours after the addition of oxometalate (P_2W_{18}, $P_2V(V)W_{17}$, $P_2V(IV)W_{17}$, $P_2V_3W_{15}$) to the media. The data is shown in Figure 7. The values shown are the mean values of three determinations, and the SD in most cases is smaller than the size of the symbol. All the compounds show very effective inhibition at the 5.0 mM level. As seen in Figure 7 no differences between the compounds can be observed at this high level. However, in comparison with the effects exhibited by the Keggin oxometalates, the

Dawson oxometalates appear to be the stronger inhibitors. An expanded study, in which lower concentrations of the Dawson oxometalates will be used, is necessary to show which of these compounds are most potent.

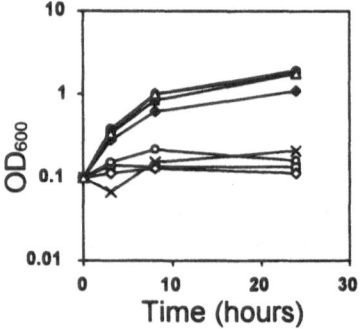

Fig. 7. Incubation of the yeast *S. cerevisiae* in the presence of Dawson oxometalates. The symbol and the pH of the growth experiments determined by the pH of compound dissolution are as indicated in parenthesis: P_2W_{18} (□, pH 4.0, 5.0 mM), $P_2V(V)W_{17}$ (◇, pH 5.0, 5.0 mM), $P_2V(IV)W_{17}$ (○, pH 4.0, 5.0 mM), $P_2V_3W_{15}$ (×, pH 4.0, 5.0 mM), Na_2WO_4 (△, pH 6.7, 5.0 mM). Buffered controls with no oxometalate addition were carried out in the entire pH range 2.5-5.5: pH 2.5 (◆), pH 3.0 (■), pH 5.5 (●).

Mechanistic Implications and Previous Studies

The studies described in this work have demonstrated that all of the tested Keggin and Dawson oxometalates are more potent than V_{10} in inhibiting yeast growth. The only compounds examined in this study that failed to inhibit growth are the buffers and other reagents used as well as the simple tungstate and molybdate salts. Previous studies with isolated enzymes and simple oxovanadates and decavanadates have documented the affinity proteins have for even the smaller oligomeric oxovanadates.[2] These results led to an unpublished study in which the affinity of a series of Keggin oxotungstates for hexokinase was determined.[22] This work was communicated at the Bielefeld symposium (1999) but not reported in the proceedings from that meeting,[46] and because of its relevance to the studies reported here, the results will be summarized briefly.

The affinity of the series of Keggin oxometalates described above for hexokinase was examined using a ^{31}P NMR assay. This assay modification was necessary given the absorption spectra of these anions precluding use of coupling assays. Studies were carried out with W_{10}, V_{10}, $PV(V)W_{11}$, $PV(IV)W_{11}$, $1,2-PV_2W_{10}$, and $1,2,3-PV_3W_9$. The results obtained by analyses of spectra were examined with regard to formation of glucose-6-phosphate (the product of the hexokinase reaction). The following order of inhibitory potency was established (most potent first): $PV^VW_{11} > PV^{IV}W_{11} = PV_2W_{10} > PV_3W_9 > W_{10} > V_{10}$. Substituting the compounds with their charges gives the following order: -4 > -5 = –5 > -6 > -4 > -6. *Increasing inhibitory potency in this series of compounds clearly does not correlate with increasing charge.* Two series of oxometalates with similar overall charges were investigated: decametalates (V_{10}, W_{10}) and Keggin structures (PV^VW_{11}, $PV^{IV}W_{11}$, PV_2W_{10}, and PV_3W_9). Since we fail to find a correlation of potency with charge (Coulombic forces) in both series, it appears that inhibitory potency may be correlated with complementarity, size, or other factors. These results are in contrast to most studies with enzymes with positively charged surfaces such as DNA and RNA processing enzymes that are governed by Coulombic forces.[2,21]   Although further studies are needed to determine if yeast growth inhibition follows the pattern observed for hexokinase or enzymes governed by Coulombic forces, the studies at the 5.0 mM level compound suggest that the latter is the case.

Studies in yeast have shown that vanadate inhibited growth, however, the experiments to determine which specific oxovanadate is inhibitory were not carried out.[23,24]   In this work we show that tetravanadate is the inhibiting species and thereby extend the selectivity of soluble isolated enzymes for V_4 to cell recognition in yeast.   Several reasonable modes of interaction are possible.   The oxometalate can bind to a transport protein or other surface relevant protein and exhibit its inhibition.   The oxometalate can also bind to protein-associated carbohydrates and in this manner exhibit its inhibition indirectly. Finally, oxometalates are large molecules extending > 10 Å, and the mode of action could simply be that they spatially cover up certain regions of the cell.   Even the smaller oxometalates such as V_{10} occupy a large volume, approximately 8 x 10 x 12 Å.   However, such a mechanism implies specific association of the oxometalate to cell surface components.   At this time we have not carried out experiments that will distinguish between the possible modes of action of inhibition.   However, studies have been conducted with some of the oxometalates that are potent inhibitors of reverse transcriptase and HIV protease, and their activity has been traced to the interaction of the oxometalate with the g-protein on the cell surface.[47]

SUMMARY

The Willsky group had previously shown that vanadate inhibits growth.   By using the combination of yeast growth and ^{51}V NMR spectroscopy we show that the inhibiting species is V_4.   Similar effects were observed when a structurally similar, chemically stable molybdate tetramer was tested.   To extend this result further, detailed studies were conducted exploring the effect of V_{10} on cell growth and it was observed that V_{10} is a poorer inhibitor than V_4.   Furthermore, it appears that the cell concentrates the absorbed vanadium into an acidic compartment where the cell-associated V_{10} will be generated. Spectroscopic studies suggest that V_{10} does enter and/or is formed inside the cell, although the mechanism is not known.   An initial screen of the inhibitory effects of a series of Keggin and Dawson oxometalate at the 5.0 mM level was conducted.   Both the Keggin and Dawson structures were found to be more potent inhibitors than V_{10}, but the Dawson complexes were found to be slightly more potent than the Keggins.   Further studies at lower concentrations are necessary to effectively compare the oxometalates' potency.

ACKNOWLEDGMENTS

DCC and GRW thank the National Institute of Health for funds characterizing the mechanisms of actions of insulin mimetic vanadium compounds.   KN thank the Ministry of Education, Science, Sports and Culture of Japan for a Grant-in Aid for Scientific Research.

REFERENCES

1. C. Jasmin, J.-C. Chermann, G. Herve, A. Teze, P. Souchay, C. Boy-Loustay, N. Raybaud, F. Simoussi, M. Raynaud, In Vivo Inhibition of Murine Leukemia and Sarcoma Viruses by the Heteropolyanion, *J. Nat. Can. Inst.* **53**, 469 (1974).

2. D. C. Crans, Enzyme Interactions with Labile Oxovanadates and Other Polyoxometalates, *Comment Inorg. Chem.* **16**, 35 (1994).

3. J. T. Rhule, C. L. Hill, D. A. Judd: Polyoxometalates in Medicine *Chem. Rev.* **98**, 327 (1998).

4. T. Douglas, M. Young: Host-Guest Encapsulation of Materials by Assembled Virus Protein Cages *Nature* **393**, 152 (1998).

5. S. Weinstein, W. Jahn, C. Glotz, F. Schlunzen, I. Levin, D. Janell, J. Harms, I. Kolln, H. A. S. Hansen, M. Gluhmann, S. S. Bennett, H. Bartels, A. Bashan, I. Agmon, M. Kessler, M. Pioletti, H. Avila, K. Anagnostopoulos, M. Peretz, T. Auerbach, F. Franceschi, A. Yonath: Metal-Compounds as Tools for the Construction and the Interpretation of Medium-Resolution Maps of Ribosomal Particles *J. Struct. Biol.* **127**, 141 (1999).

6. N. Ban, P. Nissen, J. Hansen, P. B. Moore, T. A. Steitz: The Complete Atomic-Structure of the Large Ribosomal-Subunit at 2.4 Angstrom Resolution *Science* **289**, 905 (2000).

7. A. Yonath, J. Harms, H. A. S. Hansen, A. Bashan, F. Schlünzen, I. Levin, I. Koelln, A. Tocili, I. Agmon, M. Peretz, H. Bartels, W. S. Bennett, S.; Krumbholz, D. Janell, S. Weinstein, T. Auerbach, H. Avila, M. Pioletti, S. Morlang, F. Franceschi: Crystallographic Studies on the Ribosome, a Large Macromolecular Assembly Exhibiting Severe Nonisomorphism, Extreme Beam Sensitivity and No Internal Symmetry *Acta Cryst.*, **A54**, 945 (1998).

8. J. Fu, A. L. Gnatt, D. A. Bushnell, G. J. Jensen, N. E. Thompson, R. R. Burgess, P. R. David, R. D. Kornberg, R. D.: Yeast RNA-Polymerase-II at 5 Angstrom Resolution *Cell* **98**, 799 (1999).

9. B. T. Wimberly, D. E. Brodersen, W. M. Clemons, Jr, R. J. Morgan-Warren, A. P. Carter, C. Vonrhein, T. Hartsch, V. Ramakrishnan: Structure of the 30s Ribosomal-Subunit *Nature* **407**, 327 (2000).

10. J. Lowe, D. Stock, B. Jap, P. Zwickl, W. Baumeister, R. Huber: Crystal-Structure of the 20s Proteasome from the Archaeon T-Acidophilum at 3.4-Angstrom Resolution *Science* **268**, 533 (1995).

11. R. Ladenstein, A. Bacher, R. Huber: Some observations of a correlation between the symmetry of large heavy-atom complexes and their binding sites on proteins *J. Mol. Biol.* **195**, 751 (1987).

12. D. C. Crans, M. Mahroof-Tahir, O. P. Anderson, M. M. Miller: X-Ray Structure of $(NH_4)_6(Gly-Gly)_2V_{10}O_{28}\cdot4H_2O$ - Model Studies for Polyoxometalate-Protein Interactions *Inorg. Chem.* **33**, 5586 (1994).

13. T. Yamase, M. Inoue, H. Naruke, K. Fukaya: X-Ray Structural Characterization of Molybdate-Tripeptide Complex, $Mo_4O_{12}(Glycylglycylglycine)_2\cdot9H_2O$ *Chem. Lett.* 563 (1999).

14. M. Biagioli, L. Strinna-Erre, G. Micera, A. Panzanelli, M. Zema: Tetrahydrogendecavanadate(V) and Its Binding to Glycylglycine *Inorg. Chem. Comm.* **2**, 214 (1999).

15. J. M. Messmore, R. T. Raines: Decavanadate Inhibits Catalysis by Ribonuclease-A *Arch. Boichem. Biophys.* **381**, 25 (2000).

16. S. Pluskey, M. Mahroof-Tahir, D. C. Crans, D. S. Lawrence: Vanadium Oxoanions and cAMP-Dependent Protein-Kinase - An Anti-Substrate Inhibitor *Biochem. J.* **321**, 333 (1997).

17. D. C. Crans, E. M. Willging, S. K. Butler: Vanadate Tetramer As the Inhibiting Species in Enzyme-Reactions Invitro and Invivo *J. Am. Chem. Soc.* **112**, 427 (1990).

18. D. W. Boyd, K. Kustin, M. Niwa, Do Vanadate Polyanions Inhibit Phosphotransferase Enzymes?, *Biochem. Biophys. Acta* **827**, 472 (1985).

19. C. L. Hill, M. Hartnup, M. Faraj, M. Weeks, C. M. Prosser-McCartha, R. B. Brown, Jr., M. Kadkhodayan, J.-P. Sommadossi, R. F. Schinazi: in *Advances in Chemotherapy of AIDS*, eds. R. B. Diasio, Sommadossi, J.-P., Pergamon Press Inc., New York (1990) p. 33.

20. C. Hill, M. S. Weeks, R. F. Schinzai: Anti-HIV-1 Activity, Toxicity, and Stability Studies of Representative Structural Families of Polyoxometalates *J. Med. Chem.* **33**, 2767 (1990).

21. S. G. Sarafianos, U. Kortz, M. T. Pope, M. J. Modak: Mechanism of Polyoxometalate-Mediated Inactivation of DNA-Polymerases - An Analysis with HIV-1 Reverse-Transcriptase Indicates Specificity for the DNA-Binding Cleft *Biochem. J.* **319**, 619 (1996).

22. D. C. Crans, B. Zhang, K. Nomiya, unpublished results.

23. G. R. Willsky, D. A. White, B. C. McCabe, Metabolism of Added Orthovanadate to Vanadyl and High-molecular-weight Vanadates by *Saccharomyces cerevisiae*, *J. Biol. Chem.* **259**, 13273 (1984).

24. G. R. Willsky, J. O. Leung, P. V. Offermann, Jr., E. K. Plotnick, S. F. Doesch, Isolation and Characterization of Vanadate-Resistant Mutants of *Saccharomyces cerevisiae*, *J. Bact.* **164**, 611 (1985).

25. G. K. Johnson, R. K. Murmann, Sodium and Ammonium Decavandates(V), *Inorg. Synth.* **19**, 140 (1979).

26. P. J. Domaille, Vanadium(V) Substituted Decatungstophosphates, *Inorg. Synth.* **27**, 96 (1990).

27. D. P. Smith, H. So., J. Bender, M. T. Pope, Optical and Electron Spin Resonance Spectra of the 11-Tungstovanado(IV)phosphate Anion. A Heteropoly Blue Analog, *Inorg. Chem.* **12**, 685 (1973).

28. P. J. Domaille, 1- and 2-Dimensional Tungsten-183 and Vanadium-51 NMR Characterization of Isopolymetalates and Heteropolymetalates, *J. Am. Chem. Soc.* **106**, 7677 (1984).

29. R. G. Finke, M. W. Droege, P. J. Domaille, Trivacant Heteropolytungstate Derivatives. 3. Rational Syntheses, Characterization, Two-Dimensional ^{183}W NMR, and Properties of $P_2W_{18}M_4(H_2O)_2O_{68}^{10-}$ and $P_4W_{30}M_4(H_2O)_2O_{112}^{16-}$ (M= Co, Cu, Zn), *Inorg. Chem.* **26**, 3886 (1987).

30. S. P. Harmalker, M. A. Leparulo, M. T. Pope, Mixed-Valence Chemistry of Adjacent Vanadium Centers in Heteropolytungstate Anions. 1. Synthesis and Electronic Structures of Mono-, Di-, and Trisubstituted Derivatives of α-$[P_2W_{18}O_{62}]^{6-}$, *J. Am. Chem. Soc.* **105**, 4286 (1983).

31. R. G. Finke, B. Rapko, R.J. Saxton, P. Domaille, Trisubstituted Heteropolytungstates as Soluble Metal Oxide Analogues. 3. Synthesis, Characterization, ^{31}P, ^{29}Si, ^{51}V, and 1- and 2-D ^{183}W NMR, Deprotonation, and H$^+$ Mobility Studies of Organic Solvent Soluble Forms of $HxSiW_9V_3O_{40}^{x-7}$ and $H_xP_2W_{15}V_3O_{62}^{x-9}$, *J. Am. Chem. Soc.* **108**, 2947 (1986).

32. K. Nomiya, Y. Nemoto, T. Hasegawa, S. Matsuoka: Multicenter Active-Sites of Vanadium-Substituted Polyoxometalate Catalysts on Benzene Hydroxylation with Hydrogen-Peroxide and 2 Reaction Types with and Without an Induction Period *J. Mol. Catal. A.* **152**, 55 (2000).

33. D. Rehder, A Survey of Vanadium-51 NMR Spectroscopy, *Bull. Mag. Reson.* **4**, 33 (1982).

34. D. C. Crans, C. D. Rithner, L. A Theisen: Application of Time-Resolved V-51 2D NMR for Quantitation of Kinetic Exchange Pathways Between Vanadate Monomer, Dimer, Tetramer, and Pentamer *J. Am. Chem. Soc.* **112**, 2901 (1990).

35. D. C. Crans, G. R. Willsky: *Oxovanadate and Oxomolybdate Cluster Interactions with Enzymes and Whole Cells*, 25th Steenbock Symposium, 116 (1997).

36. N. D. Chasteen, The Biochemistry of Vanadium, *Structure and Bonding* **53**, 105 (1983).

37. D. C. Crans, A. S. Tracey: *The Chemistry of Vanadium in Aqueous and Nonaqueous Solution*, ACS Symposium Series, **711**, 2 (1998).

38. G. R. Willsky, A. B. Goldfine, and P. J. Kostyniak: *Pharmacology and Toxicology of Oxovanadium Species: Oxovanadium Pharmacology*, ACS Symposium Series, **711**, 278 (1998).

39. K. Kustin, W. E. Robinson: Vanadium Transport in Animal Systems In *Metal Ions in Biological Systems*, eds. H. Sigel, A. Sigel, Marcel Dekker, Inc., New York, **31**, 511 (1995).

40. G. W. F. H. Borst-Pauwels, Ion Transport in Yeast, *Biochim. Biophys. Acta* **650**, 88 (1981).

41. S. Shigeta, S. Mori, J. Watanabe, M. Baba, A. M. Khenkin, C. L. Hill, R. F. Schinazi: In-Vitro Antimyxovirus and Anti-Human-Immunodeficiency-Virus Activities of Polyoxometalates *Antivir. Chem. Chemoth.* **6**, 114 (1995).

42. G.-S. Kim, D. A. Judd, C. L. Hill, R. F. Schinazi: Synthesis, Characterization, and Biological-Activity of a New Potent Class of Anti-HIV Agents, the Peroxoniobium-Substituted Heteropolytungstates *J. Med. Chem.* **37**, 816 (1994).

43. H. Michibata: The Mechanism of Accumulation of Vanadium by Ascidians - Some Progress Towards an Understanding of This Unusual Phenomenon *Zool. Sci.* **13**, 489 (1996).

44. A. Müller, F. Peters, M. T. Pope: Polyoxometalates - Very Large Clusters - Nanoscale Magnets *Chem. Rev.* **98**, 239 (1998).

45. G. R. Willsky: *Vanadium in the Biosphere*, in *Vanadium in Biological Systems: Physiology and Biochemistry*, ed. N. D. Chasteen, Kluwer, Boston (1990), p. 1.

46. M. T. Pope, A. Müller: *Polyoxometalates: From Topology to Self-Assembly to Applications*, Kluwer, Dodrecht in press (2000).

47. N. Yamamoto, D. Schols, E. de Clercq, Z. Debyser, R. Pauwels, J. Balzarini, H. Nakashima, M. Baba, M. Hosoya, R. Snoeck, J. Neyts, G. Andrei, B. A. Murrer, B. Theobald, G. Bossard, G. Henson, M. Abrams, D. Picker: Mechanism of Anti-Human-Immunodeficiency-Virus Action of Polyoxometalates, a Class of Broad-Spectrum Antiviral Agents *Mol. Pharm.* **42**, 1109 (1992).

SELECTIVE OXIDATION OF HYDROCARBONS WITH MOLECULAR OXYGEN CATALYZED BY TRANSITION-METAL-SUBSTITUTED SILICOTUNGSTATES

Noritaka Mizuno, Masaki Hashimoto, Yasutaka Sumida, Yoshinao Nakagawa and Keigo Kamata

Department of Applied Chemistry, School of Engineering, The University of Tokyo, Hongo, Bunkyo-ku, Tokyo 113-8656, Japan

INTRODUCTION

Catalytic oxidations of hydrocarbons are important both industrially and in organic synthesis.[1-3] Among hydrocarbons, alkanes have attracted much attention because they are abundant as resources and low in reactivities.[4-8] In particular, the selective catalytic oxidation using molecular oxygen is a desirable technology and is an area of continuous research and development.[8,9]

However, there are only a few examples of oxidations of hydrocarbons without reducing reagents or radical initiators because of the degradation of organic ligands of catalysts. $K[Ru^{III}(saloph)Cl_2]$ (saloph = *bis*(salicyaldehyde-*o*-phenylene diiminato)[10] and $[PW_9O_{37}\{Fe_{3-x}Ni_x(OAc)_3\}]^{(9+x)-}$ (x = predominantly 1)[11] are the examples of catalysts for cyclohexane oxidation. For the achievement of catalytic selective oxidation of hydrocarbons with 1 atm molecular oxygen without any additives, the high catalyst turnover numbers are needed and the information on the active structure is important.

Catalytic function of polyoxometalates has attracted much attention because their acidic and redox properties can be controlled at atomic or molecular levels.[3,4,6,12-16] The strong acidity or oxidizing property of polyoxometalates induces a lot of studies on the heterogeneous and homogeneous catalysis. The additional attractive and technologically significant aspects of polyoxometalates in catalysis are their inherent stability towards oxygen donors such as molecular oxygen and hydrogen peroxide. Therefore, polyoxometalates are useful catalysts for liquid-phase oxidations of hydrocarbons.

Here, our recent work on the synthesis and liquid-phase selective oxidation catalysis of γ-SiW$_{10}\{Fe(OH_2)\}_2O_{38}{}^{6-}$ with molecular oxygen is mainly described.[17-19]

Polyoxometalate Chemistry for Nano-Composite Design
Edited by Yamase and Pope, Kluwer Academic/Plenum Publishers, 2002

SYNTHESIS OF γ-SiW$_{10}${Fe(OH$_2$)}$_2$O$_{38}$$^{6-}$

TBA-I was synthesized by the reaction of the lacunary $[\gamma\text{-SiW}_{10}O_{36}]^{8-}$ with Fe(NO$_3$)$_3$ in an acidic aqueous solution and isolated as the *tetra-n*-butylammonium salt: Stoichiometric amounts of K$_8$[γ-SiW$_{10}$O$_{36}$]·12H$_2$O (1.0 mmol) and Fe(NO$_3$)$_3$·9H$_2$O (2.0 mmol) were mixed at pH = 3.9 and room temperature and I was isolated as the hydrophobic quaternary tetrabutylammonium salt. The yellow-orange precipitate, TBA-I, was obtained in 50% yield. The molecular formula of TBA-I was established by the elemental analysis. The thermogravimetric analysis shows the presence of 3 H$_2$O in accordance with the data of elemental analyses. The molecular weight of the monomeric polyoxometalate, SiW$_{10}$Fe$_2$O$_{38}$$^{6-}$, is 2586.26 and that of the dimeric anion, for example, [(SiW$_{10}$Fe$_2$O$_{38}$)$_2$O$_2$]$^{16-}$, is 5204.51. The negative ion FAB mass spectrum showed no intense peaks in the range of 4200 - 10000, in agreement with the monomer formulation of I.

TBA-I has been characterized by infrared, Raman, ^{183}W NMR, UV-*visible*, Mössbauer, and ESR spectroscopy, TG/DTA, FAB-mass, and magnetic susceptibility measurements, elemental analysis, and acid/base titration. The ^{183}W NMR, infrared, Raman, and UV-*visible* spectroscopy indicated that I has a γ-Keggin structure with C$_{2v}$ symmetry. For example, the ^{183}W NMR spectrum of TBA-I in acetonitrile-d_3 showed two peaks at -1334 and -1847 ppm with integrated intensities 2:1, respectively. The lower-field signal at -1334 ppm is assigned to equivalent W9-W10-W11-W12 atoms. The other signal at -1847 ppm is assigned to the W7 and W8 atoms. The signals due to W3-W4-W5-W6 atoms bound to FeO$_5$ were not observed because of the coupling with paramagnetic high-spin d^5 Fe^{3+}. The spectral pattern of TBA-I is analogous to that of [γ(1,2)-SiW$_{10}$Mn$^{II,II}_2$O$_{40}$]$^{12-}$, which showed two broad ^{183}W signals were observed at -1181 ppm and -1700 ppm with 2:1 intensities, respectively. These facts clearly indicate that the two iron atoms occupy edge-shared octahedra at 1 and 2 positions.

It was shown by the magnetic susceptibility measurement and Mössbauer, ESR, and UV-*visible* spectroscopy that TBA-I shows an antiferromagnetic coupling of the two high-spin Fe^{3+} centers. For example, the δ, ΔE_Q, and J values of TBA-I, *diferric* complexes, oxidized methane monooxygenase (abbreviated as MMO$_{ox}$), and oxidized ribonucleotide reductase (abbreviated as RR$_{ox}$) are shown in Table 1. The δ and ΔE_Q values for TBA-I are close to those for MMO$_{ox}$, **1**, **2**, and **4** with the symmetrical iron centers, and different from those for RR$_{ox}$ (δ_1 = 0.53 mms^{-1}, ΔE_{Q1} = 1.65 mms^{-1}, and δ_2 = 0.45 mms^{-1}, ΔE_{Q2} = 2.45 mms^{-1}) with asymmetrical two iron centers and **8**. The antiferromagnetic interaction (J = -5~-1 cm^{-1}) for TBA-I is close to those of **2**, **4**, **5**, hydroxo/aqua bridged MMO$_{ox}$ and far from those of **3**, **6**, **7**, **8**, **9**, and oxo/acetato bridged RR$_{ox}$ with one or two acetato bridges. Mössbauer and magnetic data for TBA-I are similar to those for **2**, **4**, and MMO$_{ox}$. The complex **4** and MMO$_{ox}$ have hydroxo-bridged structures. The complex **2** has a quinone-bridged structure, different from that of TBA-I. Therefore, Mössbauer and magnetic data suggest that the core structure of TBA-I is hydroxo-bridged. The proposed structure is shown in Figure 1.

Table 1. Comparison of magnetic and Mössbauer data for *di*-iron(III) complexes, oxidized MMO, and oxidized RR

Sample	J/cm^{-1}	Mössbauer	
		δ/mms^{-1}	$\Delta E_Q/mm$
TBA-**I**	-5∼-1 antiferromagnetic	0.32 (296 K)	0.81 (296
[Fe(salen)$_2$]O (salen = $C_{16}H_{14}N_2O_2{}^{2-}$) (**1**)	-95 antiferromagnetic	0.58 (300 K)	0.92 (300
Fe(salen)]$_2$Q (salen = $C_{16}H_{14}N_2O_2{}^{2-}$, Q = quinone moiety) (**2**)	-2.5 antiferromagnetic	0.36 (295 K)	0.75 (295
[Fe$_2$O(C$_{10}$H$_8$N$_2$)$_4$(CH$_3$CO$_2$)(ClO$_4$)$_3$] (**3**)	-119 antiferromagnetic	nr[1]	nr[1]
[Fe$_2$(C$_{16}$H$_{18}$N$_2$O$_2$)$_2$(OH)$_2$] (**4**)	-6.9 antiferromagnetic	0.49	0.56
[(CH$_3$)$_2$NC$_7$H$_2$NO$_4$(H$_2$O)Fe(OH)]$_2$ (**5**)	-11.7 antiferromagnetic	nr[1]	nr[1]
[Fe$_2$O(CH$_3$CO$_2$)Cl$_2$(bipy)$_2$] (bipy = 2,2'-bipyridine) (**6**)	-132 antiferromagnetic	nr[1]	nr[1]
[Fe$_2$O(CH$_3$CO$_2$)$_2$(Me$_3$TACN)$_2$](PF$_6$)$_2$ (Me$_3$TACN = 4,7-trimethyl-1,4,7-triazacyclononane) (**7**)	-119 antiferromagnetic	nr[1]	nr[1]
HB(pz)$_3$)Fe(OH)(CH$_3$CO$_2$)$_2$Fe(HB(pz)$_3$)]$^+$ (pz)$_3$ = hydrotris(1-pyrazolyl borate ion) (**8**)	-17 antiferromagnetic	0.47	1.50
C$_5$H$_{12}$N)$_3$[(CH$_3$CO$_2$){Fe(C$_6$H$_4$O$_2$)$_2$}$_2$] (**9**)	-134 antiferromagnetic	nr[1]	nr[1]
Fe$_2$L(OC$_2$H$_5$)$_2$Cl$_2$] (L = hexadentate Schiff base) (**10**)	-15.4 antiferromagnetic	nr[1]	nr[1]
Fe$_2$(salmp)$_2$] (salmp = $C_{21}H_{18}N_2O_3$) (**11**)	1.21 ferromagnetic	nr[1]	nr[1]
MMO$_{ox}$	-8 antiferromagnetic (d^5 high spin)	0.50	1.07
RR$_{ox}$	-108 antiferromagnetic (d^5 high spin)	0.53 0.45	1.65 2.45

[1] Not reported.

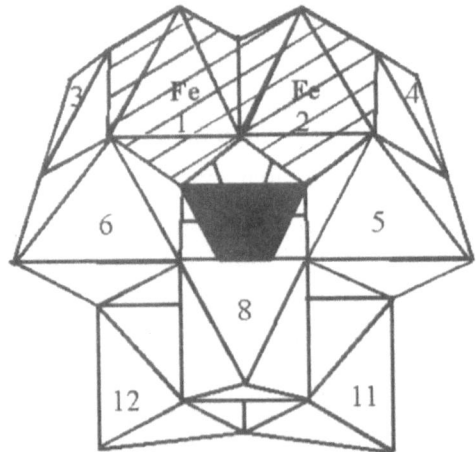

Figure 1. Polyhedral representation of **I**. The two iron atoms are represented by shaded octahedra. The WO$_6$ occupy the white octahedra, and an SiO$_4$ group is shown as the internal black tetrahedron.

OXIDATION OF ALKANES WITH MOLECULAR OXYGEN

Although the oxidation of cyclohexane has been industrialized by using cobalt or manganese acetate catalyst at pressurized molecular oxygen above 373 K,[21] it is mush more desirable that the oxidation is carried out under milder conditions. Several examples of liquid-phase oxidations of cyclohexane with 1 atm molecular oxygen and reducing reagents have been reported: Fe$_0$/pyridine (Gif system),[22] iron powder/heptanal[23] and [{Fe(HBpz$_3$)(hfacac)}$_2$O]/Zn powder (HBpz$_3$ = hydrotris-1-pyrazolylborate, hfacac = hexafluoroacetylacetone)[24] systems are the examples. On the other hand, Ishii *et al.* reported that Mn(acac)$_2$ and Co(acac)$_2$ showed high conversions of 65 and 45%, respectively, for cyclohexane oxidation in the presence of a radical initiator. *N*-hydroxyphthalimide (abbreviated as NHPI).[25,26]

TBA-**I** catalyzed the oxidation of cyclohexane with 1atm molecular oxygen at 365 K. The main products were cyclohexanol and cyclohexanone and an induction period was observed. The selectivities changed little with time. A small amount of dicyclohexyl, which is formed by the reaction of two cyclohexyl radicals, was observed. Neither acids nor oxoesters were observed. The induction period and the formation of dicyclohexyl suggest that the reaction involves a radical-chain autoxidation mechanism. The conversion was 1.1% after 96h and was increased to 2.4% by increasing the amount of catalyst by a factor of two. The turnover number of TBA-**I** reached up to 135-147 after 96 h. Table 2 compares the turnover numbers of various catalysts. The value of TBA-**I** was much higher than 18 and 5 reported for K[Ru(saloph)Cl$_2$][10] and [PW$_9$O$_{37}$\{Fe$_{3-x}$Ni$_x$(OAc)$_3$\}]$^{(9+x)-}$ (x = predominantly 1)[11], respectively, with 1 atm molecular oxygen. In addition, the value was higher than 130, 121, 90, 11 and 5 reported for Mn(acac)$_2$/NHPI,[25]

Table 2. Comparison of turnover number for cyclohexane oxidation with 1 atm molecular oxygen

Catalyst	Turnover number[1]	Refs.
TBA-I	135-147	19
K[Ru(saloph)Cl$_2$]	18	10
[PW$_9$O$_{37}${Fe$_{3-x}$Ni$_x$(OAc)$_3$}]$^{(9+x)-}$	5	11
Fe^0/PA	121	22
Iron powder/heptanal	11	23
[{Fe(HBpz$_3$)(hfacac)}$_2$O]/Zn powder	5	24
Mn(acac)$_2$/NHPI	130	25
Co(acac)$_2$/NHPI	90	26

[1] (Mol of products) / (mol of catalyst used).

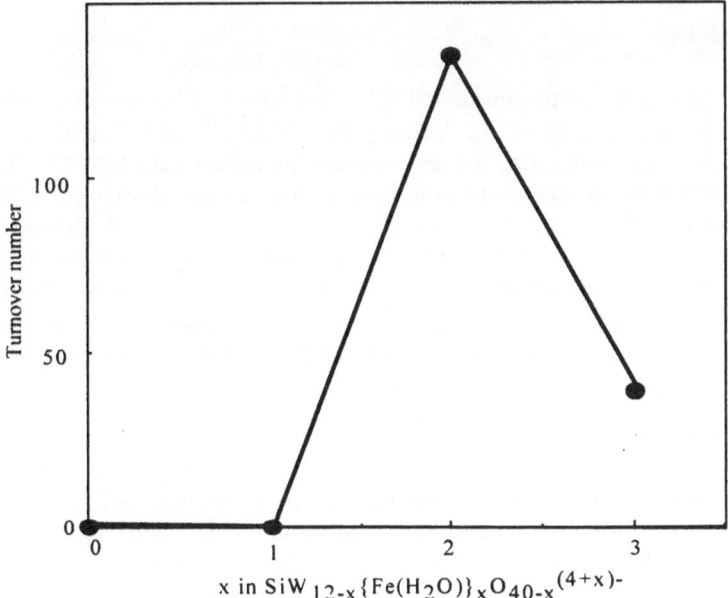

$$x \text{ in } SiW_{12-x}\{Fe(H_2O)\}_xO_{40-x}^{(4+x)-}$$

Figure 2. Comparison of turnover numbers[1] for oxidation of cyclohexane with molecular oxygen catalyzed by iron-substituted silicotungstates at 356 K.[2]
[1] Moles of products/moles of catalysts used.   [2] Reaction conditions: catalyst, 1.5 μmol; solvent, 1,2-C$_2$H$_4$Cl$_2$ (1.5 mL)/acetonitrile (0.1 mL); cyclohexane, 18.5 mmol; P(O$_2$), 1 atm; reaction time, 96 h.

Fe^0/PA,[22]   Co(acac)$_2$/NHPI,[26]   iron   powder/heptanal,[23]   and [{Fe(HBpz$_3$)(hfacac)}$_2$O]/Zn[24] systems, respectively, which work in the presence of reducing reagents or radical initiators. Radical initiator, NHPI, greatly increased the conversion from 1% to 75%. The conversion of 75% are higher than Mn(acac)$_2$+NHPI and Co(acac)$_2$+NHPI systems, which are reported to be the effective radical autoxidation system.[28,30] Adamantane was also oxidized with 1atm molecular oxygen catalyzed by γ-SiW$_{10}${Fe(OH$_2$)}$_2$O$_{38}$$^{6-}$. The conversion reached up to 3.9% after 118 h at 365 K. The main products were 1-adamantanol, 2-adamantanol and 2-adamantanone with the % selectivities of 79 : 11 : 10, respectively.   The selectivities changed little with time.

Figure 2 shows turnover numbers for cyclohexane oxidation with molecular oxygen catalyzed by iron-substituted silicotungstates.   The turnover number of TBA–I was higher than those of *non-*, *mono-* and *tri-*iron-substituted silicotungstates, showing that the *di*-iron site in TBA–I is an effective center for the oxidation.   Such a structure dependency of the catalysis is noticeable and the catalytic performance of *di*-iron-containing polyoxometalate may be related to the catalysis by methane monooxygenase.   Among *tri*-transition metal-substituted silicotungstates, α-$SiW_9\{M^{n+}(OH_2)\}_3O_{37}^{(16-3n)-}$ (M = Fe^{3+}, Ni^{2+}, Cu^{2+}), the order of turnover numbers was Fe > Ni > Cu.   In addition, *mono*-iron-substituted silicotungstate was more active than that of copper and *di*-vanadium-substituted silicotungstate was less active than TBA–I.   These facts show that iron is an effective transition metal for the selective oxidation.[27]

CONCLUSIONS

The Keggin-type *di*-iron-substituted silicotungstate, γ-$SiW_{10}\{Fe(OH_2)\}_2O_{38}^{6-}$ (I), was synthesized by the reaction of the lacunary $[\gamma$-$SiW_{10}O_{36}]^{8-}$ with $Fe(NO_3)_3$ in an acidic aqueous solution and isolated as the *tetra-n-*butylammonium salt (TBA-I).   The structure was characterized by various analyses to have an oxo-bridged *di*-iron site.   The catalytic activity greatly depended on structures of iron centers and TBA-I showed the highest turnover number of 135-147 for the oxidation of cyclohexane with 1 atm oxygen among iron- and *tri*-transition-metal-substituted silicotungstates.   Such a structure dependency of the catalysis is significant and such catalytic performance of *di*-iron-substituted polyoxometalate may be related to the catalysis by methane monooxygenase.

FUTURE OPPORTUNITIES

(1) The catalysis of *di-* or *tri*-metal-substituted polyoxometalates is interesting not only to understand the action of transition-metal-containing biomolecules but also to design an active catalyst for hydrocarbon oxidation.   (2) Further development of micro/meso porous materials containing polyoxometalates or related compounds is expected.   These materials will concentrate hydrocarbons in the pores and promote the selective oxidation catalysis.

ACKNOWLEDGMENT

This work was supported in part by a Grant-in-Aid for Scientific Research from the Ministry of Education, Science, Sports and Culture of Japan.

REFERENCES

1.   R. A. Sheldon, Homogeneous and Heterogeneous Catalytic Oxidations with Peroxide Reagents, *Top. Current Chem.*, **164**, 22 (1993).
2.   A. Sobkowiak, H. Tung, and D. T. Sawyer, Iron-and cobalt-induced activation of hydrogen peroxide and dioxygen for the selective oxidation-dehydrogenation and oxygenation of organic molecules, *Prog. Inorg. Chem.*, **40**, 291 (1992).

3. C. L. Hill, and C. M. Prosser-McCartha, Homogeneous catalysis by transition metal oxygen anion clusters, *Coord. Chem. Rev.*, **143**, 407 (1995).
4. C. L. Hill, Activation and Functionalization of Alkanes; Wiley, New York (1989), p. 243.
5. R. H. Crabtree, *Chem. Rev.*, Aspects of Methane Chemistry, **95**, 987 (1995).
6. N. Mizuno and M. Misono, Homogeneous catalysis, *Chem. Rev.*, **98**, 199(1998).
7. M. Kurioka, K. Nakata, T. Jintoku, Y. Taniguchi, K. Takaki, and Y. Fujiwara, Palladium-Catalyzed Acetic-Acid Synthesis from Methane and Carbon-Monoxide or Dioxide, *Chem. Lett.*, **1995**, 244.
8. I. Yamanaka, K. Nakagaki, T. Akimoto, and K. Otsuka, Reactivity of Active Oxygen Species Generated in the $EuCl_3$ Catalytic-System for Monooxygenation of Hydrocarbons, *J. Chem. Soc., Perkin Trans. 2*, **1996**, 2511.
9. R. Neumann and M. Dahan, A Ruthenium-Substituted Polyoxometalate as an Inorganic Dioxygenase for Activation of Molecular-Oxygen, *Nature*, **388**, 353 (1997).
10. M. M. Taqui Khan, D. Chatterjee, S. Kumar S., A. P. Rao, and N. H. Khan, Kinetics and Mechanism of the Oxidation of the Saturated-Hydrocarbons Cyclohexane and Adamantane with Molecular-Oxygen Catalyzed by Ruthenium(III) Saloph Complex, *J. Mol. Catal.*, **75**, L49(1992).
11. N. Mizuno, M. Tateishi, T. Hirose, and M. Iwamoto, *Chem. Lett.*, Oxygenation of Alkanes by Molecular-Oxygen on (Pw9O37(Fe2Ni(Oac)3))10-Heteropolyanion, **1993**, 2137.
12. I. V. Kozhevnikov, Catalysis by Heteropoly Acids and Multicomponent Polyoxometalates in Liquid-Phase Reactions, *Chem. Rev.*, **98**, 171 (1998).
13. R. Neumann, *Prog. Inorg. Chem.*, Polyoxometalate Complexes in Organic Oxidation Chemistry, **47**, 317 (1998).
14. T. Okuhara, N. Mizuno, and M. Misono, Catalytic Chemistry of Heteropoly Compounds, *Adv. Catal.*, **41**, 113 (1996).
15. M. T. Pope and A. Müller, Polyoxometalate Chemistry - An Old Field with New Dimensions in Several Disciplines, *Angew. Chem. Int. Ed. Engl.*, **30**, 34 (1991).
16. M. T. Pope, *Heteropoly and Isopoly Oxometalates* (Springer-Verlog, Berlin, 1983).
17. N. Mizuno, C. Nozaki, I. Kiyoto, and M. Misono, Highly Efficient Utilization of Hydrogen-Peroxide for Selective Oxygenation of Alkanes Catalyzed by Diiron-Substituted Polyoxometalate Precursor, *J. Am. Chem. Soc.*, **120**, 9267 (1998).
18. C. Nozaki, I. Kiyoto, Y. Minai, M. Misono, and N. Mizuno, Synthesis and Characterization of Diiron(III)-Substituted Silicotungstate, $\gamma(1,2)\text{-}SiW_{10}\{Fe(OH_2)\}_2O_{38}^{6-}$, *Inorg. Chem.*, **38**, 5724 (1999).
19. C. Nozaki, M. Misono, and N. Mizuno, Oxidation of Cyclohexane with Molecular-Oxygen Efficiently Catalyzed by di-Iron(III)-Substituted Silicotungstate, $\gamma\text{-}SiW_{10}\{Fe(OH_2)\}_2O_{38}^{6-}$, Including Radical-Chain Mechanism, *Chem. Lett.*, **1998**, 1263.
20. See refs. 40 and 44-56 in ref. 18.
21. "Kogyo Yuki Kagaku", 4th ed., edited by K. Weissermel and H.-J. Arpe, Tokyo Kagaku Dojin, Tokyo(1996), p. 270.
22. D. H. R. Barton, T. Li, and J. MacKinnon, Synergistic Oxidation of Cyclohexane and Hydrogen-Sulfide Under Gif Conditions, *Chem. Commun.*, **1997**, 557.
23. S. Murahashi, Y. Oda, and T. Naota, Iron-Catalyzed and Ruthenium-Catalyzed Oxidations of Alkanes with Molecular-Oxygen in the Presence of Aldehydes and Acids, *J. Am. Chem. Soc.*, **114**, 7913(1992).
24. N. Kitajima, M. Ito, H. Fukui, and Y. Moro-oka, Hydroxylation of Alkanes and Arenes Using Molecular-Oxygen, *J. Chem. Soc. Chem. Commun.*, **1991**, 102.
25. T. Iwahama, K. Syojyo, S. Sakaguchi, and Y. Ishii, Direct conversion of cyclohexane into adipic acid with molecular oxygen catalyzed by N-hydroxyphthalimide combined with $Mn(acac)_2$ and $Co(OAc)_2$, *Org. Proc. Res.& Dev.*, **2**, 255(1998).
26. Y. Ishii, T. Iwahama, S. Sakaguchi, K. Nakayama, and Y. Nishiyama, Alkane Oxidation with Molecular-Oxygen Using a New Efficient Catalytic-System - N-Hydroxyphthalimide (NHPI) Combined with $Co(Acac)_n$ (n=2 or 3), *J. Org. Chem.*, **61**, 4520(1996).
27. Very recently, we have reported that *di*-manganese-substituted silicotungstate, $\gamma\text{-}SiW_{10}\{Mn^{3+}(OH_2)\}_2O_{38}^{6-}$, showed the highest turnover number of 789 among various metal-substituted silicotungstates for the oxygenation of cyclohexane with 1 atm molecular oxygen, and the value was of the highest level in comparison with other catalysts so far reported.[28]
28. T. Hayashi, A. Kishida, and N. Mizuno, High Turnover Numbers of $\gamma\text{-}SiW_{10}\{Mn^{III}(OH_2)\}_2O_{38}^{6-}$ for Oxygenation of Cyclohexane with 1 Atm Molecular Oxygen, *Chem. Commun.*, **2000**, 381.

TRANSITION-METAL-SUBSTITUTED HETEROPOLY ANIONS IN NONPOLAR SOLVENTS - STRUCTURES AND INTERACTION WITH CARBON DIOXIDE

Jared Paul, Phillip Page, Philip Sauers, Katherine Ertel, Christina Pasternak, William Lin, and Mariusz Kozik

Chemistry Department
Canisius College
2001 Main Street
Buffalo, NY 14208

INTRODUCTION

This paper is a continuation of our work[1] in the area of carbon dioxide activation by transition- metal-substituted heteropoly tungstates[2], TMS HPT's, in nonpolar solvents. The idea is based on an original report by Pope et al. in 1984[3] that potassium salts of TMS HPT's can be transferred into nonpolar solvents, where water coordinated to a transition metal, Tm, can be replaced by another small molecule, Y, according to the reactions below, where X is a heteroatom, B, P, or Si.

$$[XW_{11}O_{39}Tm(H_2O)]^{n-} \rightarrow [XW_{11}O_{39}Tm(\_)]^{n-} \rightarrow [XW_{11}O_{39}Tm(Y)]^{n-}$$

Original reports by Pope's group concentrated on possible application of TMS HPT's in nonpolar solvents as *oxidation* catalysts, owing to the fact that HPT's substituted with Mn were found to complex oxygen in toluene. Our group considered the opposite, i.e. possible application of TMS HPT's in nonpolar solvents as *reduction* catalysts. The specific reduction that we have in mind is the multielectron reduction of carbon dioxide.

It was shown in 1989 by Anson et al.[4] that Fe substituted HPT's can function as catalysts for multielectron reduction of nitrite anion to ammonia, a process that requires 6 electrons.   While in the presence of many other electrocatalysts that reaction occurs via a series of one-electron steps and the products include hydroxylamine, in the case of electrocatalysis by a heteropoly tungstate, no hydroxylamine was observed and the major product of the reduction was ammonia. The first step in the electrocatalytic process observed by Anson is the complexation of nitrite by iron from the heteropoly complex. Next the reduced tungsten cage serves as a reservoir for the storage of electrons that are subsequently transferred to nitrogen in a concerted, intramolecular, multiple electron process. The unique catalytic reactivity of the iron substituted HPT in this process was

Polyoxometalate Chemistry for Nano-Composite Design
Edited by Yamase and Pope, Kluwer Academic/Plenum Publishers, 2002

attributed by Anson to the ability of the heteropoly complex to serve as both the site of substrate binding and the source of the multiple reducing electrons. Since our recent work demonstrated that TMS HPT's can complex carbon dioxide in nonpolar solvents[1], we decided to investigate if behavior analogous to that observed for nitrite reduction in water is possible for carbon dioxide in nonpolar solvents.

This paper reports our electrochemical investigation of that phenomenon. Cyclic voltammograms in dichloromethane do indeed indicate that TMS HPT's exhibit electrocatalytic activity with respect to carbon dioxide reduction. However, we also found that cyclic voltammograms of TMS HPT's in dichloromethane are irreversible, making their application as electrocatalysts impractical. We therefore investigated if it is possible to avoid reducing HPT in nonpolar solvent by carrying out the reduction process in aqueous solution, before the transfer into the nonpolar solvent. That procedure was successful, but required several new steps in order to avoid blockage of the transition metal site by an anion from the buffering solution.

During our detailed investigation of the buffer role in the phase transfer process we encountered a new and very important phenomenon. We discovered that when an aqueous solution of potassium salt of HPT and a toluene solution of tetraheptylammonium bromide are mixed, resulting in a toluene solution of tetraheptylammonium HPT and an aqueous solution of KBr, some KBr molecules become trapped inside the reverse micelles formed by THA cations and heteropoly anions. As the result, when $THA_5[PW_{11}O_{39}Co(H_2O)]$ is dried in toluene, water molecules coordinated to cobalt are replaced by bromide ions and potassium ions form ion pairs with HPT anions. This hypothesis is confirmed by several new ^{31}P NMR peaks that appear during dehydration, but disappear when water is again added to the solution.

BACKGROUND

The redox chemistry of carbon dioxide remains of great current interest owing to its potential as a C-1 feedstock and substrate for solar energy storage, as well as for its alleged role in atmospheric warming (the "greenhouse" effect).[5] Therefore, the activation of CO_2 by transition-metal complexes and its conversion into other useful chemicals are areas of great importance in inorganic chemistry. The most desirable objective is to transform carbon dioxide into its highly reduced products by a multielectron route, with methane being the ultimate goal. One approach to reach this goal is to use compounds containing transition metals that could act as both electrocatalysts and electron reservoirs, thus providing a means to simultaneously deliver the electrons necessary in multielectron reductions.

A number of transition-metal complexes, both in solution and on electrode surfaces, have been shown to be effective in the electrocatalytic reduction of carbon dioxide.[6] All of those complexes significantly decrease the overpotential for reduction of CO_2 by up to 1V (as compared to a 1-el reduction to the CO_2^- radical), and yield various multielectron reduction products. Known electrocatalysts yield primarily carbon monoxide and formate anion as the major products of the CO_2 reduction.[6] Sullivan et al. did detailed mechanistic work on a series of bipyridine complexes of transition metals, and made several suggestions concerning the design of new electrocatalysts that would be capable of reducing CO_2 past the CO and formate step.[7] Their major recommendation is to use as electrocatalysts "electron reservoir" complexes, i.e. compounds capable of storing multiple electrons.

The report by Anson and Toth[4] indicating electrocatalytic and electron-reservoir capabilities of TMS HPT's, and the report by Pope and Katsoulis[3] demonstrating reactivity of TMS HPT's in nonpolar solvents with small molecules, prompted us to investigate whether TMS HPT's in nonpolar solvents could activate CO_2 and serve as electrocatalysts in its multielectron reduction.

In 1998 we described our findings concerning reactivity of TMS HPT's with CO_2.[1] We reported that certain TMS HPT's indeed complex CO_2 in nonpolar solvents like toluene, dichloromethane, and carbon tetrachloride. The spectroscopic evidence for complex formation consisted of UV/VIS, IR, magnetic susceptibility, and ^{13}C NMR data. There was a striking similarity between the observations reported by Pope et al. for reactions of TMS HPT's with O_2[3] and our observations concerning their reactivity with CO_2. In both cases the presence of water was a requirement for complex formation and the reactions were fully reversible. The requirement for the presence of water was interpreted by hydrogen bond formation between the coordinated CO_2 molecule and oxygen atoms from the heteropoly anion. It was also suggested that there was a direct cobalt-carbon bond present and that it involved a certain amount of back bonding.

The presence of back bonding would cause the weakening of bonds within the CO_2 molecule after it is coordinated to the heteropoly anion, representing activation of CO_2. In this paper we describe our recent findings concerning CO_2 activation towards reduction.

EXPERIMENTAL

Materials

HPLC grade nonpolar solvents, toluene and dichloromethane, were purchased from Aldrich. Toluene was distilled over sodium/benzophenone, and dichloromethane was distilled over calcium hydride. All solvents were stored under nitrogen, over molecular sieves. Water content in dried solvents was typically less than 1 mM. Potassium salts of transition-metal-substituted heteropoly complexes were prepared according to the methods published previously.[8-10] The IR, UV¥VIS spectra, and Cyclic Voltammograms agreed with the data reported in the literature.[8-10] "Bone dry" grade carbon dioxide gas (Linde, 99.8%) and extra dry nitrogen gas (Linde, 99.7%) were used without further purification. For the experiments that required drier conditions, CO_2 gas was dried over P_2O_5 and distilled into the reaction vessel under vacuum. All experiments requiring dry conditions were carried out on Schlenk double manifold lines.

Phase Transfer

A slight modification of the method introduced by Katsoulis and Pope[3] was used. Typically, an unbuffered 10 mM aqueous solution of the potassium salt of heteropoly anion was shaken briefly with an equal volume of the stoichiometric amount of THABr in toluene. (The concentration of THABr solution was 10x mM, where x is the absolute value of the negative charge on the heteropoly anion, HPA). Since all the HPA's were colored, it was easy to observe the complete transfer from the aqueous to the organic layer. After the two phases had been allowed to settle for 10 minutes, they were separated. The organic layer at this stage was determined, via coulometric Karl Fisher titration, to contain between 20-25 water molecules per heteropolyanion. Next the wet toluene was removed by heating to 40° C under vacuum, and the dry solvent (toluene or dichloromethane) was added using the cannula technique on a Schlenk line. For the experiments with controlled amounts of water, water was added using gas tight syringes, or was distilled under vacuum into the reaction vessel. Phase transfer back into water was carried out using stoichiometric amount of $NaClO_4$.

Spectroscopy

UV-VIS spectra were recorded using the HP 8452A diode array spectrophotometer.

Vibrational   spectra were run using a Nicolet Impact 410 FT IR spectrometer with KBr discs. The Bruker Avance 250 MHZ spectrometer was used to record ^{31}P NMR spectra. The NMR samples were prepared using 10mm tubes with J. Young valves. The NMR spectra typically required 800 scans. The concentration of potassium ions in aqueous solutions after phase transfer back into water was determined by Spectro Flame ICP.

Electrochemistry

Electrochemical experiments were carried out on a PAR 273A potentiostat. Cyclic voltammograms in dichloromethane were performed in the presence of 0.1M TBAPF$_6$ electrolyte, using a glassy carbon working electrode, silver wire as a reference electrode, and a platinum auxiliary electrode. After cyclic voltammograms were recorded, ferrocene was added to the solution as an internal standard to confirm the peak positions.

Controlled potential electrolysis in aqueous solution was carried out with a carbon-cloth working electrode and in the presence of a Ag/AgCl double junction reference electrode, with constant nitrogen purge, as described by us previously.[11] The electrolysis was stopped when the current dropped to less than 1% of the original value.

Water Measurements

Water was determined by coulometric Karl Fisher titration using Metrohm 684 KF Coulometer filled with Aquastar coulomat single solution.

RESULTS AND DISCUSSION

Electrochemistry

Cyclic voltammograms in dichloromethane were recorded for several transition metal substituted HPT's that were found by us to react with CO_2 in nonpolar solvents [1] and for α-[$P_2W_{18}O_{62}$]$^{6-}$ ion, which was used as a "blank". CV's were always recorded first for a dry solution of the HPT, and subsequently for the solution after CO_2 gas was bubbled. Figure 1 shows both of these CV's for α-[$SiW_{11}O_{39}Co$]$^{6-}$ ion.

It can be clearly seen that there is a large change in the CV after carbon dioxide gas is bubbled through the solution. On the other hand, as seen in Figure 2, there is almost no change in the CV for α-[$P_2W_{18}O_{62}$]$^{6-}$ ion. Similarly, several other TMS HPT's that react with CO_2, also show large changes in their CV's after CO_2 is bubbled. The CV's therefore demonstrate that TMS HPT's do have electrocatalytic properties with respect to carbon dioxide reduction.

On the other hand, all CV's for TMS HPT's are irreversible, and for that reason TMS HPT's can not be used in practice as electrocatalysts for CO_2 reduction in nonpolar solvents. The irreversibility is most likely caused by the lack of protons, which are needed for the reductions. However, the CV's are reversible in aqueous solutions and we decided to investigate another route for CO_2 reduction by TMS HPT's.

The new method reverses the order of the transfer/reduction procedure of the heteropoly complex. Instead of transferring the oxidized TMS HPT into nonpolar solvent and then reducing it in that solvent, we decided to reduce the TMS HPT in aqueous solution first, and transfer the reduced HPT (heteropoly blue) into a nonpolar solvent (toluene).

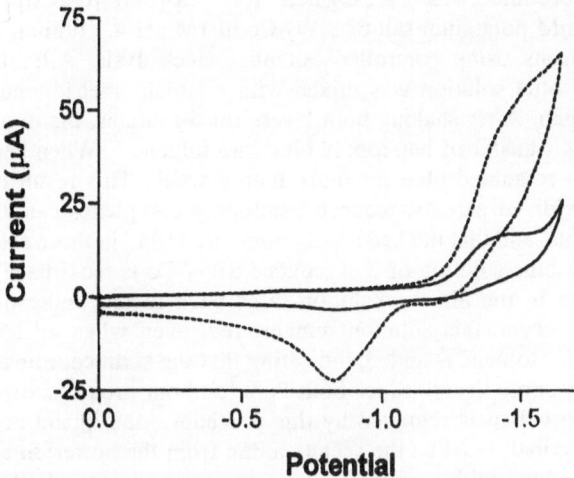

Fig.1.   Cyclic voltammograms of SiW$_{11}$Co in CH$_2$Cl$_2$ before (solid line) and after (dashed line) bubbling with CO$_2$ gas.

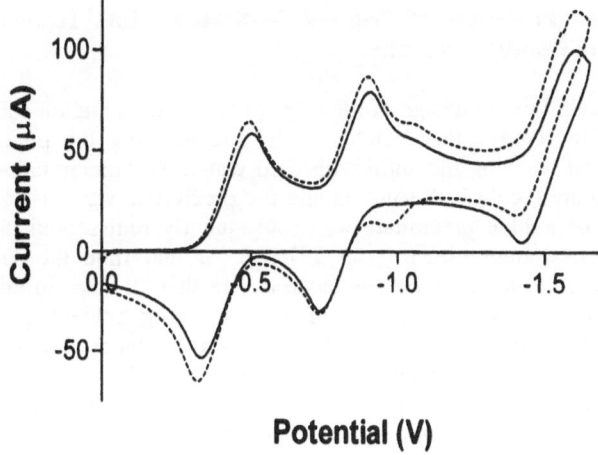

Fig.2.   Cyclic voltammograms of P$_2$W$_{18}$ in CH$_2$Cl$_2$ before (solid line) and after (dashed line) bubbling with CO$_2$ gas.

Transfer of *Reduced* TMS HPT's into Toluene

The new procedure was investigated for α-$[SiW_{11}O_{39}Co]^{6-}$ ion ($SiW_{11}Co$). Approximately 25 mM potassium salt of $SiW_{11}Co$ in 1M pH 4.5 lithium acetate buffer was reduced by 2 electrons using controlled potential electrolysis. After the reduction, the aqueous heteropoly blue solution was mixed with a stoichiometric amount of THABr in toluene under nitrogen. After shaking both layers for 5 minutes, the organic layer became dark blue, indicating transfer of heteropoly blue into toluene. When kept under nitrogen, the toluene solution remained blue for more than a week. This result demonstrates that, similarly to their oxidized parents, reduced heteropoly complexes can also be transferred into nonpolar solvents, and that the heteropoly blues are stable in those solvents.

When the dark blue solution of 2-el reduced $SiW_{11}Co$ is reoxidized by bubbling with oxygen, or left open in the air, the solution turns back to red, indicating reoxidation of heteropoly blue. However, this solution remains red, even when all solvent is removed under vacuum and dry toluene is added, indicating that the sixth coordination site on cobalt remains occupied by some ligand. Since both Pope[3] and our group[1] showed in the past that water coordinated to cobalt is removed by this procedure, the ligand occupying the sixth coordination site of cobalt must be the acetate anion from the buffer. In addition, when the red solution was bubbled with carbon dioxide, no change in the UV/VIS spectrum took place. Therefore, while the above procedure can be used to transfer the heteropoly blue into nonpolar solvent, the species that is transferred is completely useless for CO_2 reduction, owing to the blockage of the sixth coordination site on cobalt by the anion from the buffer.

Modified Transfer Procedure of *Reduced* TMS HPT's into Toluene without Buffer Blocking the Sixth Coordination Site

In order to avoid the blockage of the empty coordination site on cobalt by the acetate ion, we decided to remove the acetate ion by precipitating the potassium salt of the heteropoly blue and washing the solid with cold water. The precipitation was carried out with a nearly saturated, cold KCl solution and the precipitate was washed with cold water three times. The dark blue precipitate was subsequently redissolved in pure unbuffered water and transferred into toluene with THABr. At that time the solution in toluene remained dark blue. It needs to be emphasized here that all steps in this procedure were carried out with the complete exclusion of oxygen, using Schlenk glassware, and both water and toluene were deoxigenated by freeze thawing. After reoxidation of this solution, the removal of wet toluene by vacuum, and addition of dry toluene, the color of the solution turned green, which indicates that the procedure was successful in removing acetate ion. Most importantly, the green solution was also found to react with carbon dioxide, as demonstrated by another color change from green to burgundy red, when CO_2 gas was bubbled though the solution. Therefore, we demonstrated that it is indeed possible to transfer transition-metal-substituted heteropoly blue into toluene without blocking the sixth coordination site on the transition metal. We also found that washing less than three times was insufficient to remove the acetate ion.

Figure 3 shows visible spectra for 2-el reduced $SiW_{11}Co$ anions under three conditions: (1) in aqueous buffer solution, (2) in toluene after direct transfer from aqueous buffer, and (3) in toluene after the modified transfer involving precipitation-washing-redissolution. The presence of the same broad transition at 610nm confirms that the same reduced species is present in all three cases.

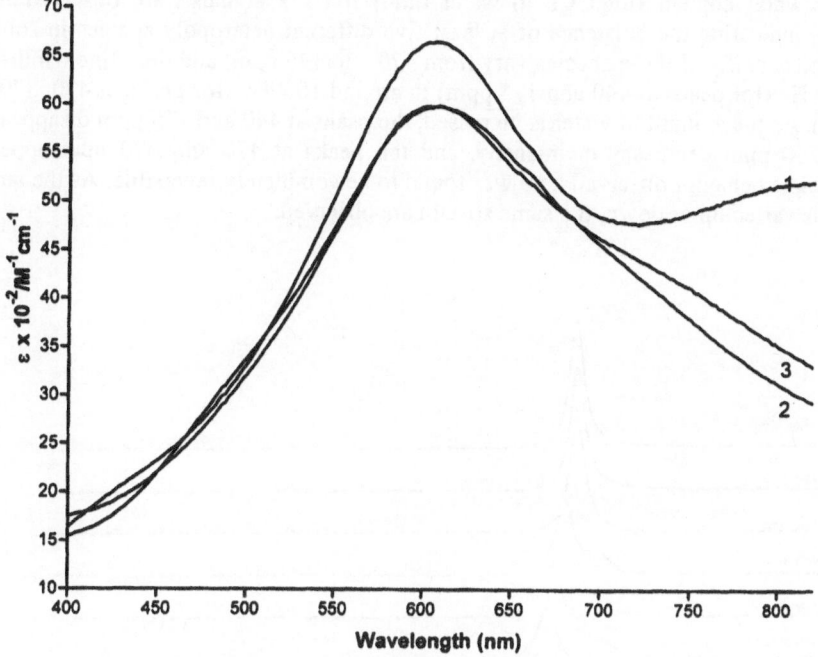

Fig. 3.   Visible spectra of 2el reduced SiW$_{11}$Co: (1) in aqueous pH=4.5 buffer, (2) THA salt in toluene after transfer from aqueous buffer, (3) THA salt in toluene after the modified transfer involving precipitation-washing-redissolution

Reactivity of 2-el Reduced SiW$_{11}$Co with CO$_2$ in Toluene

In order to determine if the heteropoly blue transferred into toluene could be used to reduce CO$_2$, we bubbled CO$_2$ gas through two toluene solutions of 2-el reduced SiW$_{11}$Co. The first solution contained 2-el blue obtained by direct transfer from aqueous buffer solution, with the sixth coordination site on cobalt blocked by the acetate ion. That solution remained dark blue after 30 minutes of bubbling and for the next several days, when it was stored under CO$_2$ gas. On the other hand, the solution obtained by the procedure involving precipitation and washing of the heteropoly blue turned immediately red as CO$_2$ gas was bubbled, demonstrating that CO$_2$ gas when coordinated to cobalt can reoxidize heteropoly blue, while being reduced itself.   When the same solution of 2-el reduced SiW$_{11}$Co was bubbled with argon, no change in color was observed. These experiments confirm our original supposition that TMS HPT's can indeed be used to activate and reduce carbon dioxide in nonpolar solvents. Our work in this area continues.

New Findings Concerning Structures of TMS HPT's in Nonpolar Solvents

The discovery that the acetate ions are transferred together with the heteropoly anions into toluene, prompted our investigation into the role that bromide ions play during the phase transfer procedure. For this project we chose α-[PW$_{11}$O$_{39}$Co]$^{5-}$ ion, referred to from now on as PW$_{11}$Co. This ion was chosen because of the presence of phosphorus atom, which is easily observable by ^{31}P NMR. Figure 4 shows ^{31}P NMR spectra of PW$_{11}$Co in toluene as a function of water content. The Co:H$_2$O ratio was varied from 3:1 to 1:12. At very low water content (high Co to water ratio) five ^{31}P signals are observed in the spectrum, indicating the existence of at least five different heteropoly species in solution. The chemical shifts of these species vary from 470 to 175 ppm, and their line widths vary from 100 Hz (for peaks at 440 and 175 ppm) to around 1000Hz (for peaks at 470, 370, and 240 ppm). As the amount of water is increased, the peaks at 440 and 175 ppm disappear, the peak at 240 ppm decreases in intensity, and the peaks at 470 and 370 ppm appear to coalesce. The behavior observed here was found to be completely reversible. As the amount of water is varied up or down, the same spectra are observed.

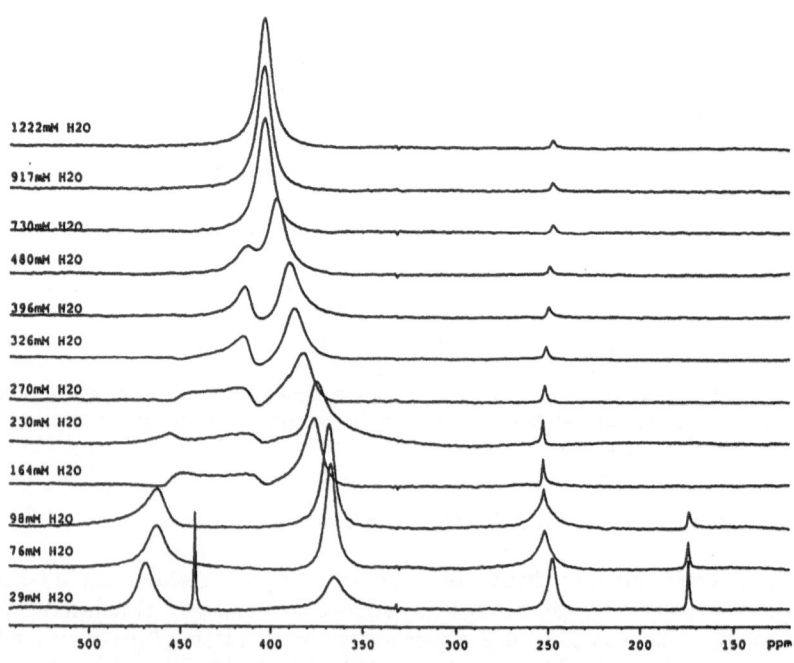

Fig.4. ^{31}P NMR spectra for 100mM solutions of THA salt of PW$_{11}$Co in toluene as water content changes from 29mM to 1222mM.

When the phase transfer is carried out in the presence of excess THABr (twice the amount required by stoichiometry), only two ^{31}P NMR signals are observed, one at 240 ppm and another at 450 ppm, with an integration ratio of 30:1, respectively. Based on these data we assign the peak at 450 ppm to the $[PW_{11}O_{39}Co(H_2O)]^{5-}$ species and the peak at 240 ppm to $[PW_{11}O_{39}CoBr]^{6-}$. The latter species contains the bromide ion coordinated to cobalt. This is an important discovery, since up to date it was believed that as PW$_{11}$Co is dried, the only two species that exist are aqua ($[PW_{11}O_{39}Co(H_2O)]^{5-}$) and pentacoordinated ($[PW_{11}O_{39}Co(\_)]^{5-}$) cobalt complexes, the latter with an empty coordination site on cobalt.

We also propose that in order to preserve the neutral charge of reverse micelles formed by THA cations and heteropoly anions, as well as the charge neutrality of aqueous KBr solution, potassium ions are also transferred into toluene solution. We do not believe that the charge is neutralized by the formation of THA$_6$[PW$_{11}$O$_{39}$CoBr], because the phase transfer process is fully stoichiometric, i.e. a five-fold excess of THABr transfers 100% of $[PW_{11}O_{39}Co(H_2O)]^{5-}$, as determined by UV/VIS spectroscopy, and any smaller amount of THABr does not transfer all of the PW$_{11}$Co. Therefore, we propose the existence of

K^+-$[PW_{11}O_{39}CoBr]^{6-}$ and K^+-$[PW_{11}O_{39}Co]^{5-}$ ion pairs in a toluene solution of $PW_{11}Co$. These ion pairs do not form at high water concentrations because at those conditions both K^+ and Br^- ions are hydrated by water molecules inside the reverse micelles formed by THA cations and HPT anions.

ICP spectroscopy was used to confirm that the potassium ions indeed are present in a toluene solution of $PW_{11}Co$. Because our ICP instrument can only be used for studying aqueous solutions, we looked for the presence of K^+ ions in the aqueous solution of $PW_{11}Co$ obtained by back transfer from toluene into water, using $NaClO_4$. Pure aqueous $NaClO_4$ showed zero concentration of K^+ ions, while the aqueous solution of the Na salt of $PW_{11}Co$ obtained by back transfer from toluene solution, showed a 1:20 ratio of K to Co. Since most likely not all K^+ ions were back transferred into aqueous solution, the real concentration of K^+ ions in toluene solution was probably higher.

At this time we do not attempt to assign all of the ^{31}P NMR peaks observed at low water concentration. Most likely the peaks that disappear first as water is added (at 175 and 440 ppm) originate from the $PW_{11}Co$ ion with the empty sixth coordination site on cobalt. One of these species could have K^+ ion paired to the heteropoly anion. We continue to investigate this phenomenon for $PW_{11}Co$ and other heteropoly tungstates.

CONCLUSIONS

Reduced transition-metal-substituted heteropoly tungstates can be transferred into nonpolar solvents by the same method as the one developed by Pope at al. for their oxidized parents, using THABr as the transfer agent. However, acetate ions from buffer solution used as an electrolyte during the reduction remain attached to the transition metal after the transfer.

It is possible to transfer reduced TMS HPT's from aqueous buffer solution into toluene without blocking the sixth coordination site on the transition metal, by the following modified procedure: reduction in aqueous buffer solution, precipitation of potassium salt of heteropoly blue, washing the solid blue with a sufficient amount of water, redissolving the washed solid in pure water, and phase transfer with THABr.

Reduced TMS HPT's transferred into toluene by the modified procedure are reoxidized with carbon dioxide, indicating CO_2 activation.

A certain amount of potassium bromide remains trapped in the reverse micelles formed by THA cations and heteropoly anions after the transfer of potassium salts of HPT's with THABr. In wet toluene solution both ions are hydrated inside the reverse micelles. As the water is removed bromide ions coordinate to transition metal on HPT and potassium ions form ion pairs with HPT's.

ACKNOWLEDGMENTS

This work was supported by grants from the National Science Foundation (CHE 932 0987, the NMR instrument grant ILI 9451327), Petroleum Research Fund (#28102-B3), Oishei Foundation, as well as Hughes fellowships to undergraduate students at Canisius College and internal undergraduate stipends at Canisius College.

REFERENCES

1. Szczepankiewicz, S.; Ippolito, C.; Santora, B.; Ippolito, G.; Van de Ven, T.; Fronckowiak, L.; Wiatrowski, F.; Power, T.; Kozik, M. Interaction of Carbon Dioxide with Transition-Metal-Substituted Heteropolyanions in Nonpolar Solvents. Spectroscopic Evidence for Complex Formation, *Inorg. Chem.* **1998**, *37*, 4344.

2. For reviews of chemistry of heteropoly complexes see: (a) Baker, L.C.W.; Glick, D.C. Present General Status of Understanding of Heteropoly Electrolytes and a Tracing of Some Major Highlights in the History of Their Elucidation *Chem. Rev.* **1998**, *98*, 3. (b) Pope, M. T.; Muller, A. Polyoxometalate Chemistry - An Old Field with New Dimensions in Several Disciplines, *Angew. Chem. Int. Ed. Engl.* **1991**, *30*, 34. (c) Day, V. W.; Klemperer, W. G., Metal Oxide Chemistry in Solution: The Early Transition Metal Polyoxoanions, *Science* **1985**, *228*, 533. (d) Pope, M. T. *Heteropoly and Isopoly Oxometalates*, Springer, New York, **1983**.

3. (a) Katsoulis, D. E.; Pope, M. T., New Chemistry for Heteropolyanions in Anhydrous Nonpolar Solvents. Coordinative Unsaturation of Surface Atoms. Polyanion Oxygen Carriers, *J. Am. Chem. Soc.*, **1984**, *106*, 2737. (b)   Katsoulis, D. E. *Doctoral Dissertation*, Georgetown University, **1985**. (c) Katsoulis, D. E.; Taush, V. S.; Pope, M. T., Interaction of Sulfur Dioxide with Heteropolyanions in Nonpolar Solvents. Evidence for Complex Formation, *Inorg. Chem.* **1987**, *26*, 215. (d) Katsoulis, D. E.; Pope, M. T. Reactions of Heteropolyanions in Non-Polar Solvents .3. Activation of Dioxygen by Manganese(II) Centers in Polytungstates - Oxidation of Hindered Phenols, *J. Chem. Soc. Dalton Trans.* **1989**, 1483.

4. Toth, J. E.; Anson F. C. Electrocatalytic Reduction of Nitrite and Nitric-Oxide to Ammonia with Iron-Substituted Polyoxotungstates, *J. Am. Chem. Soc.* **1989**, *111*, 2444.

5. (a) Gibson, D. H. The Organometallic Chemistry of Carbon-Dioxide, *Chem. Rev.* **1996**, *96*, 2063. (b) Leitner, W. Carbon-Dioxide as a Raw-Material - The Synthesis of Formic-Acid and Its Derivatives from CO_2, *Angew. Chem. Int. Ed. Engl.* **1995**, *34*, 2207. (c) Tanaka, K. Carbon dioxide fixation catalyzed by metal complexes *Adv. Inorg. Chem.* **1995**, *43*, 409. (d) Ayers, W. M. Ed. *Catalytic Activation of Carbon Dioxide, ACS Symposium Series 363*, Washington, DC; **1988**. (e) Kolomnikow, I. S.; Lysyak, T.V. Carbon-Dioxide in Coordination Chemistry and Catalysis, *Russ. Chem. Rev.* **1990**, *59*, 589. (f) Behr, A. *Carbon Dioxide Activation by Metal Complexes*, VCH, Weinheim, **1988**. (g) Darensbourg, D. J.; Kudaroski, R. A. The activation of carbon dioxide by metal complexes, *Adv. Organomet. Chem.* **1983**, *22*, 129. (h) Eisenberg, R.; Hendricksen, D. E. The binding and activation of carbon monooxide, carbon dioxide, and nitric oxide and their homogeneously catalyzed reactions, *Adv. Catal.* **1979**, *28*, 79. (i) Souter, P. F.; Andrews, L. P. F. A Spectroscopic and Theoretical-Study of the Reactions of Group-6 Metal Atoms with Carbon-Dioxide, *J. Am. Chem. Soc.* **1997**, *119*, 7350.

6. (a) Sullivan, B. P. Ed.   *Electrochemical and Electrocatalytic Reactions of CO_2*, Elsevier, **1993**. (b) Collin, J. P.; Sauvage, J. P. Electrochemical Reduction of Carbon-Dioxide Mediated by Molecular Catalysts *Coord. Chem. Rev.* **1989**, *93*, 245. (c) Sullivan, B. P. Reduction of carbon dioxide with platinum metals electrocatalysts. A potentially important route for the future production of fuels and chemicals *Platinum Metals Rev.* **1989**, *33*, 2. (d) Cheng, S. C.; Blaine, C. A.; Hill, M. G.; Mann, K. R. Electrochemical and IR Spectroelectrochemical Studies of the Electrocatalytic Reduction of Carbon Dioxide by $[Ir_2(dimen)_4]^{2+}$ (dimen = 1,8-Diisocyanomenthane), *Inorg. Chem.* **1996**, *35*, 7704. (e) Bhugun, I.; Lexa, D.; Saveant, J. M. Catalysis of the Electrochemical Reduction of Carbon-Dioxide by Iron(O) Porphyrins - Synergystic Effect of Weak Bronsted Acids *J. Am. Chem. Soc.* **1996**, *118*, 1769. (f) Haines, R. J.; Wittrig, R. E.; Kubiak, C. P. Electrocatalytic Reduction of Carbon-Dioxide by the Binuclear Copper Complex (Cu-2(6-(Diphenylphosphino)-2,2'-Bipyridyl)_2(MeCN)_2)(PF_6)_2 *Inorg. Chem.* **1994**, *21*, 4723. (g) Zhang, J. J.; Pietro, W. J.; Lever, A. B. P. Rotating-Ring-Disk Electrode Analysis of CO_2 Reduction Electrocatalyzed by a Cobalt Tetramethylpyridoporphyrazine on the Disk and Detected as Co on a Platinum Ring *J. Electroanal. Chem.* **1996**, *403*, 93. (h) Lee, Y. F.; Kirchhoff, J. R.; Berger, R. M.; Gosztola, D. Spectroelectrochemistry and Excited-State Absorption-Spectroscopy of Rhenium(I) Alpha,Alpha'-Diimine Complexes *J. Chem. Soc. Dalton Trans.* **1995**, *22*, 3677. (i) Toyohara, K.; Nagao, H.; Mizukawa, T.; Tanaka, K. Ruthenium Formyl Complexes as the Branch Point in 2-Electron and Multielectron Reductions of CO_2 *Inorg. Chem.* **1995**, *34*, 5399. (j) Steffey, B. D.; Curtis, C. J.; Dubois, D. L. Electrochemical Reduction of CO_2 Catalyzed by a Dinuclear Palladium Cooperativity *Organometallics* **1995**, *14*, 4937. (k) Ogata, T.; Yanagida, S.; Brunschwig, B. S.; Fujita, E. Mechanistic and Kinetic-Studies of Cobalt Macrocycles in a Photochemical CO_2 Reduction System - Evidence of Co-CO_2 Adducts as Intermediates *J. Am. Chem. Soc.* **1995**, *117*, 6708. (l) Arana, C.; Keshavarz, M.; Potts, K. T.; Abruna, H. D. Electrocatalytic Reduction of CO_2 and O-2 with Electropolymerized Films of Vinyl-Terpyridine Complexes of Fe, Ni and Co *Inorg. Chim. Acta* **1994**, *225*, 285.

7. Sullivan, B. P.; Bruce, M. R. M.; O'Toole, T. R.; Bolinger, C. M.; Megehee, E.; Thorp, H.; Meyer, T. J. in *Reference 5d*, Chapter 6, p.52.

8. (a) Malik, S. A.; Weakley, T. J. R., Heteropolyanions Related to $P_2W_{18}O_{62}^{6-}$ containing Heteroatoms of Two Elements, *J. Chem. Soc. Chem. Commun.* **1967**, 1094. (b) Weakley, T. J. R.; Malik, S. A., Heteropolyanions Containing Two Different Heteroatoms I., *J. Inorg. Nucl. Chem.* **1967**, *29*, 2935. (c) Malik, S. A.; Weakley, T. J. R., Heteropolyanions Containing Two Different Heteroatoms. Part

II. Anions Related to 18-Tungstodiphosphate, *J. Chem. Soc. A* **1968**, 2647. (d) Tourne, C.; Tourne, G.; Malik, S. A.; Weakley, T. J. R., Triheteropolyanions Containing Copper(II), Manganese(II), or Manganese(III), *J. Inorg. Nucl. Chem.* **1970**, *32*, 3875. (e) Weakely, T. J. R., Heteropolyanions Containing Two Different Heteroatoms. Part III. Cobalto(II)undecatungstophosphate and Related Anions, *J. Chem. Soc. Dalton* **1973**, 341. (f) Zonnevijlle, F.; Tourne, C. M.; Tourne, G. F., Preparation and Characterization of Iron(III)- and Rhodium(III)-Containing Heteropolytungstates. Identification of Novel Oxo-Bridged Iron(III) Dimers, *Inorg.Chem.* **1982**, *21*, 2751. (g) Ortega, F.; Pope, M. T., Polyoxotungstate Anions Containing High-Valent Rhenium. 1. Keggin Anion Derivatives, *Inorg.Chem.* **1984**, *23*, 3292.

9. (a) Contant, R.; Ciabrini, J-P., Preparations and Solution Properties of Some 'Defect' Heteropolyanions Related to 18-tungsto-2-phosphates (α- and β-Isomers), *J. Chem. Res. Synop.* **1977**, 222; *J. Chem. Res. Miniprint* **1977**, 2601. (b) Lyon, D. K.; Miller, W. K.; Novet, T.; Domaille, P. J.; Evitt, E.; Johnson, D. C.; Finke, R. G. Highly Oxidation Resistant Inorganic-Porphyrin Analog Polyoxometalate Oxidation Catalysts .1. The Synthesis and Characterization of Aqueous-Soluble Potassium-Salts of α-2-$P_2W_{17}O_{61}(M^{n+}.OH_2)^{n-10}$ and Organic- Solvent Soluble Tetra-Normal-Butylammonium Salts of α-2-$P_2W_{17}O_{61}(M^{n+}.Br)^{n-11}$ (M = Mn^{3+}, Fe^{3+}, Co^{2+}, Ni^{2+}, Cu^{2+}) *J. Am. Chem. Soc.* **1991**, *113*, 7209.

10. (a) Weakley, T. J. R.; Evans, H. T. Jr.; Showell, J. S.; Tourne, G. F.; Tourne, C. M., 18-Tungstotetracobalto(II)diphosphate and Related Anions: a Novel Structural Class of Heteropolyanions, *J.C.S. Chem. Commun.* **1973**, 139. (b) Evans H. T.; Tourne, C. M.; Tourne, G. F. Weakley, T. J. R., X-Ray Crystallographic and Tungsten-183 Nuclear Magnetic Resonance Structural Studies of the $[M_4(H_2O)_2(XW_9O_{34})_2]^{10-}$ Heteropolyanions (M= Co^{II} or Zn, X= P or As) *J. Chem. Soc. Dalton Trans.* **1986**, 2699.

11. Kozik, M.; Casan-Pastor, N.; Hammer, C.F.; Baker, L.C.W., Ring Currents in Wholly Inorganic Heteropoly Blue Complexes. Evaluation by a Modification of Evans's Susceptibility Method, *J. Am. Chem. Soc.* **1988**, *110*, 7697.

POLYOXOMETALATES AND SOLID STATE REACTIONS AT LOW HEATING TEMPERATURES

Su Jing, Feibo Xin, and Xinquan Xin

Coordination Chemistry Institute
Nanjing University
Nanjing 210093
P.R.China

PREFACE

Polyoxometalates are unique in their topological and electronic versatility[1,2] and are important in analytical chemistry,[3] catalysis and photocatalysis,[4-7] biochemistry,[8] medicine[9] and solid-state devices.[10,11] Thus, new classes of structures with often-unexpected reactivities and applications are still being characterized. A number of heteropolymolybdate and heteropolytungstate anions, including more than 70 different heteroelements, are known. They exhibit dozens of stoichiometries and structures.[12,13] Most of the known polyoxometalates are prepared in solution but a few compounds have been prepared by high-temperature solid-state reactions.[14]

In order to get stable materials, most chemists work with solution reactions, and material scientists with high temperature reactions. Solid-state reactions at low heating temperatures have attracted little attention, but in recent years, there has been some progress in organic solid state reactions.[15-18] Research in solid-state reactions of inorganic compounds at room temperature has also progressed.[19] In our lab, more than two hundred new cluster compounds with excellent third-order nonlinear optical properties[20] and dozens of nanomaterials have been successfully synthesized by solid state reactions at low heating temperatures. It is a simple, convenient, environmentally friendly, and cheap synthetic method for preparation of these materials[21]. In this article, polyoxometalates obtained by solid state reactions at low heating temperatures are discussed.

MECHANISM OF SOLID STATE REACTION OF $(NH_4)_6Mo_7O_{24} \cdot 4H_2O$ AND P_2O_5

There are four steps in a typical solid-state reaction: (i) diffusion, (ii) reaction, (iii)

Polyoxometalate Chemistry for Nano-Composite Design
Edited by Yamase and Pope, Kluwer Academic/Plenum Publishers, 2002

217

nucleation, and (iv) growth. As we know, the rate-controlling steps at high temperatures are usually (i) and (iv), but at low heating temperatures, any step can be the rate determining step of a solid state reaction, and product structures of different morphologies have been obtained.[21-24]

Powder X-ray diffraction (XRD) is an efficient way to identify solid phases, and is used to detect components in mixed solids. We have used XRD to trace the progress of the solid phase reaction, and to infer the reaction mechanism. In the solid-state reaction,

$$A(Solid) \ + \ B(Solid) \ \rightarrow \ C(Solid)$$

the following five processes are possible.

(1) If there is no reaction between A and B, no change will be shown in the XRD pattern with time (Fig. 1(A)).
(2) If diffusion is the rate-controlling step, with the growth of the lines from the products, those of the reactants disappear gradually until the end of the reaction (Fig. 1(B)).
(3) An intermediate phase is observed if reaction is the rate-determining step. The lines of the new phase grow first, and then disappear as the reaction proceeds (Fig. 1(C)).
(4) If nucleation is the rate-determining step, no XRD pattern of the product is observed until the crystal nuclei are produced (Fig. 1(D)).
(5) If growth is the rate-determining step, the XRD pattern (Fig. 1(E)) should be similar to Fig. 1(D). The difference is that broad signals are observed during the reaction.

We have studied the mechanism of the solid state reaction between $(NH_4)_6Mo_7O_{24} \cdot 4H_2O$ and P_2O_5. These two solids were mixed with a molar ratio Mo: P $=12:1$, and ground at room temperature. The color of the mixture changed gradually from white to light yellow, then to bright yellow. The XRD measurement showed that, as soon as the reaction began, the diffraction peaks of P_2O_5 disappeared quickly, while most of $(NH_4)_6Mo_7O_{24} \cdot 4H_2O$ still remained. This indicates that P_2O_5 molecules diffuse into the crystal lattice of $(NH_4)_6Mo_7O_{24} \cdot 4H_2O$ quickly. On further reaction, the peaks of $(NH_4)_6Mo_7O_{24} \cdot 4H_2O$ disappeared gradually. These two steps lasted about 0.5 hour. Meanwhile, no lines attributable to the product appeared. No such lines were observed even when the reaction system was heated at 80°C for 24 hours. The yellow solid product was washed in an ultrasonic bath, first with distilled water until the pH of the washed solution was 7-8, with ethanol twice, and then dried in air. The solid contained nanoparticles as shown in Fig. 4. The ^{31}P MAS NMR chemical shift of the product was similar to that for $(NH_4)_3PMo_{12}O_{40} \cdot 4H_2O$ (Fig. 2) which had been synthesized from solution . This indicates that the P atom in the amorphous product has the same environment as that in $(NH_4)_3PMo_{12}O_{40} \cdot 4H_2O$. On keeping this amorphous product at 85°C for 48 hours, a new XRD pattern appeared, which was consistent with the pattern of $(NH_4)_3PMo_{12}O_{40} \cdot 4H_2O$ (JCPDS No: 9-412).[25] So we can infer that in this reaction, velocities of the diffusion and reaction steps are faster than that of the rate-controlling step.

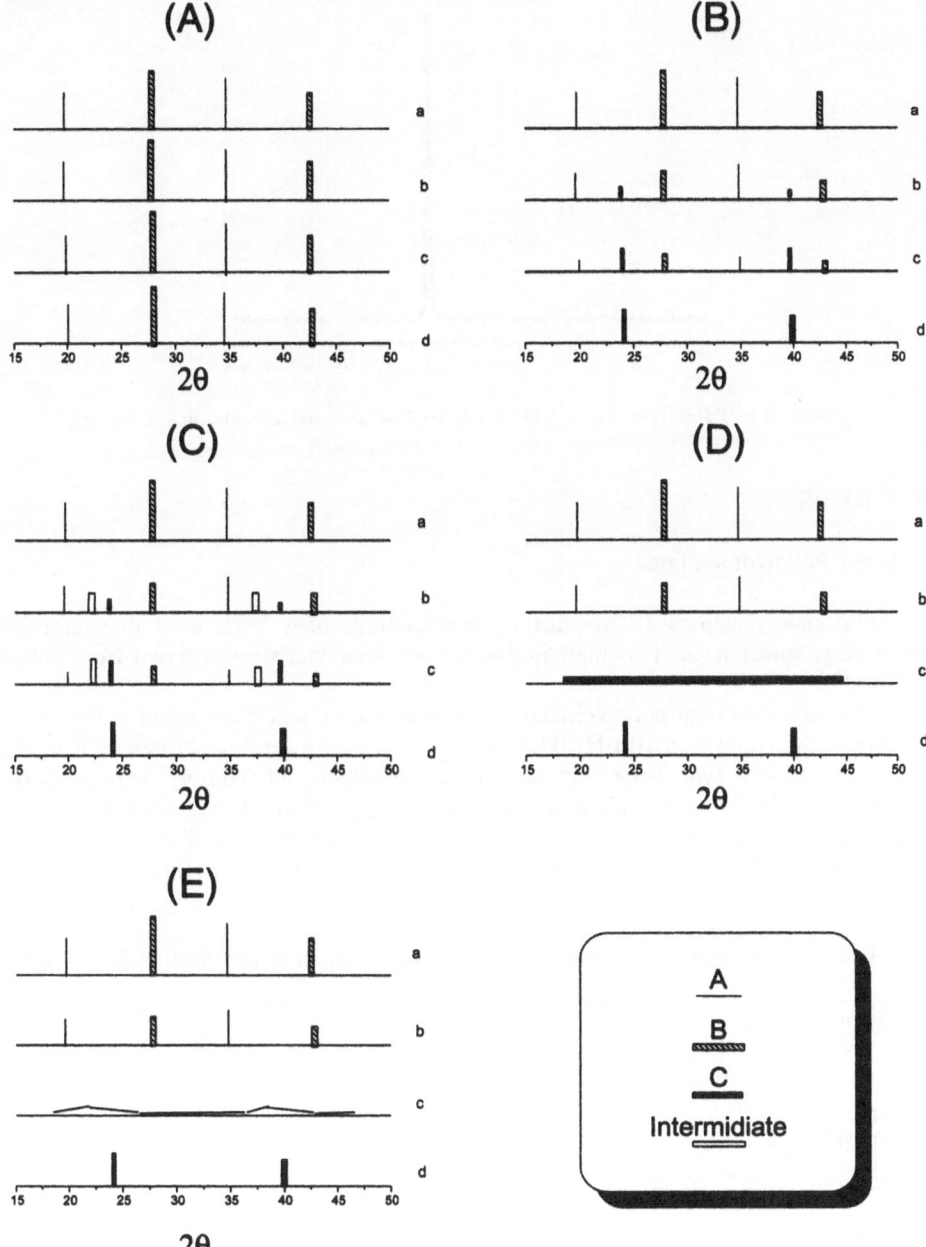

Figure 1. Variations of XRD patterns with reaction time (a→d). (A): No reaction. Rate determining step: diffusion (B); reaction (C); nucleation (D); growth (E).

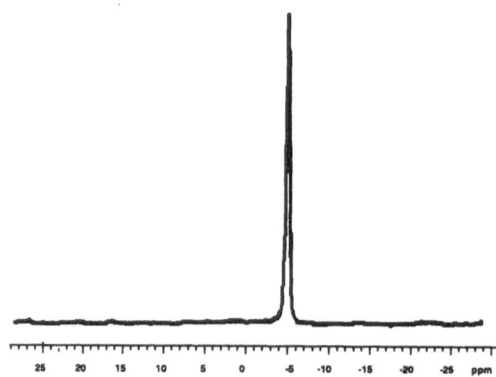

Figure 2. The ^{31}P spectrum of $(NH_4)_3PMo_{12}O_{40}\cdot4H_2O$ synthesized through solution reaction

SYNTHESIS

1. Cluster Polyoxometalates

Solid state reactions at low heating temperatures have been used in synthesizing cluster polyoxometalates. This method has its own characteristics different from solution reactions[26-33].

Some new structural polyoxometalates were obtained, which are listed in the Table 1. In Table 1, the compound $[Bu_4N]_6[H_3O]_2[Mo_{13}O_{40}]_2$ synthesized by Professor K.L. Tang was a new Keggin-type isopolyoxomolybdate, in which, the $[Bu_4N]^+$ cations, $[H_3O]^+$ cations, $[Mo_{13}O_{40}]^{4-}$ anions have different symmetry. Both anions have Keggin-type structures. The structures are based on a central $Mo(A)O_4$ slightly distorted tetrahedron surrounded by 12 $Mo(B)O_6$ octahedra.

Table 1. Polyoxometalates synthesized by solid state reactions at low heating temperatures

Compounds	References
$[(n\text{-}Bu)_4N]_4[Mo_6O_{19}][Ag_2I_4]$	34
$[(n\text{-}Bu)_4N]_4[W_6O_{19}][Ag_2I_4]$	35
$[Bu_4N]_2[Mo_2O_2(OH)_2Cl_4(C_2O_4)]$	36
$[Bu_4N]_6[H_3O]_2[Mo_{13}O_{40}]_2$	37
$H_2[n\text{-}Bu_4N]_3[As(SiW_{11}O_{39})]$	38

The cluster polyoxometalates, $[(n\text{-}Bu)_4N]_4[M_6O_{19}][Ag_2I_4]$ (M = Mo, W) synthesized in our group, are composed of three parts: $[(n\text{-}Bu)_4N]^+$, hexamolybdate $[Mo_6O_{19}]^{2-}$ (or hexatungstate $[W_6O_{19}]^{2-}$) and cluster anion $[Ag_2I_4]^{2-}$. In the Mo_6O_{19} unit having cage structure, each Mo atom is surrounded by a distorted octahedron of oxygen atoms: one central O, one terminal O and four bridging O atoms. The six Mo atoms form an octahedron whose center is occupied by O. The structure of the Mo_6O_{19} unit may be regarded as the condensation of six MoO_6 octahedra sharing a common central vertex of O. The W_6O_{19} unit is isostructural with Mo_6O_{19}. The difference between $[(n\text{-}Bu)_4N]_4[M_6O_{19}][Ag_2I_4]$ and heteropolyoxometalate compounds synthesized via reactions in solution is that the former consists of polyoxometalate, cluster anion, and organic cations, while the latter contain only heteropolyoxometalate and organic cations.[1,39] The molecular arrangement of [(n-Bu)$_4$N]$_4$[Mo$_6$O$_{19}$][Ag$_2$I$_4$] is shown in Fig. 3.

We have also synthesized several other polyoxometalates, which contain cluster cations. For example, [MS$_4$Cu$_4$(γ-MePy)$_8$][M$_6$O$_{19}$] (M= Mo, W)[40] contains M-Cu-S cluster cation and polyoxometalate anion. The compound [Ag(PPh$_3$)$_4$]$_2$[Mo$_6$O$_{19}$]·3CH$_2$Cl$_2$ is a hexamolybdate with two tetrakis(triphenylphosphine)silver cations acting as counter ions.[41]

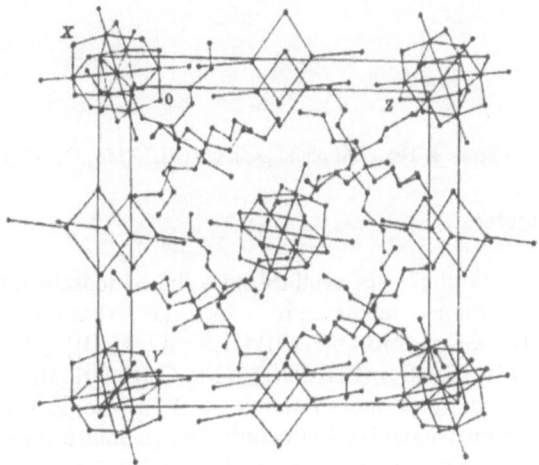

Figure 3. The molecular arrangement of [(n-Bu)$_4$N]$_4$[Mo$_6$O$_{19}$][Ag$_2$I$_4$]

2. Nano-Polyoxometalates

The development of novel methods to produce materials at the nanometer scale is of intense current interest. Nanomaterials are studied for their unique physical and chemical properties as well as novel applications. Over the past decade, much progress has been made in the preparation of nanomaterials. However, there are inherent limitations with those approaches, which include high reaction temperatures, solvent involvement, low yields and high cost, complex process control, non-uniformity, agglomeration, and aggregate formation. The solid state reaction at low heating temperature is a simple, convenient, environmentally friendly, and cheap method to synthesize nanomaterials. A dozen nanomaterials have been synthesized by solid state reactions at low heating temperatures.[21,22] Through the solid state reaction of (NH$_4$)$_6$Mo$_7$O$_{24}$·4H$_2$O and P$_2$O$_5$, the nanomaterial of (NH$_4$)$_3$PMo$_{12}$O$_{40}$·4H$_2$O with 30-nm particle size has been synthesized as shown in Fig. 4. Also, Professor Enbo Wang has prepared K$_3$PMo$_{12}$O$_{40}$·8H$_2$O as 11-nm nanoparticles with the size 11 nm by solid state reaction.[42] These particles have electric conductivity ($\rho = 1.2 \times 10^{-6}$ s/cm).

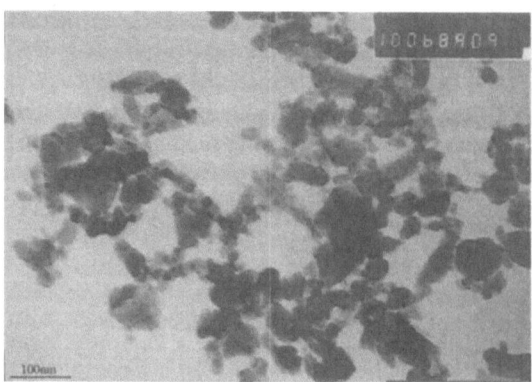

Figure 4. The TEM photograph of $(NH_4)_3PMo_{12}O_{40} \cdot 4H_2O$

3. Layer- Polyoxometalates

Professor Achim Müller[43] has synthesized a linked icosahedral compound in a solid state reaction at room temperature. In the structure of the compound $[H_4Mo_{72}Fe_{30}O_{254}(CH_3COO)_{10}\{Mo_2O_7(H_2O)\}\{H_2Mo_2O_8(H_2O)\}_3(H_2O)_{87}] \cdot 80H_2O$, every cluster unit of the $[Mo_{72}Fe_{30}O_{252}(CH_3COO)_{12}\{Mo_2O_7(H_2O)\}\{H_2Mo_2O_8(H_2O)\}(H_2O)_{91}]$ is covalently linked to four other units through Fe-O-Fe bridges (Fe-Fe = 3.79(4) Å) and thereby forms a two-dimensional layer structure. The geometrical parameters of each single cluster building unit in this compound are equal to the corresponding values found in the discrete cluster. The remarkable result of this work is the cross-linking of giant spheres in a solid state reaction at room temperature, as well as the packing of icosahedral units into a two-dimensional layer structure.

CONCLUSION

The solid state reactions at low-heating temperatures have their own characteristics. There are four steps in a typical solid state reaction: diffusion, reaction, nucleation, and growth. At low heating temperatures, any step can be the rate determining step of solid state reaction. By means of this method, many new compounds have been synthesized. In recent years, much progress has been made in the preparation of nanomaterials by solid state reactions at ambient temperatures.

REFERENCES

1. M. T. Pope, A. Müller, *Polyoxometalates: from Platonic Solids to Anti-Retroviral Activity*, Kluwer Academic Publishers, Dordrecht, The Netherlands (1994).
2. M. T. Pope, *From Simplicity to Complexity in Chemistry and Beyond* Müller, A., Dress A., Vgtle, F., EDS., Vieweg Verlag, Weisbaden, (1996).
3. E. N. Semenovskaya, Application of heteropoly compounds to the analysis of pharmaceuticals and biological samples and in medicobiological studies, *Zh. Anal. Khim.*, 41:1925 (1986).
4. T. Yamase, Photo and electrochromism of polyoxometalates and related materials, *Chem. Rev*, 98(1): 307 (1998).

5. T. Yamase, E. Ishikawa, and Y. Asai , Alkene epoxidation by hydrogen peroxide in the presence of titanium-substituted keggin-type polyoxotungstates, *J. Mol. Catal. A: Chem.*, 114(1-3): 237 (1996).

6. I. V. Kozhevnikov, Catalysis by heteropoly acids and multicomponent polyoxometalates in liquid-phase reactions, *Chem. Rev.*, 98:171 (1998).

7. N. Mizuno, M. Misono, Heterogeneous catalysis, *Chem. Rev.*, 98: 199 (1998).

8. S. G. Sarafianos, U. Kortz , M. T. Pope, Methanism of polyoxometalate-mediated inactiviation of DNA polymerases, *J. Biol. Chem.*, 226:3511 (1991).

9. J. T. Rhule, C. L. Hill, D. A. Judd, R. F. Schinazi, Polyoxometalate in medicine, *Chem. Rev.*, 98:327 (1998).

10. M. Tatsumisago, T. Minomi, Preparation of proton-conducting amorphous films containing dodecamolybdaophosphoric acid by the sol-gel method, *J. Am. Ceram. Soc.*, 72:484 (1989).

11. A. M ller, F. Peters, M. T. Pope, and D Gatteschi, Polyoxometalates: very large cluster- nanoscale magnets, *Chem. Rev.*, 239 (1998).

12. M. T. Pope, X. Y. Wei, New developments in the chemistry of heteropolytungstates of rhodium and cerium, *C. R. Acad. Sci , Ser. C: Chim.*, 1(5-6): 297 (1998).

13. M. T. Pope, A. Müller, Chemistry of polyoxometalates, *Angew. Chem. Int. Ed. Eng.*, 30:34 (1991).

14. R. E McCarley, K. H. Lii, P. A. Edwards, and L. F. Brough, New extended clusters in ternary molybdenum oxides, *J. Solid State Chem.*, 57: 17 (1985).

15. S. Wendeborn, A. Mesmaeker, A. Brill, and S. Berteina, Synthesis of diverse and complex molecules on the solid phase linkers and charge strategies in solid-phase organic, *Acc. Chem. Res.*, 33: 215 (2000).

16. F. Guillier, D. Drain, M. Bradley, Synthesis and combinational chemistry, *Chem. Rev.*, 100:2091 (2000).

17. F. Toda, Solid state organic chemistry: efficient reactions,remarkable yields and stereoselectivity, *Acc. Chem. Res.*, 28(12): 480 (1995).

18. J. O. Metzger, Solvent-free organic syntheses, *Angew. Chem. Int. Ed.*, 37:2975 (1998).

19. Y. M. Zhou, X. Q. Xin, Synthetic chemistry for solid state reaction at low-heating temperatures, *Chinese J. of Inorg. Chem.*, 15(3): 273 (1999).

20. C. Zhang, G. X. Jin, and X. Q. Xin, Studies on the solid-state synthesis and the third-order nonlinear optical properties of the Mo(W)/S/Cu(Ag) cluster compounds, *Chinese J. of Inor. Chem.*, 16(2): 229 (2000).

21. X. R. Ye, D.Z. Jia, X. H. Yu, X.Q. Xin, and Z.L. Xue, One step solid-state reactions at ambient temperatures- a novel approach to nanocrystal synthesis, *Adv. Mater.*, 11:11 (1999).

22. X. H.Yu, F. Li, X.Q. Xin, Synthesis of Ce(IV) oxide ultrafine particle by solid state reaction, *J. Am. Ceram. Soc.*, 4:964 (2000).

23. Y. M Zhou, X. R. Ye, D. H. Li, and X. Q. Xin, Studies on the Diffusion-controlled Solid-solid State Reaction by Means of XRD, *Chem. J. Chin. Univ.*, 20(3):361 (1999).

24. Z. Lai, X. Q. Xin, and H. N. Zhou, Studies on the Solid State Reactions of Coordination Compounds LXXIV, *Chin. J. Inorg. Chem.*, 13(3): (1997).

25. *Nat. Bur. Standards Circ.* , 8:539(1958)

26. Q. Miao, X. Q. Xin, and C. Hu, Solid state reactions of coordination compounds at low heating temperatures LXV, *Chinese J. Chem. Phy .*, 92: 118 (1994).

27. X. Q. Xin and L. M. Zheng, Solid state reactions of coordination compounds at low heating temperatures , *J. Solid State Chem.*, 106: 451 (1993).

28. M. D. Choen, G. M. J. Schmidt, and F. I. Sonntag, Topochemistry part1, A survey, *J. Chem. Soc.*, 1996 (1964); part2, The photochemistry of trans- cinnamic acids, *J. Chem. Soc.*, 2000 (1964); part3 The crystal chemistry of trans- cinnamic acids, *J. Chem. Soc.*, 2014 (1964).

29. H. W. Wei , X. Q. Xin, and S. Shi, M(W,V)-Cu(Ag)-S(Se) cluster compounds, *Coord. Chem. Rev.*, 153:25 (1996).

30. J. G. Li, X. Q. Xin, and Z. Y. Zhou, Synthesis and structural charaterization of a large molybdenum-copper-sulphur cluster compound, *J. Chem. Soc., Chem. Commun.*, 249 (1991).

31. L. X. Lei and X. Q. Xin, The solid state reaction of $CuCl_2$ $2H_2O$ and 8-hydroxyquinoline, *Thermochimica Acta*, 297: 61 (1997).

32. J. J. Lin, L. M. Zheng, and X. Q. Xin, Solid state reactions of coordination compounds at low heating temperatures LXX, *Chinese J. Inorg. Chem.*, 11(1): 106(1995).

33.   S. Jing, *Master Dissertation of Nanjing Univ.* (1995).

34. H. W. Hou, X. R. Ye, and X. Q. Xin, Terabutylammoium Di-μ-iodo-bis(iodoargentate), *Acta Cryst.*, C51: 2013 (1995).

35. H. W. Hou, X. Q. Xin, and K. B. Yu, Solid state synthesis and structure of two polyoxometalates, *Chin. J. Chem.*, 14(2): 123 (1996).

36. C. L. Tang, H. H. Ni, and X. L. Jin, Solid-state synthesis at low-heating temperature and crystal structure of a dinuclear molybdenum complex with oxalate ligand $[Bu_4N]_2[Mo_2O_2(OH)_2Cl_2(C_2O_4)]$, *Chineses J. Struct. Chem.*, 13(4): 300 (1994).

37. X. L. Jin, K. L. Tang, and H. H. Ni, Synthesis and crystal structure of a novel keggin-type isopolyoxomolybdate $[Bu_4N]_6[H_3O]_2[Mo_{13}O_{40}]_2$, *Polyhedron,* 13:2439 (1994).

38. J. P. Lang, *Dissertation, Nanjing University,* (1993).

39. M. T. Pope, *Heteropoly and Isopoly Oxometalates, Springer-Verlag,* Berlin, (1983).

40. M. T. Pope, J. P. Lang, X. Q. Xin, and K. B. Yu, Synthesis of a new salt containing polyoxometallate anion and M-Cu-S cluster cation $[MS_4Cu_4(\gamma\text{-MePy})_8][M_6O_{19}]$, *Chin. J. Chem.*, 13(1): 40 (1995).

41. D. L. Long, X. Q. Xin, X. M. Chen, and B. S. Kang, Synthesis and X-ray crystal structure of a polymetallate with a metal complex cation as counter ion, $[Ag(PPh_3)_4]_2Mo_6O_{19}\cdot3CH_2Cl_2$, *Polyhedron,* 16(7):1259 (1997).

42. W. S. You, Y. B. Wang, and E. B. Wang, Preparation and characterization of $K_3PMo_{12}O_{40}\cdot8H_2O$ nanoparticles with keggin structure by solid reaction at room temperature, *International Huaxia Symposium Solid State Chemistry and Synthesis Chemistry,* 4:74 (2000).

43. A. Müller, E. Krickemeyer, and K. S. Das, Linking icosahedral, strong molecular magnets $\{Mo^{VI}_{72}Fe^{II}_{30}\}$ to layers-a solid state reaction at room temperature, *Angew. Chem. Int. Ed.*, 39 (9):1612 (2000)

STRUCTURE DETERMINATION OF POLYOXOTUNGSTATES USING HIGH-ENERGY SYNCHROTRON RADIATION

Tomoji Ozeki,[1,*] Noritaka Honma,[1] Shunsuke Oike[1] and Katsuhiro Kusaka[2]

[1] Department of Chemistry and Materials Science
Tokyo Institute of Technology
2-12-1 O-okayama, Meguro-ku, Tokyo 152-8551, Japan
[2] CREST fellow, Core Research for the Evolutional Science and Technology (CREST), Japan Science and Technology Corporation (JST) and Japan Synchrotron Radiation Research Institute (JASRI)
1-1-1 Kouto, Mikazuki-cho, Sayo-gun, Hyogo 679-5198, Japan

INTRODUCTION

Single crystal X-ray diffraction methods have benefited polyoxometalate chemistry by revealing the complicated atomic connectivities in many polyoxometalates, most of which had been almost unpredictable using other techniques. However, the atomic connectivity is, of course, not the only information obtained from X-ray structure analyses. Modern X-ray diffraction experiments can provide us with detailed geometries including precise intra- and inter-molecular distances and angles. Full exploitation of this information can give us insight into the structural changes associated with chemical processes in which the polyoxometalates are involved, examples of which include protonation, introduction of excess electrons (reduction), substitution of constituent elements, partial degradation leading to lacunary species, further condensation leading to complexes with higher nuclearity, introduction of organic and organometallic moieties, formation of hydrogen bonds, *etc*. However, in spite of the current development in both instrumentation and data analysis software, structure determination of polyoxotungstates often fails to provide structural data of sufficient quality. One of the reasons can be attributed to the absorption effect that is intrinsically and inevitably associated with the X-ray diffraction experiment. In general, MoK_α radiation suffers much less absorption than CuK_α radiation. Nevertheless, heavy elements such as W still have large attenuation coefficients for MoK_α radiation. Although absorption corrections can be applied using various methods by means of widely distributed computer programs, they are in many cases of limited accuracy, especially

---

*Author to whom correspondence should be addressed. E-mail: tozeki@cms.titech.ac.jp.

Polyoxometalate Chemistry for Nano-Composite Design
Edited by Yamase and Pope, Kluwer Academic/Plenum Publishers, 2002

when it is difficult to make accurate measurements of the crystal shape and size due to the crystals not having well-developed faces or being sealed in glass capillaries. Diffraction data insufficiently corrected for absorption lead to unrealistically distorted (or often non-positive definite) anisotropic displacement parameters and poor fit between observed and calculated structure factors. The latter particularly decreases the precision of the atomic parameters of lighter atoms, in our case oxygen atoms, which accordingly makes geometric parameters involving these atoms less reliable. For this reason, the structural details involving oxygen atoms have been less often discussed for polyoxotungstates than for molybdates or vanadates.

A much better way to avoid this difficulty is, without any doubt, to diminish the absorption effect, which can be achieved either by making the crystal size smaller or by reducing the absorption coefficients of the material. The former is not practical because the crystals should be large enough for the diffraction to be observed with sufficient signal to noise ratio (it may also be worth noting that a considerable time is often needed for mounting very small crystals during which many unstable crystals may lose their crystallinity). The latter can be realized when using high-energy or short wavelength X-rays. However, diffraction experiments at wavelengths shorter than MoK$_\alpha$ radiation using conventional X-ray sources including rotating anodes are not as easy as might be expected. It should be noted that the absolute intensity of the diffracted X-ray is inversely proportional to the square of the wavelength. Thus, not only is the diffraction from the sample crystal weak, but the output from the monochromator crystal is also weak, which makes the incident X-ray intensity supplied to the sample crystal substantially lower. Another difficulty is the contraction of reciprocal space as a consequence of the use of shorter wavelength X-rays. Severe overlap of diffraction spots may then occur, which makes individual measurements of the diffraction spots very difficult. These difficulties are intrinsic to the use of high-energy X-rays, but are less pronounced by the use of synchrotron X-ray radiation, especially that from third generation storage rings. Its high brilliance conveys sufficient (or in many cases more than sufficient) flux of X-rays to the sample crystals, even after being monochromated by Si monochromator crystals, whose outputs are much smaller than those from graphite monochromator crystals. Low divergence and monochromaticity of the incident beam make diffraction spots sharp and narrow, which substantially reduces the peak overlap problem associated with the use of short wavelength X-rays. Having these advantages in mind, we initiated structure analyses of polyoxometalates using high-energy synchrotron X-ray radiation, the first results of which are described below.

SODIUM PARADODECATUNGSTATE 26-27 HYDRATE[§]

Structure Determination Using Synchrotron X-ray Radiation

In order to evaluate the effect of using high-energy synchrotron X-ray radiation, we employed sodium paradodecatungstate 26-27 hydrate as our first sample,[1] because it is one of the most easily obtained isopolyoxotungstate crystals. We carried out a structure determination with 30.748 keV X-rays (λ=0.3937Å) by using the vacuum camera installed in the BL02B1 beamline of SPring-8.[2] The absorption coefficient of this material is 5.28mm^{-1} for 30.748 keV X-rays, which is substantially smaller than that for MoK$_\alpha$ radiation (23.2mm^{-1}). 30843 data were collected up to (sinθ/λ)=0.893Å^{-1}, of which 19516 were independent. The data were not corrected for the effects of absorption. The structure was solved by direct methods without any special treatment and the structure was

---

[§]A preliminary result appears in Reference 1.

successfully refined to give final $R1$ and $wR2$ values of 0.0444 and 0.1182, respectively. The structure of the anion is shown in Figure 1, which illustrates that the anisotropic displacement parameters both for W and O atoms show no systematic distortions, even without the use of absorption corrections.

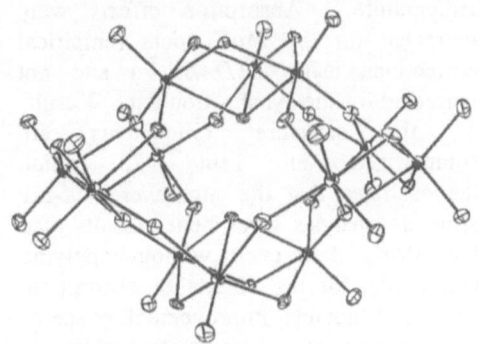

Comparison of the Structure with the Earlier Reported Structures

The first structural investigation of paradodecatungstate by X-ray diffraction was carried out by Lindqvist in 1952,[3] who pointed out that $5Na_2O \cdot 12WO_3 \cdot 28H_2O$ (or $Na_{10}[H_2W_{12}O_{42}] \cdot 27H_2O$, in modern representations) exhibits an approximately

Figure 1. Structure of the $[H_2W_{12}O_{42}]^{10-}$ anion determined by high-energy X-ray diffraction experiment.

centered cell. He reported the W atom positions corresponding to the centered cell. Cruywagen et al[4] reported in 1986 the structure of $Na_{10}[H_2W_{12}O_{42}] \cdot 26H_2O$ whose unit cell volume was reported to be approximately half of that reported by Lindqvist. They seem to have missed the weak diffractions due to the approximate cell centering. Chrissafidou et al[5] reported the structure of $Na_{10}[H_2W_{12}O_{42}] \cdot 27H_2O$ in 1995. However, no relationships between these crystal structures have been referred to.[4,5] By appropriate unit cell transformations as shown in Table 1, all these crystals can be transformed to an identical cell. The W atom positions of these crystals coincide within experimental errors.

Table 1. Comparison of the unit cell dimensions.

	Lindqvist		Cruywagen		Chrissfidou	This work
	original	transformed[a]	original	transformed[b]		
$a/\text{Å}$	11.77(3)	11.77(3)	11.811(2)	11.811	11.78(1)	11.811(1)
$b/\text{Å}$	22.19(5)	12.44(2)	12.486(2)	12.486	12.439(3)	12.557(1)
$c/\text{Å}$	12.44(2)	22.19(5)	12.206(2)	22.151	22.069[c]	22.319(3)
$\alpha/°$	86.0(3)	86.0(3)	82.29(1)	86.16	86.20(3)	86.214(4)
$\beta/°$	113.3(3)	86.0(3)	115.12(1)	86.25	86.17(2)	86.163(5)
$\gamma/°$	94.0(3)	66.6(3)	113.76(1)	66.24	66.22(2)	66.249(5)

[a] Transformation is: $a_{new}=-a_{old}$, $b_{new}=c_{old}$, $c_{new}=b_{old}$. [b] Transformation is: $a_{new}=-a_{old}$, $b_{new}=b_{old}$, $c_{new}=-a_{old}-2c_{old}$. [c] In the original publication, this value was printed as 18.76(2). However, a recalculation using the unit cell volume and the other cell constants leads to 22.069Å for c.

STRUCTURE OF THE β-DODECATUNGSTOSILICATE ANION [¶]

Given the aforementioned successful result, attempts to solve unknown structures were initiated. Among the first examples is β-dodecatungstosilicate tetrabutylammonium monohydrate. $[(n\text{-}C_4H_9)_4N]_4[\beta\text{-}SiW_{12}O_{40}] \cdot H_2O$ is orthorhombic, space group $P2_12_12_1$ with $a=19.8600(2)\text{Å}$, $b=20.5960(3)\text{Å}$, $c=25.4900(1)\text{Å}$. A crystal with dimensions of $0.3\times0.25\times0.24$ mm was measured both at the BL04B2 beamlime of SPring-8[6] and at our

---

[¶] Details will be published elsewhere.

laboratory using MoK_α radiation at ambient temperature. Absorption effects were corrected for the MoK_α data (empirical corrections using *SADABS* [7]) and not corrected for the synchrotron data. Results of the structure refinements are summarized in Table 2, which demonstrates that the high-energy X-ray structure analysis gives better results than the MoK_α data even without applying corrections for the effects of absorption. The most notable improvement concerns the precision of the parameters obtained. Standard uncertainties associated with the geometric parameters obtained from the synchrotron data are about two thirds of those obtained from the MoK_α data. The structure of the [β-SiW$_{12}$O$_{40}$]$^{4-}$ anion is shown in Figure 2, the metal-oxygen

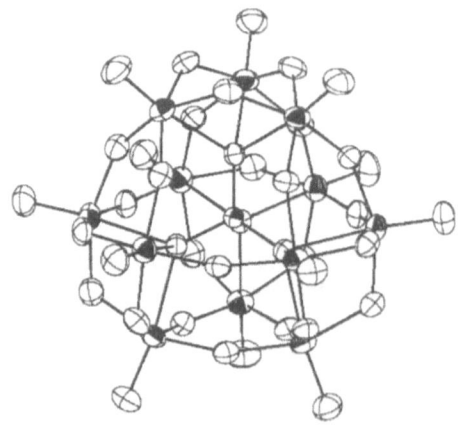

Figure 2. *ORTEP* drawing of the [β-SiW$_{12}$O$_{40}$]$^{4-}$ anion determined by high-energy X-ray diffraction experiment.

framework of which is identical to that found in β-K$_4$SiW$_{12}$O$_{40}$·9H$_2$O [8] and [(n-C$_4$H$_9$)$_4$N]$_4$[β-SiW$_{12}$O$_{40}$]. [9]

Table 2. Comparison of the structure refinements of [(n-C$_4$H$_9$)$_4$N]$_4$[β-SiW$_{12}$O$_{40}$]·H$_2$O using synchrotron and MoK_α radiations.

	Synchrotron radiation	MoK_α radiation
Diffractometer	*DIP-LABO* (imaging plate)	*SMART-CCD*
λ/Å	0.3282	0.7107
μ/mm^{-1}	1.78	13.26
Number of reflections	30470	29288
$R1(F, F_o>4\sigma(F_o))$	0.0269	0.0407
$wR2(F^2$, all data)	0.0612	0.0816
s.u. for W-W distances/Å	0.0005	0.0006-0.0008
s.u. for W-O distances/Å	0.004-0.006	0.006-0.009
s.u. for O-W-O angles/°	0.2-0.3	0.3-0.4
$\Delta\rho_{max}$/eÅ^{-3}	2.12	1.58
$\Delta\rho_{min}$/eÅ^{-3}	-0.99	-1.70

TRUE STRUCTURE OF THE β-DAWSON ANION, OR DOES WU'S A-OCTADECATUNGSTODIPHOSPHATE REALLY EXHIBIT THE β-DAWSON STRUCTURE?[†,‡]

It has long been known that octadecatungstodiphosphate has two isomers. [10] Recent work revealed that it has three isomers (and its arsenate analogue has four isomers). [11] The

---

† Extreme care must be paid when referring to the isomers of the octadecatungstodiphosphate structures. The original notation given by Wu[10] used A for the salt that precipitates first and B for the salt that precipitates afterwards. However, some authors modified Wu's notation and called A as α and B as β. [27,28] Later, Contant and Thouvenot proposed a comprehensive notation that covers all the six possible isomers and calls the structure reported by Dawson as α. [11] Thus, Wu's B (sometimes called β) salt contains anions with the α-Dawson structure. Throughout this article, the notation given by Contant and Thouvenot is employed.

‡ Details of the structure analyses will be published elsewhere.

first structure analysis of octadecatungstodiphosphate was carried out by Dawson for $K_6[P_2W_{18}O_{62}]\cdot14H_2O$. [12]   The anion structure observed here exhibits the α-Dawson structure, according to the notation given by Contant and Thouvenot.[11] This structure was also observed in Wu's B salt,[13] in a crystal with a different cation,[14] and in crystals of octadecamolybdo-diphosphate,[13, 15] -diarsenate [16] and -disulfate [17] and hexadecatungsto-divanadodisulfate.[18]   Many derivatives of the α-Dawson structure have been reported, among which are a peroxo derivative,[19] a monolacunary anion[20] and its derivatives,[20,21] a trisubstituted anion,[22] dimeric[23] and organometallic[24] derivatives of trisubstituted anions, and sandwich complexes.[25] Also, it is noteworthy that some fluorooxotungstates have a very similar metal-oxygen/fluorine framework.[26]

Given the structure revealed by Dawson, Baker and Figgis proposed a total of six possible isomers (including the original Dawson structure) for octadecatungstate anions.[27] However, observation of solid-state structures of non-α-Dawson octadecatungstate isomer (allied to Wu's A salt) is limited to only two examples: a Ba^{2+} salt of $[P_2W_{18}O_{62}]^{6-}$ reported by Matumoto and Sasaki[28] and an NH_4^+ salt of $[As_2W_{18}O_{62}]^{6-}$ reported by Neubert and Fuchs.[29]   Moreover, these two results not only contradict each other but they are also inconsistent with solution studies using ^{31}P NMR[30,11] and ^{183}W NMR[31,11] spectroscopy. Of the six possible isomeric structures proposed by Baker and Figgis, Matsumoto's structure is a staggered fusion of two α-A-PW$_9$ units[32] (α^*-P_2W_{18} structure) while Neubert's structure is a staggered fusion of two β-A-AsW$_9$ units (γ^*-As_2W_{18} structure). Solution studies ruled out the existence of the α^*-P_2W_{18} anion in aqueous solutions.[11,30,31] Only the α, β and γ isomers of $[P_2W_{18}O_{62}]^{6-}$ anion were identified.[11] As for the $[As_2W_{18}O_{62}]^{6-}$ anion, although the γ^* isomer was observed in aqueous solutions, its existence in the rhombohedral crystals was suspected.[11] To clarify this problem, we tried to investigate the solid-state structures of non-α-P_2W_{18} anions using high-energy synchrotron radiation. The crystal structures of $(NH_4)_6[P_2W_{18}O_{62}]\cdot8H_2O$ and $Ba_2Na_2[P_2W_{18}O_{62}]\cdot27H_2O$ were analyzed to reveal the true β-Dawson structure.

Structure Determinations of $(NH_4)_6[P_2W_{18}O_{62}]\cdot8H_2O$ and $Ba_2Na_2[P_2W_{18}O_{62}]\cdot27H_2O$ Using Synchrotron X-ray Radiation

Structure of Wu's A Salt, $(NH_4)_6[P_2W_{18}O_{62}]\cdot8H_2O$. Although its arsenate analogue was reported to have the space group $R\bar{3}$,[29] the structure of $(NH_4)_6[P_2W_{18}O_{62}]\cdot8H_2O$ was successfully analyzed only in the space group $R3m$. Re-determination of the structure of $(NH_4)_6[\beta$-$As_2W_{18}O_{62}]\cdot nH_2O$, which has very similar cell dimensions and is very likely to be isomorphous to Wu's A salt, may be necessary. As shown in Figure 3, the structure of the $[P_2W_{18}O_{62}]^{6-}$ anion exhibits the β-Dawson structure, which can be regarded as a fused product of α-A-PW$_9$ and β-A-PW$_9$ groups in the eclipsed arrangement of the two belt W$_6$ groups. This structure is consistent with the ^{31}P and ^{183}W NMR spectra recorded in solutions.[30,31,11] This anion has an approximate C_{3v} symmetry with one of its mirror planes coincident with the crystallographic mirror plane. The distribution of the W atoms is almost centrosymmetric with respect to the midpoint of the two P atoms and the midpoint is very close to the inversion center that exists in space group $R\bar{3}$ (this inversion center disappears in space group $R3m$). Thus, the diffraction intensity distribution approximates to the

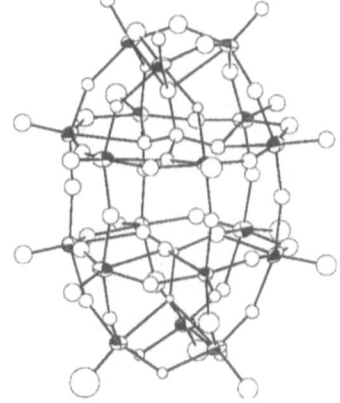

Figure 3. *ORTEP* drawing of the $[P_2W_{18}O_{62}]^{6-}$ anion in Wu's A salt, $(NH_4)_6[P_2W_{18}O_{62}]\cdot8H_2O$, determined by our high-energy X-ray diffraction experiment.

symmetry of $R3$, which makes space group assignment difficult.

Structure of $Ba_2Na_2[P_2W_{18}O_{62}] \cdot 27H_2O$. Although this compound was reported to possess orthorhombic space group $Pmn2_1$,[28] re-determination of the structure revealed that its crystal system is monoclinic with the β angle very close to 90 degrees. As monoclinic crystals with almost rectangular β angle often exhibit pseudomerohedral twinning, the unsuccessful assignment of the crystal system may have been due to twinning of the crystals. Even after successful assignment of the crystal system, there still remained the problem of space group assignment. The systematic extinction suggested the space group $P2_1/n$, in which the W atoms were successfully located but the O atom positions could not be fully determined. Thus the space group was re-examined. Of the three *translationengleiche* subgroups of $P2_1/n$, only Pn led to a successful location of the O atoms. The atomic connectivity in the $[P_2W_{18}O_{62}]^{6-}$ anion is identical with that found in $(NH_4)_6[P_2W_{18}O_{62}] \cdot 8H_2O$ (the β structure). Again, the distribution of the 18 W atoms approximates to a centrosymmetric arrangement with respect to the midpoint of the two P atoms and the midpoint coincides with the inversion center that would appear in the space group $P2_1/n$. Therefore, the diffraction from this crystal shows an intensity distribution approximating to the systematic extinction for the space group $P2_1/n$, which makes the space group assignment very confusing.

Table 3. Crystal data for the octadecatungstodiphosphate salts.

	$(NH_4)_6[P_2W_{18}O_{62}] \cdot 8H_2O$	$Ba_2Na_2[P_2W_{18}O_{62}] \cdot 27H_2O$
Chemical formula	$P_2W_{18}O_{70}N_6H_{40}$	$Ba_2Na_2P_2W_{18}O_{89}H_{54}$
Formula weight	4615.38	5170.08
Crystal system	Rhombohedral	Monoclinic
Space group	$R3m$	Pn
Unit cell dimensions	$a=37.888(5)$Å, $c=13.255(3)$Å	$a=13.027(1)$Å, $b=14.892(2)$Å $c=20.009(2)$Å, $\beta=89.042(6)°$
$\mu(37.80\text{keV})/ \mu (MoK_\alpha)$	$3.75\text{mm}^{-1}/28.29\text{mm}^{-1}$	$4.22\text{mm}^{-1}/28.22\text{mm}^{-1}$
Crystal size/mm	$0.20 \times 0.07 \times 0.07$	$0.24 \times 0.08 \times 0.06$
Temperature/K	293	95
Number of reflections	12245	18998
$R1(F, F_o>4\sigma(F_o))$	0.0390	0.0586
$wR2(F^2, \text{ all data})$	0.1263	0.1432

Difficulties Associated with the Structure Determination of the β-Dawson Structures

As Massart *et al* have already pointed out,[30] the β-Dawson structure has approximately the same W atom arrangement as the α^* isomer (and the γ^* isomer) and thus the O atom positions are very important in distinguishing these isomeric structures. However, the problem is not merely a "hunting" of oxygen atoms from the difference Fourier peaks. Since the W atoms dominate the diffraction intensities, an approximately centrosymmetric arrangement of W atoms gives rise to an approximately centrosymmetric diffraction pattern, which makes the space group assignment very difficult. Very careful measurements of the diffraction intensities, especially those for the Friedel pairs, are therefore important.

CONCLUDING REMARKS

As has been demonstrated above, high-energy X-ray structure analyses are very effective to obtain precise and accurate structures of polyoxotungstates. Elimination of the effects of absorption undoubtedly contributes to the precision and accuracy of the structures obtained.

ACKNOWLEDGEMENTS

T. O. thanks Dr. Y. Ozawa for helpful and stimulating discussions. This work was supported partly by Core Research for the Evolutional Science and Technology, Japan Science and Technology Corporation. Japan Synchrotron Radiation Research Institute is also acknowledged both for financially supporting the installation of the instrument and for the assignment of beam-time.

REFERENCES

1. T. Ozeki, Structure analysis of a sodium paradodecatungstate, *Chem. Lett.* 266 (2001).
2. Y. Noda, K. Ohshima, H. Toraya, K. Tanaka, H. Terauchi, H. Maeta and H. Konishi, First results from the crystal structure analysis beamline at SPring-8, *J. Synchrotron Rad.* 5:485 (1998).
3. I. Lindqvist, On the structure of paratungstate ion, *Acta Crystallogr.* 5:667 (1952).
4. J. J. Cruywagen, I. F. J. van der Merwe, L. R. Nassimbeni, M. L. Niven, E. A. Symonds, Crystal and molecular structure of sodium paratungstate 26 hydrate, *J. Crystallogr. Spectrosc. Res.* 16:525 (1986).
5. A. Chrissafidou, J. Fuchs, H. Hartl and R. Palm, Kristallisation und Strukturuntersuchung von Alkali-Parawolframaten, *Z. Naturforsch.* B50:217 (1995).
6. T. Ozeki, K. Kusaka, N. Honma, Y. Nakamura, S. Nakamura, S. Oike, N. Yasuda, H. Imura, H. Uekusa, M. Isshiki, C. Katayama and Y. Ohashi, Crystal structure analysis of a hexatungstate by a high-energy x-ray diffraction experiment using an imaging plate Weissenberg camera at the BL04B2 beamline of SPring-8, *Chem. Lett.* 804 (2001).
7. (a) R. H. Blessing, An empirical correction for absorption anisotropy, *Acta Crystallogr. Sect. A* 51:33 (1995). (b) G. M. Sheldrick, *SADABS absorption correction method*, University of Göttingen, Göttingen, Germany (1996).
8. K. Y. Matsumoto, A. Kobayashi and Y. Sasaki, The crystal structure of β-$K_4SiW_{12}O_{40}\cdot9H_2O$, containing an isomer of the Keggin ion, *Bull. Chem. Soc. Jpn.* 48:3164 (1975).
9. J. Fuchs, A. Thiele and R. Palm, Strukturen und Schwingungsspektrn des Tetramethylammonium-α-dodekawolframatosilikats und des Tetrabutylammonium, *Z. Naturforsch.* B36:161 (1981).
10. H. Wu, Contribution to the chemistry of phosphomolybdic acids, phosphotungstic acids, and allied substances, *J. Biol. Chem.* 43:189 (1920).
11. R. Contant and R. Thouvenot, A reinvestigation of isomerism in the Dawson structure: syntheses and ^{183}W NMR structural characterization of three new polyoxotungstates $[X_2W_{18}O_{62}]^{6-}$ ($X=P^V$, As^V), *Inorg. Chim. Acta* 212:41 (1993).
12. B. Dawson, The structure of the 9(18)-heteropoly anion in potassium 9(18)-tungstophosphate, $K_6(P_2W_{18}O_{62})\cdot14H_2O$, *Acta Crystallogr.* 6:113 (1953).
13. H. d'Amour, Vergleich der Heteropolyanionen $[PMo_9O_{31}(H_2O)_3]^{3-}$, $[P_2Mo_{18}O_{62}]^{6-}$ und $[P_2W_{18}O_{62}]^{6-}$, *Acta Crystallogr. Sect. B* 32:729 (1976).
14. E. Coronado, J. R. Galán-Mascarós, C. Giménez-Saiz, C. J. Gómez-García and V. N. Laukhin, The first radical salt of the polyoxometalate cluster $[P_2W_{18}O_{62}]^{6-}$ with bis(ethylendithio)tetrathiafulvalen (ET): $ET_{11}[P_2W_{18}O_{62}]\cdot3H_2O$, *Adv. Mater.* 8:801 (1996).
15. R. Strandberg, Multicomponent polyanions. 12. The crystal structure of $Na_6Mo_{18}P_2O_{62}(H_2O)_{24}$, a compound containing sodiumcoordinated 18-molybdodiphosphate anions, *Acta Chem. Scand. A* 29:350 (1975).
16. H. Ichida and Y. Sasaki, The structure of hexaguanidinium octadecamolybdodiarsenate enneahydrate, $(CH_6N_3)_6[As_2Mo_{18}O_{62}]\cdot9H_2O$, *Acta Crystallogr. Sect. C* 39:529 (1983).
17. T. Hori, O. Tamada and S. Himeno, The structure of 18-molybdodisulphate(VI)(4-) ion in $(NEt_4)_4S_2Mo_{18}O_{62}\cdot CH_3CN$, *J. Chem. Soc., Dalton Trans.* 1491 (1989).
18. A. Botar and J. Fuchs, Kristallstrukturen und Schwingungsspektren des Heteropolysuflats $[N(CH_3)_4]_6S_2V_2W_{16}O_{62}$ und des Vanadatowolframats $Na[N(CH_3)_4]_2VW_5O_{19}\cdot H_2O$, *Z. Naturforsch.* B37:806 (1982).

19. D. A. Judd, Q. Chen, C. F. Campana and C. L. Hill, Synthesis, solution and solid state structures, and aqueous chemistry of an unstable polyperoxo polyoxometalate: $[P_2W_{12}(NbO_2)_6O_{56}]^{12-}$, *J. Am. Chem. Soc.* 119:5461-5462 (1997).

20. T. J. R. Weakley, The crystal structures of two heteropolytungstate salts containing anions derived from α-octadecatungstodiphosphate(6-): $(NH_4)_{10}[\alpha_2\text{-}P_2W_{17}O_{61}]\cdot 8H_2O$ and $(Me_2NH_2)_8[\alpha_2\text{-}P_2Co(H_2O)W_{17}O_{61}]\cdot 11H_2O$, *Polyhedron*, 6:931 (1987).

21. (a) V. N. Molchanov, L. P. Kazanskii, E. A. Torchenkova and V. I. Simonov, Crystal structure of $[K_{16}Ce(P_2W_{17}O_{61})_2]\cdot nH_2O$ (n=50), *Kristallografiya* 24:167 (1979). (b) A. Müller, V. P. Fedin, C. Kuhlmann, H.-D. Fenske, G. Baum, H. Bögge and B. Hauptfleisch, 'Adding' stable functional complementary, nucleophiclic and electrophilic clusters: a synthetic route to $[\{(SiW_{11}O_{39})Mo_3S_4(H_2O)_3(\mu\text{-}OH)\}_2]^{10-}$ and $[\{(P_2W_{17}O_{61})Mo_3S_4(H_2O)_3(\mu\text{-}OH)\}_2]^{14-}$ as examples, *J. Chem. Soc., Chem. Commun.* 1189 (1999).

22. R. G. Finke, D. K. Lyon, K. Nomiya and T. J. R. Weakley, Structure of nonasodium α-triniobatopentadecawolframatodiphosphate-acetonitrile-water (1/2/23), $Na_9[P_2W_{15}Nb_3O_{62}]\cdot 2CH_3CN\cdot 23H_2O$, *Acta Crystallogr. Sect. C* 46:1592 (1990).

23. N. J. Crano, R. C. Chambers, V. M. Lynch and M. A. Fox, Preparation and photocatalytic studies on a novel Ti-substituted polyoxometalate, *J. Mol. Catal. A* 114:65 (1996).

24 (a) M. Pohl, Y. Lin, T. J. R. Weakley, K. Nomiya, M. Kaneko, H. Weiner and R. G. Finke, Trisubstituted heteropolytungstates as soluble metal oxide analogs. Isolation and characterization of $[(C_5Me_5)Rh\cdot P_2W_{15}Nb_3O_{62}]^{7-}$ and $[(C_6H_6)Ru\cdot P_2W_{15}Nb_3O_{62}]^{7-}$, including the first crystal structure of a Dawson-type polyoxoanion-supported organometallic complex, *Inorg. Chem.* 34:767 (1995). (b) M. Pohl, D. K. Lyon, N. Mizuno, K. Nomiya and R. G. Finke, Polyoxoanion-supported catalyst precursors. Synthesis and characterization of the iridium(I) and rhodium(I) precatalysts $[(n\text{-}C_4H_9)_4N]_5Na_3[(1,5\text{-}COD)M\cdot P_2W_{15}Nb_3O_{62}]$ (M=Ir, Rh), *Inorg. Chem.* 34:1413 (1995).

25. (a) T. J. R. Weakley and R. G. Finke, Single-crystal x-ray structures of the polyoxotungstate salts $K_{8.3}Na_{1.7}[Cu_4(H_2O)_2(PW_9O_{34})_2]\cdot 24H_2O$ and $Na_{14}Cu[Cu_4(H_2O)_2(P_2W_{15}O_{56})_2]\cdot 53H_2O$, *Inorg. Chem.* 29:1235 (1990). (b) R. G. Finke and T. J. R. Weakley, Structure of sodium bis(pentadecatungstodiphosphato)diaquatetrazincate hydrate (16:1:50), *J. Chem. Crystallogr.* 24:123 (1994). (c) C. J. Gómez-García, J. J. Borrás-Almenar, E. Coronado and L. Ouahab, Single-crystal x-ray structure and magnetic properties of the polyoxotungstate complexes $Na_{16}[M_4(H_2O)_2(P_2W_{15}O_{56})_2]\cdot nH_2O$ (M=Mn^{II}, n=53; M=Ni^{II}, n=52): an antiferromagnetic Mn^{II} tetramer and a ferromagnetic Ni^{II} tetramer, *Inorg. Chem.* 33:4016 (1994). (d) X. Zhang, Q. Chem. D. C. Duncan, C. F. Campana and C. L. Hill, Multiiron polyoxoanions. Synetheses, characterization, x-ray crystal structures, and catalysis of H_2O_2-based hydrocarbon oxidations by $[Fe^{III}_4(H_2O)_2(P_2W_{15}O_{56})_2]^{12-}$, *Inorg. Chem.* 36:4208 (1997).

26. S. H. Wasfi, C. E. Costello, A. L. Rheingold and B. S. Haggerty, Preparation and characterization of two new isomorphous heteropoly oxofluorotungstate anions: $[CoW_{17}O_{56}F_6NaH_4]^{9-}$ and $[FeW_{17}O_{56}F_6NaH_4]^{8-}$, *Inorg. Chem.* 80:1788 (1991).

27. L. C. W. Baker and J. S. Figgis, A new fundamental type of inorganic complex: hybrid between heteropoly and conventional coordination complexes. Possibilities for geometrical isomerisms in 11-, 12-, 17-, and 18-heteropoly derivatives. *J. Am. Chem. Soc.* 92:3794 (1970).

28. K. Y. Matsumoto and Y. Sasaki, Crystal structure of $\alpha\text{-}P_2W_{18}O_{62}^{6-}$ anion, *J. Chem. Soc., Chem. Comm.* 691 (1975).

29. H. Neubert and J. Fuchs, Kristallstrukturen und Schwingungsspektren zweier isomerer Oktadekawolframatodiarsenate, $(NH_4)_6As_2W_{18}O_{62}\cdot nH_2O$, *Z. Naturforsch.* B42:951 (1987).

30. R. Massart, R. Contant, J.-M. Fruchart, J.-P. Ciabrini and M. Fournier, ^{31}P NMR studies on molybdic and tungstic heteropolyanions. Correlation between structure and chemical shift, *Inorg. Chem.* 16:2916 (1977).

31. R. Acerete, S. Harmalker, C. F. Hammer, M. T. Pope and L. C. W. Baker, Concerning isomerisms and interconversions of 2:18 and 2:17 heteropoly complexes and their derivatives, *J. Chem. Soc., Chem. Comm.* 777 (1979).

32. G. Hervé and A. Tézé, Study of α- and β-enneatungstosilicates and -germanates, *Inorg. Chem.* 16:2115 (1977).

INDEX